INTRODUCTION TO
PARTIAL DIFFERENTIAL EQUATIONS

INTRODUCTION TO PARTIAL DIFFERENTIAL EQUATIONS

SECOND EDITION

GERALD B. FOLLAND

PRINCETON UNIVERSITY PRESS • PRINCETON, NEW JERSEY

Copyright © 1995 by Princeton University Press
Published by Princeton University Press, 41 William Street,
Princeton, New Jersey 08540
In the United Kingdom: Princeton University Press,
Chichester, West Sussex

Library of Congress Cataloging-in-Publication Data

Folland, G. B.
Introduction to partial differential equations / Gerald B. Folland.—
2nd ed.
p. cm.
Includes bibliographical references and indexes.

ISBN-13: 978-0-691-04361-6

ISBN-10: 0-691-04361-2

1. Differential equations, Partial. I. Title.
QA374.F54 1995   95-32308
515'.353—dc20

The publisher would like to acknowledge the author of this volume for
providing the camera-ready copy from which this book was printed

Princeton University Press books are printed on acid-free paper and
meet the guidelines for permanence and durability of the Committee
on Production Guidelines for Book Longevity of the Council on
Library Resources

Printed in the United States of America
by Princeton Academic Press

10 9 8 7 6 5 4 3 2 1

CONTENTS

PREFACE

In 1975 I gave a course in partial differential equations (PDE) at the University of Washington to an audience consisting of graduate students who had taken the standard first-year analysis courses but who had little background in PDE. Accordingly, it focused on basic classical results in PDE but aimed in the direction of the recent developments and made fairly free use of the techniques of real and complex analysis. The roughly polished notes for that course constituted the first edition of this book, which has enjoyed some success for the past two decades as a "modern" introduction to PDE. From time to time, however, my conscience has nagged me to make some revisions — to clean some things up, add more exercises, and include some material on pseudodifferential operators.

Meanwhile, in 1981 I gave another course in Fourier methods in PDE for the Programme in Applications of Mathematics at the Tata Institute for Fundamental Research in Bangalore, the notes for which were published in the Tata Lectures series under the title *Lectures on Partial Differential Equations*. They included applications of Fourier analysis to the study of constant coefficient equations (especially the Laplace, heat, and wave equations) and an introduction to pseudodifferential operators and Calderón-Zygmund singular integral operators. These notes were found useful by a number of people, but they went out of print after a few years.

Out of all this has emerged the present book. Its intended audience is the same as that of the first edition: students who are conversant with real analysis (the Lebesgue integral, L^p spaces, rudiments of Banach and Hilbert space theory), basic complex analysis (power series and contour integrals), and the big theorems of advanced calculus (the divergence theorem, the implicit function theorem, etc.). Its aim is also the same as that of the first edition: to present some basic classical results in a modern setting and to develop some aspects of the newer theory to a point where the student

will be equipped to read more advanced treatises. It consists essentially of
the union of the first edition and the Tata notes, with the omission of the
L^p theory of singular integrals (for which the reader is referred to Stein's
classic book [45]) and the addition of quite a few exercises.

Apart from the exercises, the main substantive changes from the first
edition to this one are as follows.

- §1F has been expanded to include the full Malgrange-Ehrenpreis theorem and the relation between smoothness of fundamental solutions and
  hypoellipticity, which simplifies the discussion at a few later points.

- Chapter 2 now begins with a brief new section on symmetry properties
  of the Laplacian.

- The discussion of the equation $\Delta u = f$ in §2C (formerly §2B) has been
  expanded to include the full Hölder regularity theorem (and, as a by-
  product, the continuity of singular integrals on Hölder spaces).

- The solution of the Dirichlet problem in a half-space (§2G) is now done
  in a way more closely related to the preceding sections, and the Fourier-
  analytic derivation has been moved to §4B.

- I have corrected a serious error in the treatment of the two-dimensional
  case in §3E. I am indebted to Leon Greenberg for sending me an analysis
  of the error and suggesting Proposition (3.36b) as a way to fix it.

- The discussion of functions of the Laplacian in the old §4A has been
  expanded and given its own section, §4B.

- Chapter 5 contains a new section (§5D) on the Fourier analysis of the
  wave equation.

- The first section of Chapter 6 has been split in two and expanded to
  include the interpolation theorem for operators on Sobolev spaces and
  the local coordinate invariance of Sobolev spaces.

- A new section (§6D) has been added to present Hörmander's characterization of hypoelliptic operators with constant coefficients.

- Chapter 8, on pseudodifferential operators, is entirely new.

In addition to these items, I have done a fair amount of rewriting in
order to improve the exposition. I have also made a few changes in notation
— most notably, the substitution of $\langle f \mid g \rangle$ for (f, g) to denote the Hermitian
inner product $\int f\overline{g}$, as distinguished from the bilinear pairing $\langle f, g \rangle = \int fg$.
(I have sworn off using parentheses, perhaps the most overworked symbols
in mathematics, to denote inner products.) I call the reader's attention to
the existence of an index of symbols as well as a regular index at the back
of the book.

The bias toward elliptic equations in the first edition is equally evident here. I feel a little guilty about not including more on hyperbolic equations, but that is a subject for another book by another author.

The discussions of elliptic regularity in §6C and §7F and of Gårding's inequality in §7D may look a little old-fashioned now, as the machinery of pseudodifferential operators has come to be accepted as the "right" way to obtain these results. Indeed, I rederive (and generalize) Gårding's inequality and the local regularity theorem by this method in §8F. However, I think the "low-tech" arguments in the earlier sections are also worth retaining. They provide the quickest proofs when one starts from scratch, and they show that the results are really of a fairly elementary nature.

I have revised and updated the bibliography, but it remains rather short and quite unscholarly. Wherever possible, I have preferred to give references to expository books and articles rather than to research papers, of which only a few are cited.

In the preface to the first edition I expressed my gratitude to my teachers J. J. Kohn and E. M. Stein, who influenced my point of view on much of the material contained therein. The same sentiment applies equally to the present work.

Gerald B. Folland
Seattle, March 1995

Chapter 0
PRELIMINARIES

The purpose of this chapter is to fix some terminology that will be used throughout the book, and to present a few analytical tools which are not included in the prerequisites. It is intended mainly as a reference rather than as a systematic text.

A. Notations and Definitions

Points and sets in Euclidean space

$\mathbb{R}$ will denote the real numbers, $\mathbb{C}$ the complex numbers. We will be working in $\mathbb{R}^n$, and n will always denote the dimension. Points in $\mathbb{R}^n$ will generally be denoted by x, y, ξ, η; the coordinates of x are $(x_1, \ldots, x_n)$. Occasionally $x_1, x_2, \ldots$ will denote a sequence of points in $\mathbb{R}^n$ rather than coordinates, but this will always be clear from the context. Once in a while there will be some confusion as to whether $(x_1, \ldots, x_n)$ denotes a point in $\mathbb{R}^n$ or the n-tuple of coordinate functions on $\mathbb{R}^n$. However, it would be too troublesome to adopt systematically a more precise notation; readers should consider themselves warned that this ambiguity will arise when we consider coordinate systems other than the standard one.

If U is a subset of $\mathbb{R}^n$, $\overline{U}$ will denote its closure and ∂U its boundary. The word **domain** will be used to mean an open set $\Omega \subset \mathbb{R}^n$, not necessarily connected, such that $\partial\Omega = \partial(\mathbb{R}^n \setminus \overline{\Omega})$. (That is, all the boundary points of Ω are "accessible from the outside.")

If x and y are points of $\mathbb{R}^n$ or $\mathbb{C}^n$, we set

$$x \cdot y = \sum_1^n x_j y_j,$$

so the Euclidean norm of x is given by

$$|x| = (x \cdot \bar{x})^{1/2} \quad (= (x \cdot x)^{1/2} \text{ if } x \text{ is real.})$$

We use the following notation for spheres and (open) balls: if $x \in \mathbb{R}^n$ and $r > 0$,

$$S_r(x) = \{y \in \mathbb{R}^n : |x - y| = r\},$$
$$B_r(x) = \{y \in \mathbb{R}^n : |x - y| < r\}.$$

Measures and integrals

The integral of a function f over a subset Ω of $\mathbb{R}^n$ with respect to Lebesgue measure will be denoted by $\int_\Omega f(x)\, dx$ or simply by $\int_\Omega f$. If no subscript occurs on the integral sign, the region of integration is understood to be $\mathbb{R}^n$. If S is a smooth hypersurface (see the next section), the natural Euclidean surface measure on S will be denoted by $d\sigma$; thus the integral of f over S is $\int_S f(x)\, d\sigma(x)$, or $\int_S f\, d\sigma$, or just $\int_S f$. The meaning of $d\sigma$ thus depends on S, but this will cause no confusion.

If f and g are functions whose product is integrable on $\mathbb{R}^n$, we shall sometimes write

$$\langle f, g \rangle = \int fg, \qquad \langle f \,|\, g \rangle = \int f\bar{g},$$

where $\bar{g}$ is the complex conjugate of g. The Hermitian pairing $\langle f|g \rangle$ will be used only when we are working with the Hilbert space L^2 or a variant of it, whereas the bilinear pairing $\langle f, g \rangle$ will be used more generally.

Multi-indices and derivatives

An n-tuple $\alpha = (\alpha_1, \ldots a_n)$ of nonnegative integers will be called a **multi-index**. We define

$$|\alpha| = \sum_1^n \alpha_j, \qquad \alpha! = \alpha_1! \alpha_2! \cdots \alpha_n!,$$

and for $x \in \mathbb{R}^n$,

$$x^\alpha = x_1^{\alpha_1} x_2^{\alpha_2} \cdots x_n^{\alpha_n}.$$

We will generally use the shorthand

$$\partial_j = \frac{\partial}{\partial x_j}$$

for derivatives on $\mathbb{R}^n$. Higher-order derivatives are then conveniently expressed by multi-indices:

$$\partial^\alpha = \prod_1^n \left(\frac{\partial}{\partial x_j}\right)^{\alpha_j} = \frac{\partial^{|\alpha|}}{\partial x_1^{\alpha_1} \cdots \partial x_n^{\alpha_n}}.$$

Note in particular that if $\alpha = 0$, ∂^α is the identity operator. With this notation, it would be natural to denote by ∂u the n-tuple of functions $(\partial_1 u, \ldots, \partial_n u)$ when u is a differentiable function; however, we shall use instead the more common notation

$$\nabla u = (\partial_1 u, \ldots, \partial_n u).$$

For our purposes, a **vector field** on a set $\Omega \in \mathbb{R}^n$ is simply an $\mathbb{R}^n$-valued function on Ω. If F is a vector field on an open set Ω, we define the directional derivative ∂_F by

$$\partial_F = F \cdot \nabla,$$

that is, if u is a differentiable function on Ω,

$$\partial_F u(x) = F(x) \cdot \nabla u(x) = \sum_1^n F_j(x) \partial_j u(x).$$

Function spaces

If Ω is a subset of $\mathbb{R}^n$, $C(\Omega)$ will dente the space of continuous complex-valued functions on Ω (with respect to the relative topology on Ω). If Ω is open and k is a positive integer, $C^k(\Omega)$ will denote the space of functions possessing continuous derivatives up to order k on Ω, and $C^k(\overline{\Omega})$ will denote the space of all $u \in C^k(\Omega)$ such that $\partial^\alpha u$ extends continuously to the closure $\overline{\Omega}$ for $0 \le |\alpha| \le k$. Also, we set $C^\infty(\Omega) = \bigcap_1^\infty C^k(\Omega)$ and $C^\infty(\overline{\Omega}) = \bigcap_1^\infty C^k(\overline{\Omega})$.

We next define the Hölder or Lipschitz spaces $C^\alpha(\Omega)$, where Ω is an open set and $0 < \alpha < 1$. (Here α is a real number, not a multi-index; the use of the letter "α" in both these contexts is standard.) $C^\alpha(\Omega)$ is the space of continuous functions on Ω that satisfy a locally uniform Hölder condition with exponent α. That is, $u \in C^\alpha(\Omega)$ if and only if for any compact $V \subset \Omega$ there is a constant $c > 0$ such that for all $y \in \mathbb{R}^n$ sufficiently close to 0,

$$\sup_{x \in V} |u(x + y) - u(x)| \le c|y|^\alpha.$$

(Note that $C^1(\Omega) \subset C^\alpha(\Omega)$ for all $\alpha < 1$, by the mean value theorem.) If k is a positive integer, $C^{k+\alpha}(\Omega)$ will denote the set of all $u \in C^k(\Omega)$ such that $\partial^\beta u \in C^\alpha(\Omega)$ for all muli-indices β with $|\beta| = k$ (or equivalently, with $|\beta| \leq k$; the lower-order derivatives are automatically in $C^1(\Omega) \subset C^\alpha(\Omega)$.).

The **support** of a function u, denoted by supp u, is the complement of the largest open set on which $u = 0$. If $\Omega \subset \mathbb{R}^n$, we denote by $C_c^\infty(\Omega)$ the space of all C^∞ functions on $\mathbb{R}^n$ whose support is compact and contained in Ω. (In particular, if Ω is open such functions vanish near $\partial\Omega$.)

The space $C^k(\mathbb{R}^n)$ will be denoted simply by C^k. Likewise for C^∞, $C^{k+\alpha}$, and C_c^∞.

If $\Omega \subset \mathbb{R}^n$ is open, a function $u \in C^\infty(\Omega)$ is said to be **analytic** in Ω if it can be expanded in a power series about every point of Ω. That is, u is analytic on Ω if for each $x \in \Omega$ there exists $r > 0$ such that for all $y \in B_r(x)$,

$$u(y) = \sum_{|\alpha| \geq 0} \frac{\partial^\alpha u(x)}{\alpha!}(y - x)^\alpha,$$

the series being absolutely and uniformly convergent on $B_r(x)$. When referring to complex-analytic functions, we shall always use the word **holomorphic**.

The **Schwartz class** $\mathcal{S} = \mathcal{S}(\mathbb{R}^n)$ is the space of all C^∞ functions on $\mathbb{R}^n$ which, together with all their derivatives, die out faster than any power of x at infinity. That is, $u \in \mathcal{S}$ if and only if $u \in C^\infty$ and for all multi-indices α and β,

$$\sup_{x \in \mathbb{R}^n} |x^\alpha \partial^\beta u(x)| < \infty.$$

Big O and little o

We occasionally employ the big and little o notation for orders of magnitude. Namely, when we are considering the behavior of functions in a neighborhood of a point a (which may be ∞), $O(f(x))$ denotes any function $g(x)$ such that $|g(x)| \leq C|f(x)|$ for x near a, and $o(f(x))$ denotes any function $h(x)$ such that $h(x)/f(x) \to 0$ as $x \to a$.

B. Results from Advanced Calculus

A subset S of $\mathbb{R}^n$ is called a **hypersurface of class** C^k ($1 \leq k \leq \infty$) if for every $x_0 \in S$ there is an open set $V \subset \mathbb{R}^n$ containing x_0 and a real-valued function $\phi \in C^k(V)$ such that $\nabla\phi$ is nonvanishing on $S \cap V$ and

$$S \cap V = \{x \in V : \phi(x) = 0\}.$$

In this case, by the implicit function theorem we can solve the equation $\phi(x) = 0$ near x_0 for some coordinate x_i — for convenience, say $i = n$ — to obtain

$$x_n = \psi(x_1, \ldots, x_{n-1})$$

for some C^k function ψ. A neighborhood of x_0 in S can then be mapped to a piece of the hyperplane $x_n = 0$ by the C^k transformation

$$x \rightarrow (x', x_n - \psi(x')) \qquad (x' = (x_1, \ldots, x_{n-1})).$$

This same neighborhood can also be represented in **parametric form** as the image of an open set in $\mathbb{R}^{n-1}$ (with coordinate x') under the map

$$x' \rightarrow (x', \psi(x')).$$

x' may be thought of as giving local coordinates on S near x_0.

Similar considerations apply if "C^k" is replaced by "analytic."

With S, V, ϕ as above, the vector $\nabla\phi(x)$ is perpendicular to S at x for every $x \in S \cap V$. We shall always suppose that S is **oriented**, that is, that we have made a choice of unit vector $\nu(x)$ for each $x \in S$, varying continuously with x, which is perpendicular to S at x. $\nu(x)$ will be called the **normal** to S at x; clearly on $S \cap V$ we have

$$\nu(x) = \pm\frac{\nabla\phi(x)}{|\nabla\phi(x)|}.$$

Thus ν is a C^{k-1} function on S. If S is the boundary of a domain Ω, we always choose the orientation so that ν points out of Ω.

If u is a differentiable function defined near S, we can then define the **normal derivative** of u on S by

$$\partial_\nu u = \nu \cdot \nabla u.$$

We pause to compute the normal derivative on the sphere $S_r(y)$. Since lines through the center of a sphere are perpendicular to the sphere, we have

$$(0.1) \qquad \nu(x) = \frac{x - y}{r}, \quad \partial_\nu = \frac{1}{r}\sum_1^n (x_j - y_j)\partial_j \qquad \text{on } S_r(y).$$

We will use the following proposition several times in the sequel:

(0.2) Proposition.
Let S be a compact oriented hypersurface of class C^k, $k \geq 2$. There is a neighborhood V of S in $\mathbb{R}^n$ and a number $\epsilon > 0$ such that the map

$$F(x,t) = x + t\nu(x)$$

is a C^{k-1} diffeomorphism of $S \times (-\epsilon, \epsilon)$ onto V.

Proof (sketch): F is clearly C^{k-1}. Moreover, for each $x \in S$ its Jacobian matrix (with respect to local coordinates on $S \times \mathbb{R}$) at $(x,0)$ is nonsingular since ν is normal to S. Hence by the inverse mapping theorem, F can be inverted on a neighborhood W_x of each $(x,0)$ to yield a C^{k-1} map

$$F_x^{-1} : W_x \longrightarrow (S \cap W_x) \times (-\epsilon_x, \epsilon_x)$$

for some $\epsilon_x > 0$. Since S is compact, we can choose $\{x_j\}_1^N \subset S$ such that the W_{x_j} cover S, and the maps $F_{x_j}^{-1}$ patch together to yield a C^{k-1} inverse of F from a neighborhood V of S to $S \times (-\epsilon, \epsilon)$ where $\epsilon = \min_j \epsilon_{x_j}$.   ∎

The neighborhood V in Proposition (0.2) is called a **tubular neighborhood** of S. It will be convenient to extend the definition of the normal derivative to the whole tubular neighborhood. Namely, if u is a differentiable function on V, for $x \in S$ and $-\epsilon < t < \epsilon$ we set

(0.3) $\partial_\nu u(x + t\nu(x)) = \nu(x) \cdot \nabla u(x + t\nu(x)).$

If $F = (F_1, \ldots, F_n)$ is a differentiable vector field on a subset of $\mathbb{R}^n$, its **divergence** is the function

$$\nabla \cdot F = \sum_1^n \partial_j F_j.$$

With this terminology, we can state the form of the general Stokes formula that we shall need.

(0.4) The Divergence Theorem.
Let $\Omega \subset \mathbb{R}^n$ be a bounded domain with C^1 boundary $S = \partial\Omega$, and let F be a C^1 vector field on $\overline{\Omega}$. Then

$$\int_S F(y) \cdot \nu(y) \, d\sigma(y) = \int_\Omega \nabla \cdot F(x) \, dx.$$

The proof can be found, for example, in Treves [52, §10].

Every $x \in \mathbb{R}^n \setminus \{0\}$ can be written uniquely as $x = ry$ with $r > 0$ and $y \in S_1(0)$ — namely, $r = |x|$ and $y = x/|x|$. The formula $x = ry$ is called the **polar coordinate** representation of x. Lebesgue measure is given in polar coordinates by

$$dx = r^{n-1} \, dr \, d\sigma(y),$$

where $d\sigma$ is surface measure on $S_1(0)$. (See Folland [14, Theorem (2.49)].)

For example, if $0 < a < b < \infty$ and $\lambda \in \mathbb{R}$, we have

$$\int_{a<|x|<b} |x|^\lambda \, dx = \int_{S_1(0)} \int_a^b r^{n-1+\lambda} \, dr = \begin{cases} \omega_n \dfrac{b^{n+\lambda} - a^{n+\lambda}}{n+\lambda} & \text{if } \lambda \neq -n, \\ \omega_n \log(b/a) & \text{if } \lambda = -n, \end{cases}$$

where ω_n is the area of $S_1(0)$ (which we shall compute shortly). As an immediate consequence, we have:

(0.5) Proposition.
The function $x \to |x|^\lambda$ is integrable on a neighborhood of 0 if and only if $\lambda > -n$, and it is integrable outside a neighborhood of 0 if and only if $\lambda < -n$.

As another application of polar coordinates, we can compute what is probably the most important definite integral in mathematics:

(0.6) Proposition.
$\int e^{-\pi|x|^2} \, dx = 1.$

Proof: Let $I_n = \int_{\mathbb{R}^n} e^{-\pi|x|^2} \, dx$. Since $e^{-\pi|x|^2} = \prod_1^n e^{-\pi x_j^2}$, Fubini's theorem shows that $I_n = (I_1)^n$, or equivalently that $I_n = (I_2)^{n/2}$. But in polar coordinates,

$$I_2 = \int_0^{2\pi} \int_0^\infty e^{-\pi r^2} r \, dr \, d\theta = 2\pi \int_0^\infty r e^{-r^2} \, dr = \pi \int_0^\infty e^{-\pi s} \, ds = 1. \quad \blacksquare$$

This trick works because we know that the measure of $S_1(0)$ in $\mathbb{R}^2$ is 2π. But now we can turn it around to compute the area ω_n of $S_1(0)$ in $\mathbb{R}^n$ for any n. Recall that the **gamma function** $\Gamma(s)$ is defined for $\operatorname{Re} s > 0$ by

$$\Gamma(s) = \int_0^\infty e^{-t} t^{s-1} \, dt.$$

One easily verifies that

$$\Gamma(s+1) = s\Gamma(s), \qquad \Gamma(1) = 1, \qquad \Gamma(\tfrac{1}{2}) = \sqrt{\pi}.$$

(The first formula is obtained by integration by parts, and the last one reduces to (0.6) by a change of variable.) Hence, if k is a positive integer,

$$\Gamma(k) = (k-1)!, \qquad \Gamma(k+\tfrac{1}{2}) = (k-\tfrac{1}{2})(k-\tfrac{3}{2})\cdots(\tfrac{1}{2})\sqrt{\pi}.$$

(0.7) Proposition.
The area of $S_1(0)$ in $\mathbb{R}^n$ is

$$\omega_n = \frac{2\pi^{n/2}}{\Gamma(n/2)}.$$

Proof: We integrate $e^{-\pi|x|^2}$ in polar coordinates and set $s = \pi r^2$:

$$1 = \int e^{-\pi|x|^2}\, dx = \int_{S_1(0)} \int_0^\infty e^{-\pi r^2} r^{n-1}\, dr\, d\sigma$$

$$= \omega_n \int_0^\infty e^{-\pi r^2} r^{n-1}\, dr = \frac{\omega_n}{2\pi^{n/2}} \int_0^\infty e^{-s} s^{(n/2)-1}\, ds$$

$$= \frac{\omega_n \Gamma(n/2)}{2\pi^{n/2}}. \qquad\qquad\blacksquare$$

Note that, despite appearances, ω_n is always a rational multiple of an integer power of π.

(0.8) Corollary.
The volume of $B_1(0)$ in $\mathbb{R}^n$ is

$$\frac{\omega_n}{n} = \frac{2\pi^{n/2}}{n\Gamma(n/2)}.$$

Proof: $\int_{B_1(0)} dx = \omega_n \int_0^1 r^{n-1}\, dr = \omega_n/n.$ $\blacksquare$

(0.9) Corollary.
For any $x \in \mathbb{R}^n$ and any $r > 0$, the area of $S_r(x)$ is $r^{n-1}\omega_n$ and the volume of $B_r(x)$ is $r^n \omega_n/n$.

C. Convolutions

We begin with a general theorem about integral operators on a measure space (X, μ) which deserves to be more widely known than it is. In our applications, X will be either $\mathbb{R}^n$ or a smooth hypersurface in $\mathbb{R}^n$.

(0.10) Generalized Young's Inequality.
Let (X, μ) be a σ-finite measure space, and let $1 \leq p \leq \infty$ and $C > 0$. Suppose K is a measurable function on $X \times X$ such that

$$\sup_{x \in X} \int_X |K(x, y)| \, d\mu(y) \leq C, \qquad \sup_{y \in X} \int_X |K(x, y)| \, d\mu(x) \leq C.$$

If $f \in L^p(X)$, the function Tf defined by

$$Tf(x) = \int_X K(x, y) f(y) \, d\mu(y)$$

is well-defined almost everywhere and is in $L^p(X)$, and $\|Tf\|_p \leq C\|f\|_p$.

Proof: Suppose $1 < p < \infty$, and let q be the conjugate exponent $(p^{-1} + q^{-1} = 1)$. Then by Hölder's inequality,

$$|Tf(x)| \leq \left[\int_X |K(x, y)| \, d\mu(y)\right]^{1/q} \left[\int_X |K(x, y)| |f(y)|^p \, d\mu(y)\right]^{1/p}$$

$$\leq C^{1/q} \left[\int_X |K(x, y)| |f(y)|^p \, d\mu(y)\right]^{1/p}.$$

Raising both sides to the p-th power and integrating, we see by Fubini's theorem that

$$\int_X |Tf(x)|^p \, d\mu(x) \leq C^{p/q} \int_X \int_X |K(x, y)| |f(y)|^p \, d\mu(y) \, d\mu(x)$$

$$\leq C^{(p/q)+1} \int_X |f(y)|^p \, d\mu(y),$$

or, taking pth roots,

$$\|Tf\|_p \leq C^{(1/p)+(1/q)} \|f\|_p = C\|f\|_p.$$

These estimates imply, in particular, that the integral defining $Tf(x)$ converges absolutely a.e., so the theorem is proved for the case $1 < p < \infty$. The case $p = 1$ is similar but easier and requires only the hypothesis $\int |K(x, y)| \, d\mu(x) \leq C$, and the case $p = \infty$ is trivial and requires only the hypothesis $\int |K(x, y)| \, d\mu(y) \leq C$.   ∎

In what follows, when we say L^p we shall mean $L^p(\mathbb{R}^n)$ unless another space is specified.

Let f and g be locally integrable functions on $\mathbb{R}^n$. The **convolution** $f * g$ of f and g is defined by

$$f * g(x) = \int f(x-y)g(y)\,dy = \int f(y)g(x-y)\,dy = g * f(x),$$

provided that the integrals in question exist. (The two integrals are equal by the change of variable $y \to x - y$.) The basic theorem on the existence of convolutions is the following:

(0.11) Young's Inequality.
If $f \in L^1$ and $g \in L^p$ $(1 \leq p \leq \infty)$, then $f * g \in L^p$ and $\|f*g\|_p \leq \|f\|_1\|g\|_p$.

Proof: Apply (0.10) with $X = \mathbb{R}^n$ and $K(x, y) = f(x - y)$.   ∎

Remark: It is obvious from Hölder's inequality that if $f \in L^q$ and $g \in L^p$ where $p^{-1} + q^{-1} = 1$ then $f * g \in L^\infty$ and $\|f * g\|_\infty \leq \|f\|_q\|g\|_p$. From the Riesz-Thorin interpolation theorem (see Folland [14]) one can then deduce the following generalization of Young's inequality: *Suppose $1 \leq p, q, r \leq \infty$ and $p^{-1} + q^{-1} = r^{-1} + 1$. If $f \in L^q$ and $g \in L^p$ then $f * g \in L^r$ and $\|f * g\|_r \leq \|f\|_q\|g\|_p$.*

The next theorem underlies one of the most important uses of convolutions. Before coming to it, we need a technical lemma. If f is a function on $\mathbb{R}^n$ and $x \in \mathbb{R}^n$, we define the function f_x by

$$f_x(y) = f(x + y).$$

(0.12) Lemma.
If $1 \leq p < \infty$ and $f \in L^p$, then $\lim_{x \to 0} \|f_x - f\|_p = 0$.

Proof: If g is continuous with compact support, then g is uniformly continuous, so $g_x \to g$ uniformly as $x \to 0$. Since g_x and g are supported in a common compact set for $|x| \leq 1$, it follows also that $\|g_x - g\|_p \to 0$. Now, given $f \in L^p$ and $\epsilon > 0$, choose a continuous g with compact support such that $\|f - g\|_p < \epsilon/3$. Then also $\|f_x - g_x\|_p < \epsilon/3$, so

$$\|f_x - f\|_p \leq \|f_x - g_x\|_p + \|g_x - g\|_p + \|g - f\|_p < \|g_x - g\|_p + 2\epsilon/3.$$

But for x sufficiently small, $\|g_x - g\|_p < \epsilon/3$, so $\|f_x - f\|_p < \epsilon$.   ∎

Remark: This result is false for $p = \infty$. Indeed, the condition that $\|f_x - f\|_\infty \to 0$ as $x \to 0$ means precisely that f agrees almost everywhere with a uniformly continuous function.

(0.13) Theorem.
Suppose $\phi \in L^1$ and $\int \phi(x)\, dx = a$. For each $\epsilon > 0$, define the function ϕ_ϵ by $\phi_\epsilon(x) = \epsilon^{-n}\phi(\epsilon^{-1}x)$. If $f \in L^p$, $1 \le p < \infty$, then $f * \phi_\epsilon \to af$ in the L^p norm as $\epsilon \to 0$. If $f \in L^\infty$ and f is uniformly continuous on a set V, then $f * \phi_\epsilon \to af$ uniformly on V as $\epsilon \to 0$.

Proof: By the change of variable $x \to \epsilon x$ we see that $\int \phi_\epsilon(x)\, dx = a$ for all $\epsilon > 0$. Hence,

$$f * \phi_\epsilon(x) - af(x) = \int [f(x-y) - f(x)]\phi_\epsilon(y)\, dy = \int [f(x-\epsilon y) - f(x)]\phi(y)\, dy.$$

If $f \in L^p$ and $p < \infty$, we apply the triangle inequality for integrals (Minkowski's inequality; see Folland [14]) to obtain

$$\|f * \phi_\epsilon - af\|_p \le \int \|f_{-\epsilon y} - f\|_p |\phi(y)|\, dy.$$

But $\|f_{-\epsilon y} - f\|_p$ is bounded by $2\|f\|_p$ and tends to zero as $\epsilon \to 0$ for each y, by Lemma (0.12). The desired result therefore follows from the dominated convergence theorem.

On the other hand, suppose $f \in L^\infty$ and f is uniformly continuous on V. Given $\delta > 0$, choose a compact set W so that $\int_{\mathbb{R}^n \setminus W} |\phi| < \delta$. Then

$$\sup_{x \in V} |f * \phi_\epsilon(x) - af(x)| \le \sup_{x \in V,\, y \in W} |f(x - \epsilon y) - f(x)| \int_W |\phi| + 2\|f\|_\infty \delta.$$

The first term on the right tends to zero as $\epsilon \to 0$, and δ is arbitrary, so $f * \phi_\epsilon$ tends uniformly to af on V. ∎

If $\phi \in L^1$ and $\int \phi(x)\, dx = 1$, the family of functions $\{\phi_\epsilon\}_{\epsilon > 0}$ defined in Theorem (0.13) is called an **approximation to the identity**. What makes these useful is that by choosing ϕ appropriately we can get the functions $f * \phi_\epsilon$ to have nice properties. In particular:

(0.14) Theorem.
If $f \in L^p$ $(1 \le p \le \infty)$ and ϕ is in the Schwartz class $\mathcal{S}$, then $f * \phi$ is C^∞ and $\partial^\alpha(f * \phi) = f * \partial^\alpha \phi$ for all multi-indices α.

Proof: If $\phi \in \mathcal{S}$, for every bounded set $V \subset \mathbb{R}^n$ we have

$$\sup_{x \in V} |\partial^\alpha \phi(x - y)| \le C_{\alpha,V}(1 + |y|)^{-n-1} \qquad (y \in \mathbb{R}^n).$$

The function $(1 + |y|)^{-n-1}$ is in L^q for every q by (0.5), so the integral

$$f * \partial^\alpha \phi(x) = \int f(y)\partial^\alpha \phi(x - y)\, dy$$

converges absolutely and uniformly on bounded subsets of $\mathbb{R}^n$. Differentiation can thus be interchanged with integration, and we conclude that $\partial^\alpha (f * \phi) = f * \partial^\alpha \phi$. ∎

We can get better results by taking $\phi \in C^\infty$. In that case we need only assume that f is locally integrable for $f * \phi$ to be well-defined, and the same argument as above shows that $f * \phi \in C^\infty$.

Since the existence of nonzero functions in C_c^∞ is not completely trivial, we pause for a moment to construct some. First, we define the function f on $\mathbb{R}$ by

$$f(t) = \begin{cases} e^{1/(1-t^2)} & (|t| < 1), \\ 0 & (|t| \ge 1). \end{cases}$$

Then $f \in C_c^\infty(\mathbb{R})$, so $\psi(x) = f(|x|^2)$ is a nonnegative C^∞ function on $\mathbb{R}^n$ whose support is $\overline{B_1(0)}$. In particular, $\int \psi > 0$, so $\phi = \psi / \int \psi$ is a function in $C_c^\infty(\mathbb{R}^n)$ with $\int \phi = 1$. It now follows that there are lots of functions in C_c^∞:

(0.15) Lemma.
If f is supported in V and g is supported in W, then $f * g$ is supported in $\{x + y : x \in V, \ y \in W\}$.

Proof: Exercise. ∎

(0.16) Theorem.
C_c^∞ is dense in L^p for $1 \le p < \infty$.

Proof: Choose $\phi \in C_c^\infty$ with $\int \phi = 1$, and define ϕ_ϵ as in Theorem (0.13). If $f \in L^p$ has compact support, it follows from (0.14) and (0.15) that $f * \phi_\epsilon \in C_c^\infty$ and from (0.13) that $f * \phi_\epsilon \to f$ in the L^p norm. But L^p functions with compact support are dense in L^p, so we are done. ∎

Another useful construction is the following:

(0.17) Theorem.
Suppose $V \subset \mathbb{R}^n$ is compact, $\Omega \subset \mathbb{R}^n$ is open, and $V \subset \Omega$. Then there exists $f \in C_c^\infty(\Omega)$ such that $f = 1$ on V and $0 \le f \le 1$ everywhere.

Proof: Let $\delta = \inf\{|x - y| : x \in V, \, y \notin \Omega\}$. (If $\Omega = \mathbb{R}^n$, let $\delta = 1$.) By our assumptions on V and Ω, $\delta > 0$. Let

$$U = \left\{ x : |x - y| < \tfrac{1}{2}\delta \text{ for some } y \in V \right\}.$$

Then $V \subset U$ and $\overline{U} \subset \Omega$. Let χ be the characteristic function of U, and choose a nonnegative $\phi \in C_c^\infty(B_{\delta/2}(0))$ such that $\int \phi = 1$. Then we can take $f = \chi * \phi$; the simple verification is left to the reader.   ∎

We can now prove the existence of "partitions of unity." We state the following results only for compact sets, which is all we need, but they can be generalized.

(0.18) Lemma.
Let $K \subset \mathbb{R}^n$ be compact and let $V_1, \ldots, V_N$ be open sets with $K \subset \bigcup_1^N V_j$. Then there exist open sets $W_1, \ldots, W_N$ with $\overline{W}_j \subset V_j$ and $K \subset \bigcup_1^N W_j$.

Proof: For each $\epsilon > 0$ let V_j^ϵ be the set of points in V_j whose distance from $\mathbb{R}^n \setminus V_j$ is greater than ϵ. Clearly V_j^ϵ is open and $\overline{V_j^\epsilon} \subset V_j$. We claim that $K \subset \bigcup_1^N V_j^\epsilon$ if ϵ is sufficiently small. Otherwise, for each $\epsilon > 0$ there exists $x_\epsilon \in K \setminus \bigcup_1^N V_j^\epsilon$. Since K is compact, the x_ϵ have an accumulation point $x \in K$ as $\epsilon \to 0$. But then $x \in K \setminus \bigcup_1^N V_j$, which is absurd.   ∎

(0.19) Theorem.
Let $K \subset \mathbb{R}^n$ be compact and let $V_1, \ldots, V_N$ be bounded open sets such that $K \subset \bigcup_1^N V_j$. Then there exist functions $\zeta_1, \ldots, \zeta_n$ with $\zeta_j \in C_c^\infty(V_j)$ such that $\sum_1^N \zeta_j = 1$ on K.

Proof: Let $W_1, \ldots, W_N$ be as in Lemma (0.18). By Theorem (0.17), we can choose $\phi_j \in C_c^\infty(V_j)$ with $0 \le \phi_j \le 1$ and $\phi_j = 1$ on $\overline{W}_j$. Then $\Phi = \sum_1^N \phi_j \ge 1$ on K, so we can take $\zeta_j = \phi_j/\Phi$, with the understanding that $\zeta_j = 0$ wherever $\phi_j = 0$.   ∎

The collection of functions $\{\zeta_j\}_1^N$ is called a **partition of unity on** K **subordinate to the covering** $\{V_j\}_1^N$.

D. The Fourier Transform

In this section we give a rapid introduction to the theory of the Fourier transform. For a more extensive discussion, see, e.g., Strichartz [47] or Folland [14], [17].

If $f \in L^1(\mathbb{R}^n)$, its **Fourier transform** $\hat{f}$ is a bounded function on $\mathbb{R}^n$ defined by

$$\hat{f}(\xi) = \int e^{-2\pi i x \cdot \xi} f(x)\, dx.$$

There is no universal agreement as to where to put the factors of 2π in the definition of $\hat{f}$, and we apologize if this definition is not the one the reader is used to. It has the advantage of making the Fourier transform both an isometry on L^2 and an algebra homomorphism from L^1 (with convolution) to L^∞ (with pointwise multiplication).

Clearly $\hat{f}(\xi)$ is well-defined for all ξ and $\|\hat{f}\|_\infty \leq \|f\|_1$. Moreover:

(0.20) Theorem.
If $f, g \in L^1$ then $(f * g)\hat{} = \hat{f}\hat{g}$.

Proof: This is a simple application of Fubini's theorem:

$$(f * g)\hat{}(\xi) = \iint e^{-2\pi i x \cdot \xi} f(x - y) g(y)\, dy\, dx$$

$$= \iint e^{-2\pi i (x-y)\cdot\xi} f(x-y) e^{-2\pi i y \cdot \xi} g(y)\, dx\, dy$$

$$= \hat{f}(\xi) \int e^{-2\pi i y \cdot \xi} g(y)\, dy = \hat{f}(\xi)\hat{g}(\xi). \qquad \blacksquare$$

The Fourier transform interacts in a simple way with composition by translations and linear maps:

(0.21) Proposition.
Suppose $f \in L^1(\mathbb{R}^n)$.
  a. If $f_a(x) = f(x + a)$ then $(f_a)\hat{}(\xi) = e^{2\pi i a \cdot \xi} \hat{f}(\xi)$.
  b. If T is an invertible linear transformation of $\mathbb{R}^n$, then $(f \circ T)\hat{}(\xi) = |\det T|^{-1} \hat{f}((T^{-1})^*\xi)$.
  c. If T is a rotation of $\mathbb{R}^n$, then $(f \circ T)\hat{} = \hat{f} \circ T$.

Proof: (a) and (b) are easily proved by making the substitutions $y = x + a$ and $y = Tx$ in the integrals defining $(f_a)\hat{}(\xi)$ and $(f \circ T)\hat{}(\xi)$, respectively. (c) follows from (b) since $T^* = T^{-1}$ and $|\det T| = 1$ when T is a rotation. $\qquad \blacksquare$

The easiest way to develop the other basic properties of the Fourier transform is to consider its restriction to the Schwartz class $\mathcal{S}$. In what follows, if α is a multi-index, $x^\alpha f$ denotes the function whose value at x is $x^\alpha f(x)$.

(0.22) Proposition.
Suppose $f \in \mathcal{S}$.
  a. $\widehat{f} \in C^\infty$ and $\partial^\beta \widehat{f} = [(-2\pi i x)^\beta f]\widehat{}$.
  b. $(\partial^\beta f)\widehat{} = (2\pi i \xi)^\beta \widehat{f}$.

Proof: To prove (a), just differentiate under the integral sign. To prove (b), write out the integral for $(\partial^\beta f)\widehat{}(\xi)$ and integrate by parts; the boundary terms vanish since f and its derivatives vanish at infinity. ∎

(0.23) Proposition.
If $f \in \mathcal{S}$ then $\widehat{f} \in \mathcal{S}$.

Proof: By Proposition (0.22),

$$\partial^\beta \xi^\alpha \widehat{f} = (-1)^{|\beta|} (2\pi i)^{|\beta|-|\alpha|} [x^\beta \partial^\alpha f]\widehat{},$$

so $\partial^\beta \xi^\alpha \widehat{f}$ is bounded for all α, β. It then follows by the product rule for derivatives and induction on β that $\xi^\alpha \partial^\beta \widehat{f}$ is bounded for all α, β, that is, $\widehat{f} \in \mathcal{S}$. ∎

(0.24) The Riemann-Lebesgue Lemma.
If $f \in L^1$ then $\widehat{f}$ is continuous and tends to zero at infinity.

Proof: This is true by Proposition (0.23) if f lies in the dense subspace $\mathcal{S}$ of L^1. But if $\{f_j\} \subset \mathcal{S}$ and $f_j \to f$ in L^1, then $\widehat{f_j} \to \widehat{f}$ uniformly (because $\|\widehat{f_j} - \widehat{f}\|_\infty \le \|f_j - f\|_1$), and the result follows immediately. ∎

(0.25) Theorem.
Let $f(x) = e^{-\pi a |x|^2}$ where $a > 0$. Then

$$\widehat{f}(\xi) = a^{-n/2} e^{-\pi |\xi|^2 / a}.$$

Proof: By making the change of variable $x \to a^{-1/2} x$ we may assume $a = 1$. Since the exponential function converts sums into products, by Fubini's theorem we have

$$\widehat{f}(\xi) = \int e^{-2\pi i x \cdot \xi - \pi |x|^2} \, dx = \prod_1^n \int e^{-2\pi i x_j \xi_j - \pi x_j^2} \, dx_j,$$

and it suffices to show that the jth factor in the product is $e^{-\pi\xi_j^2}$, i.e., to prove the theorem for $n = 1$. Now when $n = 1$ we have

$$\int e^{-2\pi i x \xi - \pi x^2}\, dx = e^{-\pi\xi^2} \int e^{-\pi(x+i\xi)^2}\, dx.$$

But $f(z) = e^{-\pi z^2}$ is an entire holomorphic function of $z \in \mathbb{C}$ which dies out rapidly as $|\operatorname{Re} z| \to \infty$ when $|\operatorname{Im} z|$ remains bounded. Hence by Cauchy's theorem we can shift the contour of integration from $\operatorname{Im} z = 0$ to $\operatorname{Im} z = -\xi$, which together with (0.6) yields

$$e^{-\pi\xi^2} \int e^{-\pi(x+i\xi)^2}\, dx = e^{-\pi\xi^2} \int e^{-\pi x^2}\, dx = e^{-\pi\xi^2}. \qquad \blacksquare$$

(0.26) Theorem.
If $f, g \in \mathcal{S}$ then $\int f\hat{g} = \int \hat{f}g$.

Proof: By Fubini's theorem,

$$\int f\hat{g} = \iint f(x)g(y)e^{-2\pi i x \cdot y}\, dy\, dx = \int \hat{f}g. \qquad \blacksquare$$

For $f \in L^1$, define the function $f^\vee$ by

$$f^\vee(x) = \int e^{2\pi i x \cdot \xi} f(\xi)\, d\xi = \hat{f}(-x).$$

(0.27) The Fourier Inversion Theorem.
If $f \in \mathcal{S}$, $(\hat{f})^\vee = f$.

Proof: Given $\epsilon > 0$ and $x \in \mathbb{R}^n$, set $\phi(\xi) = e^{2\pi i x \cdot \xi - \pi\epsilon^2 |\xi|^2}$. Then by Theorem (0.25),

$$\hat{\phi}(y) = \int e^{-2\pi i(y-x)\cdot\xi} e^{-\pi\epsilon^2|\xi|^2}\, d\xi = \epsilon^{-n} e^{-\pi|x-y|^2/\epsilon^2}.$$

Thus,

$$\hat{\phi}(y) = \epsilon^{-n} g(\epsilon^{-1}(x-y)) = g_\epsilon(x-y) \text{ where } g(x) = e^{-\pi|x|^2}.$$

By (0.26), then,

$$\int e^{-\pi\epsilon^2|\xi|^2} e^{2\pi i x \cdot \xi} \hat{f}(\xi)\, d\xi = \int \hat{f}\phi = \int f\hat{\phi} = \int f(x)g_\epsilon(x-y)\, dy = f * g_\epsilon(x).$$

By (0.6) and (0.14), $f * g_\epsilon \to f$ uniformly as $\epsilon \to 0$ since functions in $\mathcal{S}$ are uniformly continuous. But clearly, for each x,

$$\int e^{-\pi\epsilon^2|\xi|^2} e^{2\pi i x \cdot \xi} \hat{f}(\xi)\, d\xi \to \int e^{2\pi i x \cdot \xi} \hat{f}(\xi)\, d\xi = (\hat{f})^\vee(x). \qquad \blacksquare$$

(0.28) Corollary.
The Fourier transform is an isomorphism of $\mathcal{S}$ onto itself.

(0.29) The Plancherel Theorem.
The Fourier transform on $\mathcal{S}$ extends uniquely to a unitary isomorphism of L^2 onto itself.

Proof: Since $\mathcal{S}$ is dense in L^2 (Theorem (0.16)), by Corollary (0.28) it suffices to show that $\|\widehat{f}\|_2 = \|f\|_2$ for $f \in \mathcal{S}$. If $f \in \mathcal{S}$, set $g(x) = \overline{f(-x)}$. One easily checks that $\widehat{g} = \overline{\widehat{f}}$. Hence by Theorems (0.20) and (0.27),

$$\|f\|_2^2 = \int f(x)\overline{f(x)}\,dx = f * g(0) = \int (f*g)\check{}(\xi)\,d\xi = \int \widehat{f}(\xi)\overline{\widehat{f}(\xi)}\,d\xi$$
$$= \|\widehat{f}\|_2^2. \qquad\blacksquare$$

The results (0.20)–(0.29) are the fundamental properties of the Fourier transform which we shall use repeatedly. We shall also sometimes need the Fourier transform as an operator on tempered distributions, to be discussed in the next section, and the following result.

(0.30) Proposition.
If $f \in L^1$ has compact support, then $\widehat{f}$ extends to an entire holomorphic function on $\mathbb{C}^n$. If $f \in C_c^\infty$, then $\widehat{f}(\xi)$ is rapidly decaying as $|\operatorname{Re}\xi| \to \infty$ when $|\operatorname{Im}\xi|$ remains bounded.

Proof: The integral $\widehat{f}(\xi) = \int e^{-2\pi i x \cdot \xi} f(x)\,dx$ converges for every $\xi \in \mathbb{C}^n$, and $e^{-2\pi i x \cdot \xi}$ is an entire function of $\xi \in \mathbb{C}^n$. Hence one can take complex derivatives of $\widehat{f}$ simply by differentiating under the integral. Moreover, if $f \in C_c^\infty$ and f is supported in $\{x : |x| \le K\}$, for any multi-index α we have

$$|(2\pi i\xi)^\alpha \widehat{f}(\xi)| = \left| \int e^{-2\pi i x \cdot \xi}\partial^\alpha f(x)\,dx \right| \le e^{K|\operatorname{Im}\xi|}\|\partial^\alpha f\|_1,$$

which yields the second assertion. $\blacksquare$

E. Distributions

We now outline the elements of the theory of distributions. The material sketched here is covered in more detail in Folland [14] and Rudin [41], and a more extensive treatment at an elementary level can be found in Strichartz [47]. See also Treves [49] and Hörmander [27, vol. I] for a deeper study of distributions.

Let Ω be an open set in $\mathbb{R}^n$. We begin by defining a notion of sequential convergence in $C_c^\infty(\Omega)$. Namely, we say that $\phi_j \to \phi$ in $C_c^\infty(\Omega)$ if the ϕ_j's are all supported in a common compact subset of Ω and $\partial^\alpha \phi_j \to \partial^\alpha \phi$ uniformly for every multi-index α. (This notion of convergence comes from a locally convex topology on $C_c^\infty(\Omega)$, whose precise description we shall not need. See Rudin [41] or Treves [49].)

If u is a linear functional on the space $C_c^\infty(\Omega)$, we denote the number obtained by applying u to $\phi \in C_c^\infty(\Omega)$ by $\langle u, \phi \rangle$ (or sometimes by $\langle \phi, u \rangle$: it is convenient to maintain this flexibility). A **distribution** on Ω is a linear functional u on $C_c^\infty(\Omega)$ that is continuous in the sense that if $\phi_j \to \phi$ in $C_c^\infty(\Omega)$ then $\langle u, \phi_j \rangle \to \langle u, \phi \rangle$. A bit of functional analysis (cf. Folland [14, Prop. (5.15)]) shows that this notion of continuity is equivalent to the following condition: for every compact set $K \subset \Omega$ there is a constant C_K and an integer N_K such that for all $\phi \in C_c^\infty(K)$,

$$(0.31) \qquad |\langle u, \phi \rangle| \le C_K \sum_{|\alpha| \le N_K} \|\partial^\alpha \phi\|_\infty.$$

The space of distributions on Ω is denoted by $\mathcal{D}'(\Omega)$, and we set $\mathcal{D}' = \mathcal{D}'(\mathbb{R}^n)$. We put the weak topology on $\mathcal{D}'(\Omega)$; that is, $u_j \to u$ in $\mathcal{D}'(\Omega)$ if and only if $\langle u_j, \phi \rangle \to \langle u, \phi \rangle$ for every $\phi \in C_c^\infty(\Omega)$.

Every locally integrable function u on Ω can be regarded as a distribution by the formula $\langle u, \phi \rangle = \int u\phi$, which accords with the notation introduced earlier. (The continuity follows from the Lebesgue dominated convergence theorem.) This correspondence is one-to-one if we regard two functions as the same if they are equal almost everywhere. Thus distributions can be regarded as "generalized functions." Indeed, we shall often pretend that distributions are functions and write $\langle u, \phi \rangle$ as $\int u(x)\phi(x)\,dx$; this is a useful fiction that makes certain operations involving distributions more transparent.

Every locally finite measure μ on Ω defines a distribution by the formula $\langle \mu, \phi \rangle = \int \phi\,d\mu$. In particular, if we take μ to be the point mass at 0, we obtain the graddaddy of all distributions, the **Dirac δ-function** $\delta \in \mathcal{D}'$ defined by $\langle \delta, \phi \rangle = \phi(0)$. Theorem (0.13) implies that if $u \in L^1$, $\int u = a$, and $u_\epsilon(x) = \epsilon^{-n} u(\epsilon^{-1} x)$, then $u_\epsilon \to a\delta$ in $\mathcal{D}'$ when $\epsilon \to 0$.

If $u, v \in \mathcal{D}'(\Omega)$, we say that $u = v$ on an open set $V \subset \Omega$ if $\langle u, \phi \rangle = \langle v, \phi \rangle$ for all $\phi \in C_c^\infty(V)$. The **support** of a distribution u is the complement of the largest open set on which $u = 0$. (To see that this is well-defined, one needs to know that if $\{V_\alpha\}_{\alpha \in A}$ is a collection of open sets and $u = 0$ on each V_α, then $u = 0$ on $\bigcup V_\alpha$. But if $\phi \in C_c^\infty(\bigcup V_\alpha)$, supp ϕ is covered by finitely many V_α's. By means of a partition of unity on supp ϕ subordinate to this covering, one can write $\phi = \sum_1^N \phi_j$ where each ϕ_j is supported in some V_α. It follows that $\langle u, \phi \rangle = \sum \langle u, \phi_j \rangle = 0$, as desired.)

The space of distributions on $\mathbb{R}^n$ whose support is a compact subset of the open set Ω is denoted by $\mathcal{E}'(\Omega)$, and we set $\mathcal{E}' = \mathcal{E}'(\mathbb{R}^n)$.

Suppose $u \in \mathcal{E}'$. Let Ω be a bounded open set such that supp $u \subset \Omega$, and choose $\psi \in C_c^\infty(\Omega)$ with $\psi = 1$ on a neighborhood of supp u (by Theorem (0.17)). Then for any $\phi \in C_c^\infty$ we have

$$\langle u, \phi \rangle = \langle u, \psi\phi \rangle.$$

This has two consequences. First, u is of "finite order": indeed, by (0.31) with $K = \overline{\Omega}$,

$$|\langle u, \phi \rangle| \leq C_{\overline{\Omega}} \sum_{|\alpha| \leq N_{\overline{\Omega}}} \|\partial^\alpha(\psi\phi)\|_\infty.$$

Expanding $\partial^\alpha(\psi\phi)$ by the product rule, we see that

(0.32)
$$|\langle u, \phi \rangle| \leq C \sum_{|\alpha| \leq N} \sup_{x \in \Omega} |\partial^\alpha \phi(x)|,$$

where $N = N_{\overline{\Omega}}$ and C depends only on $C_{\overline{\Omega}}$ and the constants $\|\partial^\beta \psi\|_\infty$, $|\beta| \leq N$. Second, $\langle u, \psi\phi \rangle$ makes sense for all $\phi \in C^\infty$, compactly supported or not, so if we define $\langle u, \phi \rangle$ to be $\langle u, \psi\phi \rangle$ for all $\phi \in C^\infty$, we have an extension of u to a linear functional on C^∞. This extension is clearly independent of the choice of ψ, and it is unique subject to the condition that $\langle u, \phi \rangle = 0$ whenever supp ϕ and supp u are disjoint. Thus distributions with compact support can be regarded as linear functionals on C^∞ that satisfy estimates of the form (0.32). Conversely, the restriction to C_c^∞ of any linear functional on C^∞ satisfying (0.32) is clearly a distribution supported in $\overline{\Omega}$.

The general philosophy for extending operations from functions to distributions is the following. Let T be a linear operator on $C_c^\infty(\Omega)$ that is continuous in the sense that if $\phi_j \to \phi$ in $C_c^\infty(\Omega)$ then $T\phi_j \to T\phi$ in $C_c^\infty(\Omega)$. Suppose there is another such operator T' such that $\int (T\phi)\psi = \int \phi(T'\psi)$ for all $\phi, \psi \in C_c^\infty(\Omega)$. (We call T' the **dual** or **transpose** of T.) We can then extend T to act on distributions by the formula

$$\langle Tu, \phi \rangle = \langle u, T'\phi \rangle.$$

The linear functional Tu on $C_c^\infty(\Omega)$ defined in this way is continuous on $C_c^\infty(\Omega)$ since T' is assumed continuous. The most important examples are the following; in all of them the verification of continuity is left as a simple exercise.

1. Let T be multiplication by the function $f \in C^\infty(\Omega)$. Then $T' = T$, so we can multiply any distribution u by $f \in C^\infty(\Omega)$ by the formula $\langle fu, \phi \rangle = \langle u, f\phi \rangle$.

2. Let $T = \partial^\alpha$. By integration by parts, $T' = (-1)^{|\alpha|}\partial^\alpha$. Hence we can differentiate any distribution as often as we please to obtain other distributions by the formula $\langle \partial^\alpha u, \phi \rangle = (-1)^{|\alpha|}\langle u, \partial^\alpha \phi \rangle$.

3. We can combine (1) and (2). Let $T = \sum_{|\alpha| \le k} a_\alpha \partial^\alpha$ be a differential operator of order k with C^∞ coefficients a_α. Integration by parts shows that the dual operator T' is given by $T'\phi = \sum_{|\alpha| \le k}(-1)^{|\alpha|}\partial^\alpha(a_\alpha \phi)$. For any distribution u, then, we define Tu by $\langle Tu, \phi \rangle = \langle u, T'\phi \rangle$.

Clearly, if $u \in C^k(\Omega)$, the distribution derivatives of u of order $\le k$ are just the pointwise derivatives. The converse is also true:

(0.33) Proposition.
If $u \in C(\Omega)$ and the distribution derivatives $\partial^\alpha u$ are in $C(\Omega)$ for $|\alpha| \le k$ then $u \in C^k(\Omega)$.

Proof: By induction it suffices to assume that $k = 1$. Since the conclusion is of a local nature, moreover, it suffices to assume that Ω is a cube, say $\Omega = \{x : \max |x_j - y_j| \le r\}$ for some $y \in \mathbb{R}^n$. For $x \in \Omega$, set

$$v(x) = \int_{y_1}^{x_1} \partial_1 u(t, x_2, \ldots, x_n)\, dt + u(y_1, x_2, \ldots, x_n).$$

It is easily checked that v and u agree as distributions on Ω, hence $v = u$ as functions on Ω. But $\partial_1 u$ is clearly a pointwise derivative of v. Likewise for $\partial_2 u, \ldots, \partial_n u$; thus $u \in C^1(\Omega)$.   ∎

We now continue our list of operations on distributions. In all of the following, we take $\Omega = \mathbb{R}^n$.

4. Given $x \in \mathbb{R}^n$, let $T\phi = \phi_x$, where $\phi_x(y) = \phi(x + y)$. Then $T'\phi = \phi_{-x}$. Thus for any distribution u, we define its translate u_x by $\langle u_x, \phi \rangle = \langle u, \phi_{-x} \rangle$.

5. Let $T\phi = \tilde{\phi}$, where $\tilde{\phi}(x) = \phi(-x)$. Then $T' = T$, so for any distribution u we define its reflection in the origin $\tilde{u}$ by $\langle \tilde{u}, \phi \rangle = \langle u, \tilde{\phi} \rangle$.

6. Given $\psi \in C_c^\infty$, define $T\phi = \phi * \psi$, which is in C_c^∞ by (0.14) and (0.15). It is easy to check that $T'\phi = \phi * \tilde{\psi}$, where $\tilde{\psi}$ is defined as in (5). Thus, if u is a distribution, we can define the distribution $u * \psi$ by $\langle u * \psi, \phi \rangle = \langle u, \phi * \tilde{\psi} \rangle$. On the other hand, notice that $\phi * \psi(x) = \langle \phi, (\psi_x)^\smallfrown \rangle$, so we can also define $u * \psi$ pointwise as a continuous function by $u * \psi(x) = \langle u, (\psi_x)^\smallfrown \rangle$.

In fact, these two definitions agree. To see this, let $\phi \in C_c^\infty$, let K be a compact set containing $\text{supp}(\psi_x)^\smallfrown$ for all $x \in \text{supp}\,\phi$, and let N_K be as in (0.31). From the relation $\phi * \tilde{\psi}(y) = \int \phi(x)(\psi_x)^\smallfrown(y)\,dx$ it is is not hard to see that there is a sequence of Riemann sums $\sum \phi(x_j)(\psi_{x_j})^\smallfrown \Delta x_j$ that converge uniformly to $\phi * \tilde{\psi}$ together with their derivatives of order $\le N_K$. But then (0.31) implies that if $u * \psi$ is defined as a continuous function, we have

$$\langle u * \psi, \phi \rangle = \lim \sum u * \psi(x_j)\phi(x_j)\,\Delta x_j$$
$$= \lim \sum \langle u, (\psi_{x_j})^\smallfrown \rangle \phi(x_j)\,\Delta x_j = \langle u, \phi * \tilde{\psi} \rangle,$$

which is the action of the distribution $u * \psi$ on ϕ.

Moreover, by (2) and integration by parts, we see that the distribution $\partial^\alpha(u * \psi)$ is given by

$$\langle \partial^\alpha(u * \psi), \phi \rangle = (-1)^{|\alpha|}\langle u, (\partial^\alpha \phi) * \tilde{\psi} \rangle = \langle u, \phi * (\partial^\alpha \psi)^\smallfrown \rangle = \langle u * \partial^\alpha \psi, \phi \rangle,$$

so $\partial^\alpha(u * \psi) = u * \partial^\alpha \psi$ is a continuous function. Hence $u * \psi$ is actually a C^∞ function.

7. The same considerations apply when $u \in \mathcal{E}'$ and $\psi \in C^\infty$. That is, we can define $u * \psi$ either as a distribution by $\langle u * \psi, \phi \rangle = \langle u, \phi * \tilde{\psi} \rangle$, or as a C^∞ function by $u * \psi(x) = \langle u, (\psi_x)^\smallfrown \rangle$.

8. If $u \in \mathcal{E}'$ and $\psi \in C^\infty$, as in (0.15) we see that $u * \psi \in C_c^\infty$. Hence we can consider the operator $T\psi = u * \psi$ on C^∞, whose dual is clearly $T'\psi = \tilde{u} * \psi$. It follows that if $u \in \mathcal{E}'$ and $v \in \mathcal{D}'$, $u * v$ can be defined as a distribution by the formula $\langle u * v, \psi \rangle = \langle v, \tilde{u} * \psi \rangle$. We leave it as an exercise to verify that for any multi-index α we have $\partial^\alpha(u * v) = (\partial^\alpha u) * v = u * (\partial^\alpha v)$.

We shall also need to consider the class of "tempered distributions." We endow the Schwartz class $\mathcal{S}$ with the Fréchet space topology defined by the family of norms $\|\phi\|_{(\alpha,\beta)} = \|x^\alpha \partial^\beta \phi\|_\infty$. That is, $\phi_j \to \phi$ in $\mathcal{S}$ if and only if

$$\sup_x |x^\alpha \partial^\beta (\phi_j - \phi)(x)| \to 0 \quad \text{for all } \alpha, \beta.$$

A **tempered distribution** is a continuous linear functional on $\mathcal{S}$; the space of tempered distributions is denoted by $\mathcal{S}'$. Since C_c^∞ is a dense subspace of $\mathcal{S}$ in the topology of $\mathcal{S}$, and the topology on C_c^∞ is stronger than the topology on $\mathcal{S}$, the restriction of every tempered distribution to C_c^∞ is a distribution, and this restriction map is one-to-one. Hence, every tempered distribution "is" a distribution. On the other hand, Proposition (0.32) shows that every distribution with compact support is tempered. Roughly speaking, the tempered distributions are those which "grow at most polynomially at infinity." For example, every polynomial is a tempered distribution, but $u(x) = e^{|x|}$ is not. (Exercise: prove this.)

One can define operations on tempered distributions as above, simply by replacing C_c^∞ by $\mathcal{S}$. For example, if $u \in \mathcal{S}'$, then:

1. $\partial^\alpha u$ is a tempered distribution for all multi-indices α;

2. fu is a tempered distribution for all $f \in C^\infty$ such that $\partial^\alpha f$ grows at most polynomially at infinity for all α;

3. $u * \phi$ is a tempered distribution, and also a C^∞ function, for any $\phi \in \mathcal{S}$.

The importance of tempered distributions lies in the fact that they have Fourier transforms. Indeed, since the Fourier transform maps $\mathcal{S}$ continuously onto itself and is self-dual by Proposition (0.26), for any $u \in \mathcal{S}'$ we can define $\widehat{u} \in \mathcal{S}'$ by $\langle \widehat{u}, \phi \rangle = \langle u, \widehat{\phi} \rangle$ ($\phi \in \mathcal{S}$), which is consistent with the definition for functions. It is easy to see that Propositions (0.21) and (0.22) are still valid when $f \in \mathcal{S}'$, as is the Fourier inversion theorem (0.27), provided $f^\vee$ is defined by $\langle f^\vee, \phi \rangle = \langle f, \phi^\vee \rangle$. Also, if $u \in \mathcal{S}'$ and $\phi \in \mathcal{S}$, we have $(u * \phi)^\widehat{} = \widehat{\phi}\widehat{u}$; the proof is left as an easy exercise.

For example, the Fourier transform of the Dirac δ-function is given by $\langle \widehat{\delta}, \phi \rangle = \langle \delta, \widehat{\phi} \rangle = \widehat{\phi}(0) = \int \phi(x)\,dx = \langle 1, \phi \rangle$, so $\widehat{\delta}$ is the constant function 1. It then follows from the inversion theorem that $\widehat{1} = \delta$, and from Proposition (0.22) that $(\partial^\alpha \delta)^\widehat{} = (2\pi i)^{|\alpha|}\xi^\alpha$ and that $(x^\alpha)^\widehat{} = (i/2\pi)^{|\alpha|}\partial^\alpha \delta$.

F. Compact Operators

Let $\mathcal{X}$ be a Banach space and let T be a bounded linear operator on $\mathcal{X}$. We denote the nullspace and range of T by $\mathcal{N}(T)$ and $\mathcal{R}(T)$. T is called **compact** if whenever $\{x_j\}$ is a bounded sequence in $\mathcal{X}$, the sequence $\{Tx_j\}$ has a convergent subsequence. Equivalently, T is compact if it maps bounded sets into sets with compact closure. T is said to be of **finite rank** if $\mathcal{R}(T)$ is finite-dimensional. Clearly every bounded operator of finite rank is compact.

(0.34) Theorem.

The set of compact operators on $\mathcal{X}$ is a closed two-sided ideal in the algebra of bounded operators on $\mathcal{X}$ with the norm topology.

Proof: Suppose T_1 and T_2 are compact, and $\{x_j\} \subset \mathcal{X}$ is bounded. We can choose a subsequence $\{y_j\}$ of $\{x_j\}$ such that $\{T_1 y_j\}$ converges, and then choose a subsequence $\{z_j\}$ of $\{y_j\}$ such that $\{T_2 z_j\}$ converges. It follows that $a_1 T_1 + a_2 T_2$ is compact for all $a_1, a_2 \in \mathbb{C}$. Also, it is clear that if T is compact and S is bounded, then TS and ST are compact. Thus the set of compact operators is a two-sided ideal.

Suppose $\{T_m\}$ is a sequence of compact operators converging to a limit T in the norm topology. Given a seqence $\{x_j\} \subset \mathcal{X}$ with $\|x_j\| \le C$ for all j, choose a subsequence $\{x_{1j}\}$ such that $\{T_1 x_{1j}\}$ converges. Proceeding inductively, for $m = 2, 3, 4, \ldots$, choose a subsequence $\{x_{mj}\}$ of $\{x_{(m-1)j}\}$ such that $\{T_m x_{mj}\}$ converges. Setting $y_j = x_{jj}$, one easily sees that $\{T_m y_j\}$ converges for all m. But then

$$\|Ty_j - Ty_k\| \le \|(T - T_m)y_j\| + \|T_m(y_j - y_k)\| + \|(T_m - T)y_k\|$$
$$\le 2C\|T - T_m\| + \|T_m y_j - T_m y_k\|.$$

Given $\epsilon > 0$, we can choose m so large that $\|T - T_m\| \le \epsilon/4C$, and then with this choice of m we have $\|T_m y_j - T_m y_k\| < \epsilon/2$ when j and k are sufficiently large. Thus $\{Ty_j\}$ is convergent, so T is compact. ∎

(0.35) Corollary.

If T is a bounded operator on $\mathcal{X}$ and there is a sequence $\{T_m\}$ of operators of finite rank such that $\|T_m - T\| \to 0$, then T is compact.

In case $\mathcal{X}$ is a Hilbert space, this corollary has a converse.

(0.36) Theorem.

If T is a compact operator on a Hilbert space $\mathcal{X}$, then T is the norm limit of operators of finite rank.

Proof: Suppose $\epsilon > 0$, and let B be the unit ball in $\mathcal{X}$. Since $T(B)$ has compact closure, it is totally bounded: there is a finite set $y_1, \ldots, y_n$ of elements of $T(B)$ such that every $y \in T(B)$ satisfies $\|y - y_j\| < \epsilon$ for some j. Let P_ϵ be the orthogonal projection onto the space spanned by $y_1, \ldots, y_n$, and set $T_\epsilon = P_\epsilon T$. Then T_ϵ is of finite rank. Also, since $T_\epsilon x$ is the element closest to Tx in $\mathcal{R}(P_\epsilon)$, for $x \in B$ we have

$$\|Tx - T_\epsilon x\| \le \min_{1 \le j \le n} \|Tx - y_j\| < \epsilon.$$

In other words, $\|T - T_\epsilon\| < \epsilon$, so $T_\epsilon \to T$ as $\epsilon \to 0$. ∎

Remark: For many years it was an open question whether Theorem (0.36) were true for general Banach spaces. The answer is negative even for some separable, reflexive Banach spaces; see Enflo [12].

(0.37) Theorem.
The operator T on the Banach space $\mathcal{X}$ is compact if and only if the dual operator T^* on the dual space $\mathcal{X}^*$ is compact.

Proof: Let B and B^* be the unit balls in $\mathcal{X}$ and $\mathcal{X}^*$. Suppose T is compact, and let $\{f_j\}$ be a bounded sequence in $\mathcal{X}^*$. Multiplying the f_j's by a small constant, we may assume $\{f_j\} \subset B^*$. The functions $f_j : \mathcal{X} \to \mathbb{C}$ are equicontinuous and uniformly bounded on bounded sets, so by the Arzelà-Ascoli theorem there is a subsequence (still denoted by $\{f_j\}$) which converges uniformly on the compact set $\overline{T(B)}$. Thus $T^* f_j(x) = f_j(Tx)$ converges uniformly for $x \in B$, so $\{T^* f_j\}$ is Cauchy in the norm topology of $\mathcal{X}^*$. Hence T^* is compact.

Likewise, if T^* is compact then T^{**} is compact on $\mathcal{X}^{**}$. But $\mathcal{X}$ is isometrically embedded in $\mathcal{X}^{**}$, and T is the restriction of T^{**} to $\mathcal{X}$, so T is compact. ∎

We now present the main structure theorem for compact operators. This theorem was first proved by I. Fredholm (by different methods) for certain integral operators on L^2 spaces. In the abstract Hilbert space setting it is due to F. Riesz, and it was later extended to arbitrary Banach spaces by J. Schauder. For this reason it is sometimes called the *Riesz-Schauder theory.* We shall restrict attention to the Hilbert space case, which is all we shall need, and for which the proof is easier; see Rudin [41] for the general case.

(0.38) Fredholm's Theorem.
Let T be a compact operator on a Hilbert space $\mathcal{X}$ with inner product $\langle \cdot \mid \cdot \rangle$. For each $\lambda \in \mathbb{C}$, let

$$\mathcal{V}_\lambda = \{x \in \mathcal{X} : Tx = \lambda x\}, \qquad \mathcal{W}_\lambda = \{x \in \mathcal{X} : T^* x = \lambda x\}.$$

Then:
a. The set of $\lambda \in \mathbb{C}$ for which $\mathcal{V}_\lambda \neq \{0\}$ is finite or countable, and in the latter case its only accumulation point is 0. Moreover, $\dim(\mathcal{V}_\lambda) < \infty$ for all $\lambda \neq 0$.
b. If $\lambda \neq 0$, $\dim(\mathcal{V}_\lambda) = \dim(\mathcal{W}_{\overline{\lambda}})$.
c. If $\lambda \neq 0$, $\mathcal{R}(\lambda I - T)$ is closed.

Proof: (a) is equivalent to the following statement: For any $\epsilon > 0$, the linear span of the spaces $\mathcal{V}_\lambda$ with $|\lambda| \geq \epsilon$ is finite-dimensional. Suppose to the contrary that there exist $\epsilon > 0$ and an infinite sequence $\{x_j\} \subset \mathcal{X}$ of linearly independent elements such that $Tx_j = \lambda_j x_j$ with $|\lambda_j| \geq \epsilon$ for all j. Since $|\lambda_j| \leq \|T\|$, by passing to a subsequence we can assume that $\{\lambda_j\}$ is a Cauchy sequence. Let $\mathcal{X}_m$ be the linear span of $x_1, \ldots, x_m$. For each m, choose $y_m \in \mathcal{X}_m$ with $\|y_m\| = 1$ and $y_m \perp \mathcal{X}_{m-1}$. Then $y_m = \sum_1^m c_{mj} x_j$ for some scalars c_{mj}, so

$$\lambda_m^{-1} T y_m = c_{mm} x_m + \sum_1^{m-1} c_{mj} \lambda_j \lambda_m^{-1} x_j = y_m + \sum_1^{m-1} c_{mj} (\lambda_j \lambda_m^{-1} - 1) x_j$$
$$= y_m \pmod{\mathcal{X}_{m-1}}.$$

If $n < m$, then,

$$\lambda_m^{-1} T y_m - \lambda_n^{-1} T y_n = y_m \pmod{\mathcal{X}_{m-1}}.$$

Therefore, since $y_m \perp \mathcal{X}_{m-1}$, the Pythagorean theorem yields

$$\|\lambda_m^{-1} T y_m - \lambda_n^{-1} T y_n\| \geq 1.$$

But then

$$1 \leq |\lambda_m^{-1}| \|T y_m - T y_n\| + |\lambda_m^{-1} - \lambda_n^{-1}| \|T y_n\|,$$

or

$$\|T y_n - T y_m\| \geq |\lambda_m| - |1 - \lambda_m \lambda_n^{-1}| \|T y_n\|.$$

As $m, n \to \infty$ the second term on the right tends to zero since $\|T y_n\| \leq \|T\|$ and $\lambda_m \lambda_n^{-1} \to 1$, and the first term is bounded below by ϵ. Thus $\{T y_m\}$ has no convergent subsequence, contradicting compactness.

Now consider (b). Given $\lambda \neq 0$, by Theorem (0.36) we can write $T = T_0 + T_1$ where T_0 has finite rank and $\|T_1\| < |\lambda|$. The operator $\lambda I - T_1 = \lambda(I - \lambda^{-1} T_1)$ is invertible (the inverse being given by the convergent geometric series $\sum_0^\infty \lambda^{-k-1} T_1^k$), and we have

$$(0.39) \quad (\lambda I - T_1)^{-1}(\lambda I - T) = (\lambda I - T_1)^{-1}(\lambda I - T_1 - T_0) = I - (\lambda I - T_1)^{-1} T_0.$$

Set $T_2 = (\lambda I - T_1)^{-1} T_0$. Then clearly $x \in \mathcal{V}_\lambda$ if and only if $x - T_2 x = 0$. On the other hand, taking the adjoint of both sides of (0.39), we have

$$(\overline{\lambda} I - T^*)(\overline{\lambda} I - T_1^*)^{-1} = I - T_2^*,$$

so $y = (\bar{\lambda}I - T_1^*)^{-1}x$ is in $W_{\bar{\lambda}}$ if and only if $x - T_2^* x = 0$. We must therefore show that the equations $x - T_2 x = 0$ and $x - T_2^* x = 0$ have the same number of independent solutions.

Since T_0 has finite rank, so does T_2. Let $u_1, \ldots, u_N$ be an orthonormal basis for $\mathcal{R}(T_2)$. Then for any $x \in X$ we have $T_2 x = \sum_1^N f_j(x)u_j$ where $\sum_1^N |f_j(x)|^2 = \|T_2 x\|^2$. It follows that $x \rightarrow f_j(x)$ is a bounded linear functional on X, so there exist $v_1, \ldots, v_N$ such that

$$T_2 x = \sum_1^N \langle x \mid v_j \rangle u_j \qquad (x \in X).$$

Set $\beta_{jk} = \langle u_j \mid v_k \rangle$, and given $x \in X$, set $\alpha_j = \langle x \mid v_j \rangle$. If $x - T_2 x = 0$, then $x = \sum_1^n \alpha_j u_j$, and we see by taking the scalar product with v_k that

$$(0.40) \qquad \alpha_k - \sum_j \beta_{jk}\alpha_j = 0, \qquad k = 1, \ldots, N.$$

Conversely, if $\alpha_1, \ldots, \alpha_n$ satisfy (0.40), then $x = \sum \alpha_j u_j$ satisfies $x - T_2 x = 0$. On the other hand, one easily verifies that $T_2^* x = \sum \langle x \mid u_j \rangle v_j$, so by the same reasoning, $x - T_2^* x = 0$ if and only if $x = \sum_1^N \alpha_j v_j$, where

$$(0.41) \qquad \alpha_k - \sum_j \overline{\beta}_{kj}\alpha_j = 0, \qquad k = 1, \ldots, N.$$

But the matrices $(\delta_{jk} - \beta_{jk})$ and $(\delta_{jk} - \overline{\beta}_{kj})$ are adjoints of each other and so have the same rank. Thus (0.40) and (0.41) have the same number of independent solutions.

Finally, we prove (c). Suppose we have a sequence $\{y_j\} \subset \mathcal{R}(\lambda I - T)$ which converges to an element $y \in X$. We can write $y_j = (\lambda I - T)x_j$ for some $x_j \in X$; if we set $x_j = u_j + v_j$ where $u_j \in V_\lambda$ and $v_j \perp V_\lambda$, we have $y_j = (\lambda I - T)v_j$. We claim that $\{v_j\}$ is a bounded sequence. Otherwise, by passing to a subsequence we may assume $\|v_j\| \rightarrow \infty$. Set $w_j = v_j/\|v_j\|$; then by passing to another subsequence we may assume that $\{Tw_j\}$ converges to a limit z. Since the y_j's are bounded and $\|v_j\| \rightarrow \infty$,

$$\lambda w_j = Tw_j + \frac{y_j}{\|v_j\|} \rightarrow z, \qquad (j \rightarrow \infty).$$

Thus $z \perp V_\lambda$ and $\|z\| = |\lambda|$, but also

$$(\lambda I - T)z = \lim(\lambda I - T)\lambda w_j = \lim \frac{\lambda y_j}{\|v_j\|} = 0,$$

so $z \in \mathcal{V}_\lambda$. This is a contradiction since we assume $\lambda \neq 0$.

Now, since $\{v_j\}$ is a bounded sequence, by passing to a subsequence we may assume that $\{Tv_j\}$ converges to a limit x. But then

$$v_j = \lambda^{-1}(y_j + Tv_j) \to \lambda^{-1}(y + x),$$

so

$$y = \lim(\lambda I - T)v_j = (\lambda I - T)\lambda^{-1}(y + x).$$

Thus $y \in \mathcal{R}(\lambda I - T)$, and the proof is complete. ∎

(0.42) Corollary.
Suppose $\lambda \neq 0$. Then:
 i. The equation $(\lambda I - T)x = y$ has a solution if and only if $y \perp \mathcal{W}_{\overline{\lambda}}$.
 ii. $\lambda I - T$ is surjective if and only if it is injective.

Proof: (i) follows from part (c) of the theorem and the fact that $\overline{\mathcal{R}(S)} = \mathcal{N}(S^*)^\perp$ for any bounded operator S. (ii) then follows from (i) and part (b) of the theorem. ∎

In general it may happen that the spaces $\mathcal{V}_\lambda$ are all trivial. (It is easy to construct an example from a weighted shift operator on l^2.) However, if T is self-adjoint, there are lots of eigenvectors.

(0.43) Lemma.
If T is a compact self-adjoint operator on a Hilbert space $\mathcal{X}$, then either $\|T\|$ or $-\|T\|$ is an eigenvalue for T.

Proof: Clearly we may assume $T \neq 0$. Let $c = \|T\|$ (so $c > 0$), and consider the operator $A = c^2 I - T^2$. For all $x \in \mathcal{X}$ we have

$$\langle Ax \mid x \rangle = c^2 \|x\|^2 - \|Tx\|^2 \geq 0.$$

Choose a sequence $\{x_j\} \subset \mathcal{X}$ with $\|x_j\| = 1$ and $\|Tx_j\| \to c$. Then $\langle Ax_j \mid x_j \rangle \to 0$, so applying the Schwarz inequality to the nonnegative Hermitian form $(u, v) \to \langle Au \mid v \rangle$, we see that

$$\|Ax_j\|^2 = \langle Ax_j \mid Ax_j \rangle \leq \langle Ax_j \mid x_j \rangle^{1/2} \langle A^2 x_j \mid Ax_j \rangle^{1/2}$$
$$\leq \langle Ax_j \mid x_j \rangle^{1/2} \|A^2 x_j\|^{1/2} \|Ax_j\|^{1/2} \leq \|A\|^{3/2} \langle Ax_j \mid x_j \rangle^{1/2} \to 0,$$

so $Ax_j \to 0$. By passing to a subsequence we may assume that $\{Tx_j\}$ converges to a limit y, which satisfies

$$\|y\| = \lim \|Tx_j\| = c > 0, \qquad Ay = \lim ATx_j = \lim TAx_j = 0.$$

In other words,

$$y \neq 0 \quad \text{and} \quad Ay = (cI + T)(cI - T)y = 0.$$

Thus either $Ty = cy$ or $cy - Ty = z \neq 0$ and $Tz = -cz$.   ∎

(0.44) The Spectral Theorem.
If T is a compact self-adjoint operator on a Hilbert space $\mathcal{X}$, then $\mathcal{X}$ has an orthonormal basis consisting of eigenvectors for T.

Proof: It is a simple consequence of the self-adjointness of T that (i) eigenvectors for different eigenvalues are orthogonal to each other, and (ii) if $\mathcal{Y}$ is a subspace of $\mathcal{X}$ such that $T(\mathcal{Y}) \subset \mathcal{Y}$, then also $T(\mathcal{Y}^\perp) \subset \mathcal{Y}^\perp$. In particular, let $\mathcal{Y}$ be the closed linear span of all the eigenvectors of T. If we pick an orthonormal basis for each eigenspace of T and take their union, by (i) we obtain an orthonormal basis for $\mathcal{Y}$. By (ii), $T|\mathcal{Y}^\perp$ is a compact operator on $\mathcal{Y}^\perp$, and it has no eigenvectors since all the eigenvectors of T belong to $\mathcal{Y}$. But this is impossible by Lemma (0.43) unless $\mathcal{Y}^\perp = \{0\}$, so $\mathcal{Y} = \mathcal{X}$.   ∎

We conclude by constructing a useful class of compact operators on $L^2(\mu)$, where μ is a σ-finite measure on a space S. To simplify the argument a bit, we shall make the (inessential) assumption that $L^2(\mu)$ is separable. If K is a measurable function on $S \times S$, we formally define the operator T_K on functions on S by

$$T_K f(x) = \int K(x, y) f(y) \, d\mu(y).$$

If $K \in L^2(\mu \times \mu)$, K is called a **Hilbert-Schmidt kernel**.

(0.45) Theorem.
Let K be a Hilbert-Schmidt kernel. Then T_K is a compact operator on $L^2(\mu)$, and $\|T_K\| \leq \|K\|_2$.

Proof: First we show that T_K is well-defined on $L^2(\mu)$ and bounded by $\|K\|_2$. By the Schwarz inequality,

$$|T_K f(x)| \leq \int |K(x, y)||f(y)| \, d\mu(y) \leq \left[\int |K(x, y)|^2 \, d\mu(y) \right]^{1/2} \|f\|_2.$$

This shows that $T_K f$ is finite almost everywhere, and moreover

$$\|T_K f\|_2^2 = \int |T_K f(x)|^2 \, d\mu(x) \le \|f\|_2^2 \iint |K(x,y)|^2 \, d\mu(y) \, d\mu(x)$$
$$= \|K\|_2^2 \|f\|_2^2,$$

so $\|T_K\| \le \|K\|_2$.

Now let $\{\phi_j\}_1^\infty$ be an orthonormal basis for $L^2(\mu)$. It is an easy consequence of Fubini's theorem that if $\psi_{ij}(x,y) = \phi_i(x)\phi_j(y)$, then $\{\psi_{ij}\}_{i,j=1}^\infty$ is an orthonormal basis for $L^2(\mu \times \mu)$. Hence we can write $K = \sum a_{ij}\psi_{ij}$. For $N = 1, 2, \ldots$, let

$$K_N(x,y) = \sum_{i+j \le N} a_{ij}\psi_{ij}(x,y) = \sum_{i+j \le N} a_{ij}\phi_i(x)\phi_j(y).$$

Then $\mathcal{R}(T_{K_N})$ lies in the span of $\phi_1, \ldots, \phi_N$, so T_{K_N} has finite rank. On the other hand,

$$\|K - K_N\|_2^2 = \sum_{i+j > N} |a_{ij}|^2 \to 0 \text{ as } N \to \infty,$$

so by the previous remarks,

$$\|T_K - T_{K_N}\| \le \|K - K_N\|_2 \to 0 \text{ as } N \to \infty.$$

By Corollary (0.35), then, T_K is compact. ∎

Chapter 1
LOCAL EXISTENCE THEORY

In this chapter we set up the ideas and terminology with which we shall be working throughout the book, and we prove some basic results about local existence of solutions to partial differential equations.

A. Basic Concepts

Roughly speaking, a **partial differential equation of order** k is an equation of the form

$$F(x, u, \partial_1 u, \ldots, \partial_n u, \partial_1^2 u, \ldots, \partial_n^k u) = 0$$

relating a function u of the variable $x \in \mathbb{R}^n$ and its derivatives of order $\leq k$. In order to keep the notation manageable, we introduce the following terminology. Let us order the set of multi-indices by saying that α comes before β if $|\alpha| < |\beta|$ or if $|\alpha| = |\beta|$ and $\alpha_i < \beta_i$ where i is the largest number with $\alpha_i \neq \beta_i$. Given complex numbers a_α ($|\alpha| \leq k$), we denote by $(a_\alpha)_{|\alpha| \leq k}$ the element of $\mathbb{C}^{N(k)}$ given by ordering the α's in this fashion, where $N(k)$ is the cardinality of $\{\alpha : |\alpha| \leq k\}$. Similarly, if $S \subset \{\alpha : |\alpha| \leq k\}$, we can consider the ordered (card S)-tuple $(a_\alpha)_{\alpha \in S}$.

Now let Ω be an open set in $\mathbb{R}^n$, and let F be a function of the variables $x \in \Omega$ and $(u_\alpha)_{|\alpha| \leq k} \in \mathbb{C}^{N(k)}$. Then we can form the partial differential equation

$$(1.1) \qquad F\left(x, (\partial^\alpha u)_{|\alpha| \leq k}\right) = 0.$$

A (complex-valued) function $u = u(x)$ on Ω is a **classical solution** of this equation if the derivatives $\partial^\alpha u$ occurring in F exist on Ω, and

$$F\left(x, (\partial^\alpha u(x))_{|\alpha| \leq k}\right) = 0 \text{ for all } x \in \Omega.$$

We are really being too vague. The general formula (1.1) includes absurd equations such as $\exp(\partial_1 u) = 0$ which have no solutions for trivial reasons, as well as equations like $(\partial_1 u)^2 - 4 = 0$, which is really the disjunction of the two equations $\partial_1 u = 2$ and $\partial_1 u = -2$. It also allows us to think of a kth order equation as a $(k + m)$th order equation for any $m > 0$. However, we shall not formulate precise restrictions on F until we consider more specific problems.

The equation (1.1) is called **linear** if F is an affine-linear function of the vector variable $(u_\alpha)_{|\alpha| \le k}$, that is, if (1.1) can be rewritten as

$$(1.2) \qquad \sum_{|\alpha| \le k} a_\alpha(x) \partial^\alpha u = f(x).$$

In this case we speak of the **differential operator** $L = \sum_{|\alpha| \le k} a_\alpha \partial^\alpha$ and write (1.2) simply as $Lu = f$. If the coefficients a_α are C^∞ on Ω we can apply the operator L to any distribution u on Ω, and u is called a **distribution solution** or **weak solution** of (1.2) if (1.2) holds in the sense of distributions, namely,

$$\sum_{|\alpha| \le k} (-1)^{|\alpha|} \langle u, \partial^\alpha(a_\alpha \phi) \rangle = \langle f, \phi \rangle \qquad (\phi \in C_c^\infty(\Omega)).$$

More general than the linear equations are the **quasi-linear equations**, namely, those equations (1.1) where F is an affine-linear function of $(u_\alpha)_{|\alpha|=k}$. Such equations can be written as

$$(1.3) \qquad \sum_{|\alpha|=k} a_\alpha\big(x, (\partial^\beta u)_{|\beta| \le k-1}\big) \partial^\alpha u = b\big(x, (\partial^\beta u)_{|\beta| \le k-1}\big).$$

The general questions with which we shall be concerned are the following:

Existence. Can we find a solution u of (1.1), perhaps satisfying some preassigned conditions, in some neighborhood of a given point, or in some given domain? How can we construct solutions?

Uniqueness. In general (1.1) will have many solutions. What kinds of boundary conditions on u will guarantee uniqueness?

Behavior of solutions. What are the qualitative properties of solutions — in particular, differentiability properties? Do the solutions depend smoothly on the boundary conditions? In the linear case (1.2), does u depend smoothly on f?

The answers to these questions will depend on the nature of the constraints imposed on the solution u by the equation (1.1), and these in turn depend very strongly on the nature of the equation under consideration. Let us look at some simple examples in $\mathbb{R}^2$:

i. The general solution of the equation $\partial_1 u = 0$ is clearly $u(x_1, x_2) = f(x_2)$ where f is arbitrary. Thus the equation $\partial_1 u = 0$ gives complete information about the behavior of u as a function of x_1 (it must be constant), but none at all about its dependence on x_2.

ii. The general solution of $\partial_1 \partial_2 u = 0$ is easily seen to be $u(x_1, x_2) = f(x_1) + g(x_2)$ where f and g are arbitrary. Thus the equation $\partial_1 \partial_2 u = 0$ gives no information about the dependence of u on either x_1 or x_2 when the other is held fixed, but merely says that the dependences are "uncoupled" in a certain sense.

iii. Any solution of the Cauchy-Riemann equation $\partial_1 u + i \partial_2 u = 0$ is a holomorphic function of the complex variable $z = x_1 + i x_2$, and in particular is C^∞. This equation imposes very strong restrictions on all the derivatives of the solution u.

In the linear case, a simple measure of the "strength" of a differential operator in a certain direction is provided by the notion of characteristics. If $L = \sum_{|\alpha| \leq k} a_\alpha \partial^\alpha$ is a linear differential operator of order k on $\Omega \subset \mathbb{R}^n$, its **characteristic form** at $x \in \Omega$ is the homogeneous polynomial of degree k on $\mathbb{R}^n$ defined by

$$\chi_L(x, \xi) = \sum_{|\alpha| = k} a_\alpha(x) \xi^\alpha \qquad (\xi \in \mathbb{R}^n).$$

A nonzero vector ξ is called **characteristic** for L at x if $\chi_L(x, \xi) = 0$, and the set of all such ξ is called the **characteristic variety** of L at x and is denoted by $\mathrm{char}_x(L)$:

$$\mathrm{char}_x(L) = \left\{ \xi \neq 0 : \chi_L(x, \xi) = 0 \right\}.$$

To see more clearly the meaning of the characteristic variety we must consider the effect of a change of coordinates. Let F be a smooth one-to-one mapping of Ω onto some open set Ω', and set $y = F(x)$. Assume that the Jacobian matrix $J_x = [(\partial y_i / \partial x_j)(x)]$ is nonsingular for $x \in \Omega$, so that $y_1, \ldots, y_n$ is a coordinate system on Ω. We have

$$\frac{\partial}{\partial x_j} = \sum \frac{\partial y_i}{\partial x_j} \frac{\partial}{\partial y_i},$$

which we can write symbolically as $\partial_x = J_x^t \partial_y$, where J_x^t is the transpose of J_x. The operator L is then transformed into the operator

$$L' = \sum_{|\alpha| \le k} a_\alpha (F^{-1}(y)) \left(J_{F^{-1}(y)}^t \partial_y \right)^\alpha$$

on Ω'. When this is expanded out, the expression will involve derivatives of $J_{F^{-1}(y)}^t$; but these derivatives are only formed by "using up" some of the ∂_y's on $J_{F^{-1}(y)}^t$, so they do not enter into the highest order terms. One then sees that

$$\chi_{L'}(y, \xi) = \sum_{|\alpha|=k} a_\alpha(F^{-1}(y)) \left(J_{F^{-1}(y)}^t \xi \right)^\alpha.$$

Since $F^{-1}(y) = x$, on comparing with the expression

$$\chi_L(x, \xi) = \sum_{|\alpha|=k} a_\alpha(x) \xi^\alpha$$

we see that $\mathrm{char}_x(L)$ is just the image of $\mathrm{char}_y(L')$ under the linear map J_x^t.

Another way of looking at this is the following. If we associate to $\xi \in \mathbb{R}^n$ the differential form $d\xi = \sum_1^n \xi_j \, dy_j$ and use the chain rule $dy_j = \sum (\partial y_j / \partial x_i) \, dx_i$, we have

$$d\xi = \sum_j \left[\sum_i \xi_i \frac{\partial y_i}{\partial x_j} \right] dx_j.$$

Thus in the x-coordinates, $d\xi$ corresponds to the vector $J_x^t \xi$. In the terminology of differential geometry, $\mathrm{char}_x(L)$ is intrinsically defined as a subset of the cotangent space at x.

Now, note that if $\xi \ne 0$ is a vector in the x_j-direction (i.e., $\xi_i = 0$ for $i \ne j$), then $\xi \in \mathrm{char}_x(L)$ if and only if the coefficient of ∂_j^k in L vanishes at x. Moreover, given any $\xi \ne 0$, by a rotation of coordinates we can arrange for ξ to lie in a coordinate direction. Thus the condition $\xi \in \mathrm{char}_x(L)$ means that, in some sense, L fails to be "genuinely kth order" in the ξ direction at x.

L is said to be **elliptic at** x if $\mathrm{char}_x(L) = \varnothing$ and **elliptic on** Ω if it is elliptic at every $x \in \Omega$. At least on the formal level, an elliptic operator of order k exerts control on all derivatives of order k. We shall see in Chapter 6 that this formal statement is also valid analytically.

Here are some examples. The first three are the ones discussed above, and the second three are the basic operators of mathematical physics which will be studied in detail in Chapters 2–5.

i. $L = \partial_1$: $\text{char}_x(L) = \{\xi \neq 0 : \xi_1 = 0\}$.

ii. $L = \partial_1\partial_2$: $\text{char}_x(L) = \{\xi \neq 0 : \xi_1 = 0 \text{ or } \xi_2 = 0\}$.

iii. $L = \frac{1}{2}(\partial_1 + i\partial_2)$ (the Cauchy-Riemann operator* on $\mathbb{R}^2$): L is elliptic on $\mathbb{R}^2$.

iv. $L = \sum_1^n \partial_j^2$ (the Laplace operator): L is elliptic on $\mathbb{R}^n$.

v. $L = \partial_1 - \sum_2^n \partial_j^2$ (the heat operator): $\text{char}_x(L) = \{\xi \neq 0 : \xi_j = 0 \text{ for } j \geq 2\}$.

vi. $L = \partial_1^2 - \sum_2^n \partial_j^2$ (the wave operator): $\text{char}_x(L) = \{\xi \neq 0 : \xi_1^2 = \sum_2^n \xi_j^2\}$.

One last definition. A hypersurface S is called **characteristic** for L at $x \in S$ if the normal vector $\nu(x)$ to S at x is in $\text{char}_x(L)$, and S is called **non-characteristic** if it is not characteristic at any point. We remark that $\nu(x)$ can be defined in a coordinate-free way as a cotangent vector at x, i.e., it transforms under coordinate changes by the same rule as $\text{char}_x(L)$. (In fact, $\nu(x)$ is one of the two unit cotangent vectors at x that annihilate the tangent space to S at x; the choice of orientation is immaterial since $\chi_L(x, -\xi) = (-1)^k \chi_L(x, \xi)$.) Hence the condition "$S$ is non-characteristic" is independent of the choice of coordinates.

The notion of characteristics can be extended to the nonlinear theory, but there the situation is more complicated, as we shall see.

B. Real First Order Equations

Recall that the basic boundary value problem for first order ordinary differential equations is the initial value problem: given a function F on $\mathbb{R}^3$ and $(t_0, u_0) \in \mathbb{R}^2$, find a function $u(t)$ defined in a neighborhood of t_0 such that $F(t, u, u') = 0$ and $u(t_0) = u_0$. In this section we shall consider the corresponding initial value problem for real first order partial differential equations. Namely, given an equation $F(x, (\partial^\alpha u)_{|\alpha| \leq 1}) = 0$ (where F and u are assumed to be real-valued), a hypersurface S, and a (real-valued) function ϕ on S, find a solution u defined on a neighborhood of S such that $u = \phi$ on S. This theory has a pleasantly geometric flavor.

---

* Question: Why the factor of $\frac{1}{2}$? Answer: The operators $\partial_z = \frac{1}{2}(\partial_x - i\partial_y)$ and $\partial_{\bar{z}} = \frac{1}{2}(\partial_x + i\partial_y)$ are the result of formally applying the chain rule to the change of variables $z = x + iy$, $\bar{z} = x - iy$. When applied to holomorphic functions, ∂_z is just the complex derivative.

Let us first consider the linear equation

(1.4) $$Lu = \sum a_j \partial_j u + bu = f,$$

where a_j, b, and f are assumed to be C^1 functions of x. If we denote by A the vector field $(a_1, \ldots, a_n)$, we have

$$\text{char}_x(L) = \{\xi \neq 0 : A(x) \cdot \xi = 0\},$$

that is, $\text{char}_x(L) \cup \{0\}$ is the hyperplane orthogonal to $A(x)$. From this it is clear that a hypersurface S is characteristic at x if and only if $A(x)$ is tangent to S at x.

Suppose we wish to find a solution u of (1.4) with given initial values $u = \phi$ on the hypersurface S. If S is characteristic at x_0, the quantity $\sum a_j(x_0) \partial_j u(x_0)$ is completely determined as a certain directional derivative of ϕ along S at x_0, and it may be impossible to make it equal to $f(x_0) - b(x_0)u(x_0)$. (For example, if the equation is $\partial_1 u = 0$ and S is the hyperplane $x_n = 0$, we cannot have $u = \phi$ on S unless $\partial_1 \phi = 0$.) To make the initial value problem well-behaved, then, we must asume that S is non-characteristic, and we do so henceforth.

It is natural to look at the integral curves of the vector field A (sometimes called the **characteristic curves** of the equation (1.4)), that is, the parametrized curves $x(t)$ that satisfy the system of ordinary differential equations

(1.5) $$\frac{dx}{dt} = A(x), \quad \text{i.e.,} \quad \frac{dx_j}{dt} = a_j(x) \quad (j = 1, \ldots, n).$$

Along one of those curves a solution u of (1.4) must satisfy

(1.6) $$\frac{du}{dt} = \sum \partial_j u \frac{dx_j}{dt} = \sum a_j \partial_j u = f - bu.$$

By the fundamental existence and uniqueness theorem for ordinary differential equations (see, e.g., Coddington and Levinson [9]), through each point x_0 of S there passes a unique integral curve $x(t)$ of A, namely the solution of (1.5) with $x(0) = x_0$. Along this curve the solution u of (1.4) must be the solution of the ordinary differential equation (1.6) with $u(0) = \phi(x_0)$. Moreover, since A is non-characteristic, $x(t) \notin S$ for $t \neq 0$, at least for $|t|$ small, and the curves $x(t)$ fill out a neighborhood of S. Thus the initial value problem for (1.4) is reduced to the initial value problem for the ordinary differential equations (1.5) and (1.6), and we have:

(1.7) Theorem.
*Assume that S is a hypersurface of class C^1 which is non-characteristic
for (1.4), and that the functions a_j, b, f, and ϕ are C^1 and real-valued.
Then for any sufficiently small neighborhood Ω of S in $\mathbb{R}^n$ there is a unique
solution $u \in C^1$ of (1.4) on Ω that satisfies $u = \phi$ on S.*

This theorem is a special case of the corresponding result for quasi-
linear equations, which we shall prove in detail below.

What happens if we allow the quantities in (1.4) to be complex-valued?
If we set $a_j = a_j^1 + ia_j^2$, and similarly for b, f, and u, the real and imaginary
parts of (1.4) are

$$\sum a_j^1 \partial_j u^1 - \sum a_j^2 \partial_j u^2 + b^1 u^1 - b^2 u^2 = f^1,$$
$$\sum a_j^2 \partial_j u^1 + \sum a_j^1 \partial_j u^2 + b^2 u^1 + b^1 u^2 = f^2.$$

This is a system of two equations for the two real unknowns u^1 and u^2. If
the a_j's and b are real, this system is uncoupled: the first equation involves
only u^1 and the second only u^2, so we can solve them separately and obtain
the solution $u = u^1 + iu^2$ of (1.4). However, if the a_j's and b are complex,
this system may possess no solutions, as we shall see in §1E.

Now let us generalize to the quasi-linear equation

$$(1.8) \qquad\qquad \sum a_j(x, u)\partial_j u = b(x, u).$$

If u is a function of x, the normal to the graph of u in $\mathbb{R}^{n+1}$ is proportional
to $(\partial_1 u, \ldots, \partial_n u, -1)$, so (1.8) just says that the vector field

$$\mathbf{A}(x, y) = \big(a_1(x, y), \ldots, a_n(x, y), b(x, y)\big)$$

is tangent to the graph $y = u(x)$ at any point. This suggests that we look
at the integral curves of the vector field $\mathbf{A}$ in $\mathbb{R}^{n+1}$ given by solving the
ordinary differential equations

$$\frac{dx_j}{dt} = a_j(x, y), \qquad \frac{dy}{dt} = b(x, y).$$

It is clear that any graph $y = u(x)$ in $\mathbb{R}^{n+1}$ which is the union of an
$(n-1)$-parameter family of these integral curves will define a solution of
(1.8). Conversely, suppose u is a solution of (1.8). If we solve the equations

$$\frac{dx_j}{dt} = a_j(x, u(x)), \qquad x_j(0) = (x_0)_j$$

to obtain a curve $x(t)$ passing through x_0, and then set $y(t) = u(x(t))$, we have

$$\frac{dy}{dt} = \sum \partial_j u \frac{dx_j}{dt} = \sum a_j(x, u)\partial_j u = b(x, u).$$

Thus if the graph $y = u(x)$ intersects an integral curve of $\mathbf{A}$ in one point $(x_0, u(x_0))$, it contains the whole curve.

Suppose we are given initial data $u = \phi$ on a hypersurface S. If we form the submanifold

$$S^* = \{(x, \phi(x)) : x \in S\}$$

of $\mathbb{R}^{n+1}$, the graph of the solution should be the hypersurface generated by the integral curves of $\mathbf{A}$ passing through S^*. Again, we need to assume that S is non-characteristic in some sense. This is more complicated than in the linear case because the coefficients a_j depend on u as well as x, but the geometric interpretation is exactly the same: for $x \in S$, the vector

$$(a_1(x, \phi(x)), \ldots, a_n(x, \phi(x)))$$

should not be tangent to S at x. (Note that this condition involves ϕ as well as S.) If S is represented parametrically by a mapping $g : \mathbb{R}^{n-1} \to \mathbb{R}^n$ and we take coordinates $s_1, \ldots, s_{n-1}$ on $\mathbb{R}^{n-1}$, this condition is just

$$(1.9) \qquad \det \begin{pmatrix} \partial g_1/\partial s_1 & \cdots & \partial g_1/\partial s_{n-1} & a_1(g(s), \phi(g(s))) \\ \vdots & & \vdots & \vdots \\ \partial g_n/\partial s_1 & \cdots & \partial g_n/\partial s_{n-1} & a_n(g(s), \phi(g(s))) \end{pmatrix} \neq 0.$$

(1.10) Theorem.
Suppose S is a hypersurface of class C^1 in $\mathbb{R}^n$ and a_j, b, and ϕ are C^1 real-valued functions. Suppose also that the vector

$$(a_1(x, \phi(x)), \ldots, a_n(x, \phi(x)))$$

is not tangent to S at any $x \in S$. Then for any sufficiently small neighborhood Ω of S in $\mathbb{R}^n$ there is a unique solution $u \in C^1$ of (1.8) on Ω that satisfies $u = \phi$ on S.

Proof: The uniqueness follows from the preceding discussion: the graph of u must be the union of the integral curves of $\mathbf{A}$ in Ω passing through S^*. (Ω must be taken small enough so that these curves are all distinct. That is, it can happen that an integral curve of $\mathbf{A}$ intersects S^*

in more than one point, but in between these points of intersection it must leave Ω.) Now, any hypersurface S can be covered by open sets on which it admits a parametric representation $x = g(s)$. If we solve the problem on each such set, by uniqueness the local solutions agree on their common domains and hence patch together to give a solution for all of S. It therefore suffices to assume that S is represented parametrically by $x = g(s)$.

For each $s \in \mathbb{R}^{n-1}$, consider the initial value problem

$$\frac{\partial x_j}{\partial t}(s,t) = a_j(x,y), \qquad \frac{\partial y}{\partial t}(s,t) = b(x,y),$$
$$x_j(s,0) = g_j(s), \qquad y(s,0) = \phi(g(s)).$$

Here s is just a parameter, so we have a system of ordinary differential equations in t. By the fundamental theorem of ordinary differential equations (Coddington and Levinson [9]), there is a unique solution (x,y) defined for small t, and (x,y) is a C^1 function of s and t jointly. By the non-characteristic condition (1.9) and the inverse mapping theorem, the mapping $(s,t) \to x(s,t)$ is invertible on a neighborhood Ω of S, yielding s and t as C^1 functions of x on Ω such that $t(x) = 0$ and $g(s(x)) = x$ when $x \in S$. Now set

$$u(x) = y(s(x), t(x)).$$

Clearly $u = \phi$ on S, and we claim that u satisfies (1.8). Indeed, by the chain rule,

$$\sum_{j=1}^{n} a_j \partial_j u = \sum_{j=1}^{n} a_j \left(\sum_{k=1}^{n-1} \frac{\partial u}{\partial s_k} \frac{\partial s_k}{\partial x_j} + \frac{\partial u}{\partial t} \frac{\partial t}{\partial x_j} \right)$$
$$= \sum_{k=1}^{n-1} \frac{\partial u}{\partial s_k} \sum_{j=1}^{n} a_j \frac{\partial s_k}{\partial x_j} + \frac{\partial u}{\partial t} \sum_{j=1}^{n} a_j \frac{\partial t}{\partial x_j}$$
$$= \sum_{k=1}^{n-1} \frac{\partial u}{\partial s_k} \sum_{j=1}^{n} \frac{\partial s_k}{\partial x_j} \frac{\partial x_j}{\partial t} + \frac{\partial u}{\partial t} \sum_{j=1}^{n} \frac{\partial x_j}{\partial t} \frac{\partial t}{\partial x_j}$$
$$= \sum_{k=1}^{n-1} \frac{\partial u}{\partial s_k} \frac{\partial s_k}{\partial t} + \frac{\partial u}{\partial t} \frac{\partial t}{\partial t}$$
$$= 0 + \frac{\partial u}{\partial t} = b,$$

since s_k and t are functionally independent. This completes the proof. ∎

Let us see a couple of examples; others will be found in the exercises.

Example 1.

In $\mathbb{R}^3$, solve $x_1\partial_1 u + 2x_2\partial_2 u + \partial_3 u = 3u$ with $u = \phi(x_1, x_2)$ on the plane $x_3 = 0$.

Solution: We solve the ordinary differential equations

$$\frac{dx_1}{dt} = x_1, \qquad \frac{dx_2}{dt} = 2x_2, \qquad \frac{dx_3}{dt} = 1, \qquad \frac{du}{dt} = 3u$$

with initial conditions

$$(x_1, x_2, x_3, u)|_{t=0} = (s_1, s_2, 0, \phi(s_1, s_2)),$$

obtaining

$$x_1 = s_1 e^t, \qquad x_2 = s_2 e^{2t}, \qquad x_3 = t, \qquad u = \phi(s_1, s_2)e^{3t}.$$

We solve the first three equations for s_1, s_2, and t, obtaining

$$t = x_3, \qquad s_1 = x_1 e^{-x_3}, \qquad s_2 = x_2 e^{-2x_3}.$$

The solution is then

$$u = \phi(x_1 e^{-x_3}, x_2 e^{-2x_3})e^{3x_3}.$$

Example 2.

In $\mathbb{R}^2$, solve $u\partial_1 u + \partial_2 u = 1$ with $u = \frac{1}{2}s$ on the segment $x_1 = x_2 = s$, $0 < s < 1$.

Solution: First, (1.9) is satisfied, for

$$\det \begin{pmatrix} \partial x_1/\partial s & a_1(s, s, \frac{1}{2}s) \\ \partial x_2/\partial s & a_2(s, s, \frac{1}{2}s) \end{pmatrix} = \det \begin{pmatrix} 1 & \frac{1}{2}s \\ 1 & 1 \end{pmatrix} = 1 - \frac{1}{2}s \neq 0 \text{ for } 0 < s < 1.$$

We solve the equations

$$\frac{dx_1}{dt} = u, \qquad \frac{dx_2}{dt} = 1, \qquad \frac{du}{dt} = 1$$

with initial conditions

$$(x_1, x_2, u)|_{t=0} = (s, s, \tfrac{1}{2}s),$$

obtaining

$$u = t + \tfrac{1}{2}s, \qquad x_2 = t + s, \qquad x_1 = \tfrac{1}{2}t^2 + \tfrac{1}{2}st + s.$$

Since $x_2 - x_1 = \frac{1}{2}t(2 - t - s) = \frac{1}{2}t(2 - x_2)$, we can easily eliminate s and t from these equations to obtain

$$u = \frac{4x_2 - 2x_1 - x_2^2}{2(2 - x_2)}.$$

The initial value problem for a general real first order equation can also be reduced to the initial value problem for a system of ordinary differential equations. We shall derive these equations by assuming that a solution exists and working backwards.

Suppose, then, that we are given a real-valued function F of the $2n+1$ variables $x_1, \ldots, x_n, y, z_1, \ldots, z_n$. (The appropriate smoothness assumption here is $F \in C^2$.) Suppose that we have a function u of x such that when we set $y = u(x)$ and $z_j = \partial_j u(x)$, the equation

$$(1.11) \qquad F(x_1, \ldots, x_n, y, z_1, \ldots, z_n) = 0$$

is satisfied identically in x, that is, u is a solution of

$$(1.12) \qquad F(x_1, \ldots, x_n, u, \partial_1 u, \ldots, \partial_n u) = 0.$$

In what follows we stipulate that the function F and its derivatives are to be evaluated at $y = u(x)$, $z_j = \partial_j u(x)$.

First, consider the integral curves of the vector field $\nabla_z F$, i.e., the solution curves of the equations

$$(1.13) \qquad \frac{dx_j}{dt} = \frac{\partial F}{\partial z_j}.$$

Along these curves the quantity y must satisfy

$$(1.14) \qquad \frac{dy}{dt} = \sum \frac{\partial u}{\partial x_j} \frac{dx_j}{dt} = \sum z_j \frac{\partial F}{\partial z_j}.$$

In the quasi-linear case (1.8), $\partial F / \partial z_j = a_j$ and $\sum z_j(\partial F/\partial z_j) = b$ are independent of the z_j's, so (1.13) and (1.14) form a determined system in the variables $x_1, \ldots, x_n, y$ — in fact, the system in used in the proof of Theorem (1.10). For the general case we also need equations for the variables z_j, which can be obtained as follows. From (1.13), we have

$$\frac{dz_j}{dt} = \sum \frac{\partial z_j}{\partial x_k} \frac{dx_k}{dt} = \sum \frac{\partial^2 u}{\partial x_j \partial x_k} \frac{\partial F}{\partial z_k} = \sum \frac{\partial z_k}{\partial x_j} \frac{\partial F}{\partial z_k}.$$

Also, differentiating (1.11) with respect to x_j,

$$0 = \frac{\partial F}{\partial x_j} + \frac{\partial F}{\partial y} \frac{\partial y}{\partial x_j} + \sum \frac{\partial F}{\partial z_k} \frac{\partial z_k}{\partial x_j} = \frac{\partial F}{\partial x_j} + z_j \frac{\partial F}{\partial y} + \sum \frac{\partial z_k}{\partial x_j} \frac{\partial F}{\partial z_k}.$$

Combining these equations, we have

$$(1.15) \qquad \frac{dz_j}{dt} = -\frac{\partial F}{\partial x_j} - z_j \frac{\partial F}{\partial y}.$$

Dropping now the assumption that a solution u is given, we see that (1.13), (1.14), and (1.15) form a system of $2n + 1$ ordinary differential equations for the $2n + 1$ unknowns x_j, y, and z_j.

As for the initial conditions, we assume given a C^1 hypersurface S represented parametrically as $x = g(s)$ and a C^1 function ϕ on S. Then the initial conditions for (1.13) and (1.14) are

$$(1.16) \qquad x(s, 0) = g(s), \qquad y(s, 0) = \phi(g(s)).$$

Moreover, on S we must have

$$(1.17) \qquad \frac{\partial y}{\partial s_j} = \frac{\partial \phi(g(s))}{\partial s_j} = \sum \frac{\partial y}{\partial x_k} \frac{\partial g_k}{\partial s_j} = \sum z_k \frac{\partial g_k}{\partial s_j} \quad (1 \le j \le n - 1),$$
$$F\big(g(s),\, \phi(g(s)),\, z_j(g(s))\big) = 0.$$

These provide n equations in the n unknowns $z_1, \ldots, z_n$ on S. In the quasi-linear case these equations are linear, and the non-characteristic condition (1.9) ensures the existence of a unique solution. In general we have neither existence nor uniqueness (for example, a square root may appear). We must therefore assume given C^1 solutions $z_1, \ldots, z_n$ of (1.17) on S which satisfy the non-characteristic condition

$$\det \begin{pmatrix} \frac{\partial g_1}{\partial s_1} & \cdots & \frac{\partial g_1}{\partial s_{n-1}} & \frac{\partial F}{\partial z_1}\big(g(s), \phi(g(s)), z_j(g(s))\big) \\ \vdots & & \vdots & \vdots \\ \frac{\partial g_n}{\partial s_1} & \cdots & \frac{\partial g_n}{\partial s_{n-1}} & \frac{\partial F}{\partial z_n}\big(g(s), \phi(g(s)), z_j(g(s))\big) \end{pmatrix} \ne 0.$$

We can then proceed as before. We solve the system (1.13–15) with initial conditions (1.16–17) to obtain x_j, y, z_j as functions of s and t. The non-characteristic condition ensures that the mapping $(s, t) \to x(s, t)$ can be inverted to yield s and t as C^1 functions of x. We then claim that $u(x) = y(s(x), t(x))$ satisfies (1.12): it clearly satisfies $u = \phi$ on S because of (1.16). To prove the claim, one uses the uniqueness theorem for ordinary differential equations and the chain rule to verify that

$$F\big(x_1, \ldots, x_n, u(x), z_1(s(x), t(x)), \ldots, z_n(s(x), t(x))\big) = 0,$$
$$z_j(s(x), t(x)) = \partial_j u(x).$$

There is much more to be said about fully nonlinear first order equations. More complete discussions, including geometric interpretations and some applications, can be found in Courant and Hilbert [10, vol. II] and John [30].

EXERCISES

The following problems deal with differential equations on $\mathbb{R}^2$. We write x, y and ∂_x, ∂_y rather than x_1, x_2 and $\partial_{x_1}, \partial_{x_2}$.

1. Solve $\partial_x u + \partial_y u = u$ with $u = \cos x$ when $y = 0$. (Answer: $u = e^y \cos(x - y)$.)

2. Solve $x^2 \partial_x u + y^2 \partial_y u = u^2$ with $u = 1$ when $y = 2x$. (Answer: $u = xy/(xy - y + 2x)$.)

3. Show by geometric considerations that the general solution of $x\partial_x u - y\partial_y u = 0$ is $u = f(xy)$. Find the solution whose graph contains the line $u = x = y$. (Answer: $u = \sqrt{xy}$.) What happens to the initial value problem when S is the curve $y = x^{-1}$?

4. Solve $u\partial_x u + \partial_y u = 1$ with $u = 0$ when $y = x$. (Answer: $u = 1 - (1 + 2x - 2y)^{1/2}$.) Something nasty happens if we replace the initial condition $u = 0$ by $u = 1$. What?

C. The General Cauchy Problem

We now return to the general kth order equation

$$(1.18) \qquad\qquad F\big(x, (\partial^\alpha u)_{|\alpha|\leq k}\big) = 0,$$

where F will always be assumed to be (at least) C^1. The most natural generalization of the initial value problem for first order equations, which corresponds to the initial value problem for higher order ordinary differential equations, is the Cauchy problem.

Let S be a hypersurface of class C^k. If u is a C^{k-1} function defined near S, the quantities $u, \partial_\nu u, \ldots, \partial_\nu^{k-1} u$ on S are called the **Cauchy data** of u on S. (Recall that ∂_ν is the normal derivative on [a neighborhood of] S: see (0.3).) The **Cauchy problem** is to solve (1.18) when the Cauchy data of u on S are preassigned.

In this section all our considerations will be restricted to a neighborhood of a given point on S. We may therefore assume that a change of coordinates has been made so that S contains the origin and, near the origin, coincides with the hyperplane $x_n = 0$. With this in mind, it will be convenient to make a slight change of notation. We shall consider $\mathbb{R}^n$ as $\mathbb{R}^{n-1} \times \mathbb{R}$ and denote the coordinates by (x, t) where $x = (x_1, \ldots, x_{n-1})$. Derivatives with respect to the x variables will be denoted by ∂_{x_j} or ∂_x^α, where $\alpha = (\alpha_1, \ldots, \alpha_{n-1})$, and derivatives with respect to t will be denoted by ∂_t^j. We can then restate the Cauchy problem as follows: Given functions $\phi_0, \ldots, \phi_{k-1}$ of x, solve

$$(1.19) \qquad \begin{aligned} F\big(x, t, (\partial_x^\alpha \partial_t^j u)_{|\alpha|+j \le k}\big) &= 0, \\ \partial_t^j u(x, 0) &= \phi_j(x) \qquad (0 \le j < k). \end{aligned}$$

We observe to begin with that if u is a function of class C^r with $r \ge k$, then the Cauchy data $\{\phi_j\}$ determine all derivatives $\partial_x^\alpha \partial_t^j u$ on S with $j < k$ and $|\alpha| + j \le r$; in fact,

$$\partial_x^\alpha \partial_t^j u(x, 0) = \partial_x^\alpha \phi_j(x).$$

Hence the only quantity in the differential equation in (1.19) which is unknown on S is $\partial_t^k u$. In order for the Cauchy problem to be well-behaved, we must assume that the equation $F = 0$ can be solved for $\partial_t^k u$.

In the linear case,

$$F\big(x, t, (\partial_x^\alpha \partial_t^j u)_{|\alpha|+j \le k}\big) = \sum_{|\alpha|+j \le k} a_{\alpha j}(x, t) \partial_x^\alpha \partial_j^t u - f(x, t),$$

this assumption just means that S is non-characteristic. Indeed, "S is non-characteristic" means that $a_{0k}(x, 0) \ne 0$, hence by continuity $a_{0k}(x, t) \ne 0$ for small t, and we can solve for $\partial_t^k u$:

$$\partial_t^k u = (a_{0k})^{-1} \left[\sum_{|\alpha|, j \le k, \; j < k} a_{\alpha j} \partial_x^\alpha \partial_t^j u - f \right].$$

Here are some examples of the bad behavior that can occur when this condition is not satisfied:

i. The line $t = 0$ is characteristic for the equation $\partial_x \partial_t u = 0$ in $\mathbb{R}^2$. If u is a solution of this equation with $u(x, 0) = \phi_0(x)$ and $\partial_t u(x, 0) = \phi_1(x)$, we must have $\partial_x \phi_1 = 0$, i.e., ϕ_1 is constant. Thus the Cauchy problem is not solvable in general. On the other hand, if ϕ_1 is constant, then there is no uniqueness: we can take $u(x, t) = \phi_0(x) + f(t)$ where f is any function with $f'(0) = \phi_1$.

ii. The line $t = 0$ is characteristic for the equation $\partial_x^2 u - \partial_t u = 0$ in $\mathbb{R}^2$. Here if we are given that u is a solution with $u(x, 0) = \phi_0(x)$, then $\partial_t u(x, 0)$ is already completely determined: $\partial_t u(x, 0) = \phi_0''(x)$.

In the quasi-linear case,

$$F\big(x,t,(\partial_x^\alpha \partial_t^j u)_{|\alpha|+j\leq k}\big) = \sum_{|\alpha|+j=k} a_{\alpha j}\big(x,t,(\partial_x^\beta \partial_t^i u)_{|\beta|+i\leq k-1}\big)\partial_\alpha^x \partial_t^j u$$
$$- b\big(x,t,(\partial_x^\beta \partial_t^i u)_{|\beta|+i\leq k-1}\big),$$

we say that the Cauchy problem (1.19) is non-characteristic if

$$a_{0k}\big(x,0,(\partial_x^\beta \phi_i(x))_{|\beta|+i\leq k-1}\big) \neq 0$$

for all x; again, this allows us to solve for the derivative $\partial_t^k u$.

In the general case, the equation

$$F\big(x,0,(\partial_x^\alpha \phi_j(x))_{|\alpha|+j\leq k,\ j<k},u_{0k}\big) = 0$$

will usually not determine u_{0k} uniquely as a function of x on S. Therefore, we phrase the non-characteristic condition as follows: The quantity u_{0k} can be determined as a C^1 function of x on S so that

$$F\big(x,0,(\partial_x^\alpha \phi_j(x))_{|\alpha|+j\leq k,\ j<k},u_{0k}(x)\big) = 0,$$
$$\frac{\partial F}{\partial u_{0k}}\big(x,0,(\partial_x^\alpha \phi_j(x))_{|\alpha|+j\leq k,\ j<k},u_{0k}(x)\big) \neq 0$$

for all x. In this case, we can solve the equation $F = 0$ for u_{0k} as a C^1 function G of the remaining variables near S, by the implicit function theorem, and write the differential equation in the normal form:

$$(1.20) \qquad \partial_t^k u = G\big(x,t,(\partial_x^\alpha \partial_t^j u)_{|\alpha|+j\leq k,\ j<k}\big).$$

The Cauchy data $\{\phi_j\}$ together with (1.20) determine all derivatives of u of order $\leq k$ on S. If G is sufficiently smooth, we can also determine higher derivatives of u. Namely, differentiating (1.20) with respect to t,

$$\partial_t^{k+1} u = \frac{\partial G}{\partial t} + \sum_{|\alpha|+j\leq k,\ j<k} \frac{\partial G}{\partial u_{\alpha j}}\partial_x^\alpha \partial_t^{j+1} u.$$

All the quantities on the right are known on S, so $\partial_t^{k+1} u$ is also; hence we know all derivatives of u of order $\leq k+1$ on S. Applying ∂_t more times, we obtain higher derivatives. In particular, we have:

(1.21) Proposition.
Suppose that $G, \phi_0, \ldots, \phi_{k-1}$ are analytic functions. Then there is at most one analytic function u satisfying (1.20) such that $\partial_t^j u(x,0) = \phi_j(x)$ for $0 \leq j < k$.

Proof: By Taylor's formula, an analytic function is completely determined by the values of its derivatives at one point. ∎

Actually, the smoothness of G is unnecessary for the unique determination of the derivatives of u on S.

(1.22) Proposition.
Suppose G is continuous and there is a constant $C > 0$ such that for all $x, t \in \mathbb{R}^n$ and all vectors $(u_{\alpha j})$, $(v_{\alpha j})$ $(0 \le |\alpha| + j \le k, \, j < k)$,

$$\left| G(x, t, (u_{\alpha j})) - G(x, t, (v_{\alpha j})) \right| \le C \sum_{\alpha, j} |u_{\alpha j} - v_{\alpha j}|.$$

If u and v are two solutions of (1.20) with the same Cauchy data on S, and the derivatives $\partial_x^\alpha \partial_t^j u$ and $\partial_x^\alpha \partial_t^j v$ exist for $|\alpha| \le q$ and $j \le r$ $(q, r \ge k)$, then these derivatives agree on S.

Proof: Let $w = u - v$. It suffices to show that $\partial_t^m w = 0$ on S for $m \le r$, as then the x-derivatives of these functions also vanish on S. We proceed by induction on m, the cases $m < k$ being true by assumption. Suppose then that $m \ge k$ and $\partial_t^i w = 0$ on S for $i < m$. By Taylor's theorem,

$$(1.23) \qquad \partial_t^k w(x, t) = \frac{t^{m-k}}{(m-k)!} \partial_t^m w(x, 0) + o(t^{m-k}) \qquad (t \to 0),$$

and for $j < k$ and $|\alpha| + j \le k$,

$$\partial_x^\alpha \partial_t^j w(x, t) = \frac{t^{m-j}}{(m-j)!} \partial_x^\alpha \partial_t^m w(x, 0) + o(t^{m-j}) = o(t^{m-k}) \qquad (t \to 0).$$

(See Folland [16] for the form of Taylor's theorem used here, which is omitted from most calculus books.) Therefore, by the assumption on G,

$$\begin{aligned}
\left| \partial_t^k w(x, t) \right| &= \left| G(x, t, (\partial_x^\alpha \partial_t^j u(x, t))) - G(x, t, (\partial_x^\alpha \partial_t^j v(x, t))) \right| \\
&\le C \sum \left| \partial_x^\alpha \partial_t^j u(x, t) - \partial_x^\alpha \partial_t^j v(x, t) \right| \\
&= C \sum \left| \partial_x^\alpha \partial_t^j w(x, t) \right| = o(t^{m-k}) \qquad (t \to 0).
\end{aligned}$$

Hence, by (1.23),

$$\frac{t^{m-k}}{(m-k)!} \partial_t^m w(x, 0) = o(t^{m-k}) \qquad (t \to 0),$$

which forces $\partial_t^m w(x, 0)$ to be 0. ∎

Although the Cauchy problem is good from the point of view of determining a unique solution, existence is another matter, especially if we want a solution in a specified domain and not just in some neighborhood of the initial hypersurface S. In fact, the Cauchy problem tends to be overdetermined except in certain special situations. As we shall see, the appropriate boundary conditions for a differential equation depend strongly on the particular form of the equation.

D. The Cauchy-Kowalevski Theorem

In this section we consider the Cauchy problem

$$F\big(x, (\partial^\alpha u)_{|\alpha| \le k}\big) = 0, \qquad \partial_\nu^j u = \phi_j \text{ on } S \quad (0 \le j < k),$$

where the functions $F, \phi_0, \ldots, \phi_{k-1}$ and the hypersurface S are assumed to be analytic, and we look for solutions defined in a neighborhood of some point $x_0 \in S$. As in §1C, we can make an analytic change of coordinates from $\mathbb{R}^n$ to $\mathbb{R}^{n-1} \times \mathbb{R}$ so that x_0 is mapped to $(0,0)$ and a neighborhood of x_0 in S is mapped into the hyperplane $t = 0$. This transformation will of course change the function F, but the result will still be analytic. We assume the non-characteristic condition in its analytic form, that is, that the equation $F = 0$ can be solved for $\partial_t^k u$ to yield $\partial_t^k u$ as an analytic function G of the remaining variables. The Cauchy problem then takes the form

$$(1.24) \qquad \begin{aligned} &\partial_t^k u = G\big(x, t, (\partial_x^\alpha \partial_t^j u)_{|\alpha| + j \le k, \, j < k}\big), \\ &\partial_t^j u(x, 0) = \phi_j(x) \qquad (0 \le j < k). \end{aligned}$$

Our main result is the following fundamental existence theorem.

(1.25) The Cauchy-Kowalevski* Theorem.

If G, $\phi_0, \ldots, \phi_{k-1}$ are analytic near the origin, there is a neighborhood of the origin on which the Cauchy problem (1.24) has a unique analytic solution.

---

* The problem of how to spell this name is vexed not only by the usual lack of a canonical scheme for transliterating from the Cyrillic alphabet to the Latin one but also by the question of whether to use the feminine ending (-skaia instead of -ski). The spelling used here is the one preferred by Kowalevski herself in her scientific works.

We remark that G, ϕ_j, and u can be complex-valued here, or, for that matter, vector-valued: the arguments in this section work equally well for determined systems of equations. Before proceeding to the proof, we recall some properties of power series in several variables. (The proofs are the same as in the one-variable case.)

(1.26)   If f is analytic near $x^0 \in \mathbb{R}^n$, there is a cube $\Omega = \{x : |x_j - x_j^0| < r$ for $1 \le j \le n\}$ on which the Taylor series

$$\sum_\alpha \frac{\partial^\alpha f(x^0)}{\alpha!}(x - x^0)^\alpha$$

converges to $f(x)$. The convergence is absolute and uniform on compact subsets of Ω, and the series can be differentiated termwise.

(1.27)   Let $f(x) = \sum a_\alpha (x - x^0)^\alpha$ be convergent near $x = x^0 \in \mathbb{R}^n$. Moreover, let x be an analytic function of $\xi \in \mathbb{R}^m$: $x = \sum b_\beta (\xi - \xi^0)^\beta$ where $b_\beta \in \mathbb{R}^n$ and $x(\xi^0) = b_0 = x^0$. Then the composite function $F(\xi) = f(x(\xi))$ is analytic at ξ^0, and its power series expansion about ξ^0 is obtained by substituting $\sum_{\beta \neq 0} b_\beta (\xi - \xi^0)^\beta$ for $x - x^0$ in the series $\sum a_\alpha (x - x^0)^\alpha$ and multiplying out. Thus $F(\xi) = \sum c_\gamma (\xi - \xi^0)^\gamma$ where $c_\gamma = P_\gamma(a_\alpha$'s, b_β's) and P_γ is a universal polynomial (independent of the particular series involved) in those coefficients a_α and b_β for which $\alpha_j \le \gamma_j$ and $\beta_j \le \gamma_j$ for all j. Moreover, P_γ has nonnegative coefficients since only addition and multiplication are involved in expanding out the substitution.

(1.28)   Given $M, r > 0$, the function

$$f(x) = \frac{Mr}{r - (x_1 + \ldots + x_n)}$$

is analytic in the rectangle $\{x : \max |x_j| < r/n\}$. By the geometric series formula and the multinomial theorem, its Taylor series is

$$f(x) = M \sum_0^\infty \frac{(x_1 + \cdots + x_n)^k}{r^k} = M \sum_{|\alpha| \ge 0} \frac{|\alpha|!}{\alpha! r^{|\alpha|}} x^\alpha,$$

which converges absolutely for $\max |x_j| < r/n$.

(1.29)   A power series $\sum a_\alpha (x - x^0)^\alpha$ with nonnegative coefficients is said to **majorize** another series $\sum b_\alpha (x - x^0)^\alpha$ if $a_\alpha \ge |b_\alpha|$ for all α. In this case it is clear that $\sum b_\alpha (x - x^0)^\alpha$ converges absolutely whenever $\sum a_\alpha (x - x^0)^\alpha$ does.

(1.30)   Suppose $\sum a_\alpha x^\alpha$ is convergent in a rectangle $\{x : \max |x_j| < R\}$. Then a geometric series as in (1.28) can be found that majorizes $\sum a_\alpha x^\alpha$. Indeed, let us fix r with $0 < r < R$. Setting $x = (r, r, \ldots, r)$, we see that $\sum a_\alpha r^{|\alpha|}$ converges, so there is a constant M such that $|a_\alpha r^{|\alpha|}| \leq M$ for all α. Hence

$$|a_\alpha| \leq \frac{M}{r^{|\alpha|}} \leq \frac{M|\alpha|!}{\alpha! r^{|\alpha|}}.$$

We now return to the Cauchy-Kowalevski theorem. We have proved uniqueness of the solution in Proposition (1.21), and the discussion there suggests how to prove existence: determine all the derivatives of u at the origin by differentiating (1.24) and plug the results into Taylor's formula. The problem is to show that the resulting power series converges. To this end, it is technically convenient to replace our kth order equation by a first order system.

(1.31) Theorem.
The Cauchy problem (1.24) is equivalent to the Cauchy problem for a certain first order quasi-linear system,

(1.32)
$$\partial_t Y = \sum_1^{n-1} A_j(x, t, Y)\partial_{x_j} Y + B(x, t, Y),$$

$$Y(x, 0) = \Phi(x),$$

in the sense that a solution to one problem can be read off from a solution to the other. Here Y, B, and Φ are vector-valued functions, the A_j's are matrix-valued functions, and A_j, B, and Φ are explicitly determined by the functions in (1.24).

Proof: The vector Y is to have components $(y_{\alpha j})$, $0 \leq |\alpha| + j \leq k$. In what follows it is understood that $\partial_x^\alpha \partial_t^j u$ is to be replaced by $y_{\alpha j}$ as an independent variable in G. Also, if α is a nonzero multi-index, $i = i(\alpha)$ will denote the smallest number such that $\alpha_i \neq 0$, and 1_i will denote the multi-index with 1 in the ith place and 0 elsewhere. The first order system to be solved is

(1.33)

(a) $\partial_t y_{\alpha j} = y_{\alpha(j+1)}$ $(|\alpha| + j < k)$,

(b) $\partial_t y_{\alpha j} = \partial_{x_i} y_{(\alpha - 1_i)(j+1)}$ $(|\alpha| + j = k,\ j < k)$,

(c) $\partial_t y_{0k} = \dfrac{\partial G}{\partial t} + \displaystyle\sum_{|\alpha| + j < k} \dfrac{\partial G}{\partial y_{\alpha j}} y_{\alpha(j+1)}$

$$+ \sum_{|\alpha| + j = k,\ j < k} \frac{\partial G}{\partial y_{\alpha j}} \partial_{x_i} y_{(\alpha - 1_i)(j+1)},$$

and the initial conditions are

(1.34)
$$\begin{array}{ll}
\text{(a)} & y_{\alpha j}(x,0) = \partial_x^\alpha \phi_j(x) \qquad (j < k), \\
\text{(b)} & y_{0k}(x,0) = G\big(x, 0, (\partial_x^\alpha \phi_j(x))_{|\alpha|+j \le k,\, j < k}\big).
\end{array}$$

It is clear that if u is a solution of (1.24), then the functions $y_{\alpha j} = \partial_x^\alpha \partial_t^j u$ satisfy (1.33) and (1.34). Conversely, if the $y_{\alpha j}$'s satisfy (1.33) and (1.34), we claim that $u = y_{00}$ satisfies (1.24). This is nontrivial and involves the initial conditions in an essential way.

First, (1.33a) implies that

(1.35)
$$y_{\alpha(j+l)} = \partial_t^l y_{\alpha j} \qquad (j + l \le k),$$

and (1.33b) then implies that for $|\alpha| + j = k$ and $j < k$,

$$\partial_t y_{\alpha j} = \partial_t \partial_{x_i} y_{(\alpha - 1_i)j},$$

so that

$$y_{\alpha j}(x,t) = \partial_{x_i} y_{(\alpha - 1_i)j}(x,t) + c_{\alpha j}(x)$$

for some function $c_{\alpha j}$. But by (1.34a),

$$y_{\alpha j}(x,0) = \partial_x^\alpha \phi_j(x) = \partial_{x_i} \partial_x^{\alpha - 1_i} \phi_j(x) = \partial_{x_i} y_{(\alpha - 1_i)j}(x,0),$$

so that $c_{\alpha j} = 0$, and we have

(1.36)
$$y_{\alpha j} = \partial_{x_i} y_{(\alpha - 1_i)j} \qquad (|\alpha| + j = k,\ j < k).$$

Next, by (1.33c), (1.35), and (1.36),

$$\partial_t y_{0k} = \frac{\partial G}{\partial t} + \sum_{|\alpha|+j \le k,\, j < k} \frac{\partial G}{\partial y_{\alpha j}} \frac{\partial y_{\alpha j}}{\partial t} = \frac{\partial}{\partial t}\big[G(x,t,(y_{\alpha j}))\big],$$

so that

$$y_{0k}(x,t) = G\big(x,t,(y_{\alpha j}(x,t))\big) + c_{0k}(x)$$

for some function c_{0k}. But by (1.34),

$$y_{0k}(x,0) = G\big(x,0,(\partial_x^\alpha \phi_j(x))\big) = G\big(x,0,(y_{\alpha j}(x,0))\big),$$

so again $c_{0k} = 0$, and we have

(1.37)
$$y_{0k} = G\big(x,t,(y_{\alpha j})_{|\alpha|+j \le k,\, j < k}\big).$$

Finally, we propose to show that

(1.38) $$y_{\alpha j} = \partial_{x_i} y_{(\alpha-1_i)j} \qquad (\alpha \neq 0).$$

The proof proceeds by induction on $k - j - |\alpha|$, the initial step $k = j + |\alpha|$ being established in (1.36). By (1.33a), (1.35), and the inductive hypothesis,

$$\partial_t y_{\alpha j} = y_{\alpha(j+1)} = \partial_{x_i} y_{(\alpha-1_i)(j+1)} = \partial_t \partial_{x_i} y_{(\alpha-1_i)j},$$

so that

$$y_{\alpha j}(x,t) = \partial_{x_i} y_{(\alpha-1_i)j}(x,t) + c_{\alpha j}(x).$$

But by (1.34a),

$$y_{\alpha j}(x,0) = \partial_x^\alpha \phi_j(x) = \partial_{x_i} \partial_x^{\alpha-1_i} \phi_j(x) = \partial_{x_i} y_{(\alpha-1_i)j}(x,0).$$

Hence $c_{\alpha j} = 0$, and (1.38) is established.

Now we are done. Applying (1.35) and (1.38) repeatedly, we find that

(1.39) $$y_{\alpha j} = \partial_x^\alpha \partial_t^j y_{00},$$

and then by (1.39), (1.37), and (1.34a), $u = y_{00}$ satisfies (1.24).   ∎

We need one further minor simplification.

(1.40) Theorem.
The Cauchy problem (1.32) is equivalent to another problem of the same form in which $\Phi = 0$ and $A_1, \ldots, A_{n-1}, B$ do not depend on t.

Proof: To eliminate Φ we simply set $U(x,t) = Y(x,t) - \Phi(x)$. Then Y satisfies (1.32) if and only if U satisfies

$$\partial_t U = \sum_1^{n-1} \tilde{A}_i(x,t,U)\partial_{x_i}U + \tilde{B}(x,t,U), \qquad U(x,0) = 0,$$

where

$$\tilde{A}_i(x,t,U) = A_i(x,\,t,\,U+\Phi),$$

$$\tilde{B}(x,t,U) = B(x,\,t,\,U+\Phi) + \sum_1^{n-1} A_i(x,\,t,\,U+\Phi)\partial_{x_i}\Phi.$$

Next, to eliminate t from $\tilde{A}_i$ and $\tilde{B}$ we merely add on extra component u^0 to U satisfying $\partial_t u^0 = 1$ and $u^0(x,0) = 0$. Then $u^0 = t$, and we can replace t by u^0 in $\tilde{A}_i$ and $\tilde{B}$ by adding the extra equation and initial condition.   ∎

In Theorems (1.31) and (1.40) there is no harm in assuming that all functions in question are real-valued, as a $\mathbb{C}^N$-valued function can be regarded as an $\mathbb{R}^{2N}$-valued function. Since the constructions in these theorems clearly preserve analyticity, we have reduced the Cauchy-Kowalevski theorem to the following:

(1.41) Theorem.
Suppose that B is an analytic $\mathbb{R}^N$-valued function and $A_1, \ldots, A_{n-1}$ are analytic $N \times N$ real matrix-valued functions defined on a neighborhood of the origin in $\mathbb{R}^n \times \mathbb{R}^N$. Then there is a neighborhood of the origin in $\mathbb{R}^n$ on which the Cauchy problem

$$\partial_t Y = \sum_{1}^{n-1} A_i(x, Y)\partial_{x_i} Y + B(x, Y),$$

(1.42)

$$Y(x, 0) = 0$$

has a unique analytic solution.

Proof: Let $Y = (y_1, \ldots, y_N)$, $B = (b_1, \ldots, b_N)$, $A_i = (a^i_{ml})^N_{m,l=1}$. We seek solutions $y_m = \sum c^{\alpha j}_m x^\alpha t^j$ $(m = 1, \ldots, N)$ of (1.42). The Cauchy data tell us that $c^{\alpha 0}_m = 0$ for all α, m. To determine $c^{\alpha j}_m$ for $j > 0$, we substitute these power series into the differential equations

(1.43) $$\partial_t y_m = \sum_{i,l} a^i_{ml}(x, y_1, \ldots, y_N)\partial_{x_i} y_l + b_m(x, y_1, \ldots, y_N).$$

By (1.27), substituting the series for the y_k's into a^i_{ml} yields a power series in x and t whose coefficients are polynomials with nonnegative coefficients in the $c^{\alpha j}_k$ and the coefficients of the Taylor series of a^i_{ml}. Moreover, the coefficients of the terms in which t occurs to the jth power only involve the $c^{\alpha l}_k$ with $l \le j$. The same is true of the series obtained from b_m and that obtained from $\partial_{x_i} y_l$, and multiplying a^i_{ml} by $\partial_{x_i} y_l$ still preserves these properties.

In short, on the right side of (1.43) we obtain an expression of the form

$$\sum_{\alpha j} P^{\alpha j}_m\left((c^{\beta l}_k)_{l \le j}, \text{ coeff. of } A_i \text{ and } B\right) x^\alpha t^j,$$

where $P^{\alpha j}_m$ is a polynomial with nonnegative coefficients. On the left side, we have

$$\sum_{\alpha j} (j+1)c^{\alpha(j+1)}_m x^\alpha t^j.$$

Hence,

$$c_m^{\alpha(j+1)} = (j+1)^{-1} P_m^{\alpha j}\big((c_k^{\beta l})_{l \le j}\big), \text{ coeff. of } A_i \text{ and } B),$$

so if we know that $c_k^{\beta l}$ with $l \le j$ we can determine the $c_k^{\beta l}$ with $l = j+1$. Proceeding inductively, we determine all the $c_k^{\beta l}$, and more precisely we find that

$$c_m^{\alpha j} = Q_m^{\alpha j}(\text{coeff. of } A_i \text{ and } B),$$

where $Q_m^{\alpha j}$ is again a polynomial with nonnegative coefficients.

Now suppose we have another Cauchy problem

$$\partial_t \widetilde{Y} = \sum_1^{n-1} \widetilde{A}_i(x, \widetilde{Y})\partial_{x_i}\widetilde{Y} + \widetilde{B}(x, \widetilde{Y}),$$

$$\widetilde{Y}(x, 0) = 0$$

($\widetilde{Y}$ again being $\mathbb{R}^N$-valued) for which
(a) we know that an analytic solution exists near $(0, 0)$;
(b) the Taylor series of $\widetilde{A}_i$ and $\widetilde{B}$ majorize those of A_i and B.
Then the solution $\widetilde{Y} = (\widetilde{y}_1, \ldots, \widetilde{y}_N)$ of this problem is given by $\widetilde{y}_m = \sum \widetilde{c}_m^{\alpha j} x^\alpha t^j$, where

$$\widetilde{c}_m^{\alpha j} = Q_m^{\alpha j}(\text{coeff. of } \widetilde{A}_i \text{ and } \widetilde{B}),$$

$Q_m^{\alpha j}$ being the same polynomial as before. Since $Q_m^{\alpha j}$ has nonnegative coefficients,

$$|c_m^{\alpha j}| = |Q_m^{\alpha j}(\text{coeff. of } A_i \text{ and } B)| \le Q_m^{\alpha j}(\text{coeff. of } \widetilde{A}_i \text{ and } \widetilde{B}) = \widetilde{c}_m^{\alpha j}.$$

Hence the series for $\widetilde{Y}$ majorizes the series for Y, and the latter therefore converges in some neighborhood of $(0, 0)$.

It is easy to construct such a majorizing system. Indeed, if $M > 0$ is sufficiently large and $r > 0$ is sufficiently small, by (1.30) the series for A_i and B are all majorized by the series for

$$\frac{Mr}{r - (x_1 + \ldots + x_{n-1}) - (y_1 + \ldots + y_N)}.$$

Thus we consider the following Cauchy problem: for $m = 1, \ldots, N$,

$$(1.44) \qquad \partial_t y_m = \frac{Mr}{r - \sum x_j - \sum y_j}\left[\sum_i \sum_j \partial_{x_i} y_j + 1\right],$$

$$y_m(x, 0) = 0.$$

A solution to this problem is readily found. In fact, if we solve the Cauchy problem

(1.45)
$$\partial_t u = \frac{Mr}{r - s - Nu}[N(n-1)\partial_s u + 1],$$
$$u(s,0) = 0$$

for one (scalar) unknown in two (scalar) variables s and t, and then set

$$y_j(x,t) = u(x_1 + \ldots + x_{n-1}, t) \qquad (j = 1, \ldots, N),$$

then $Y = (y_1, \ldots, y_N)$ will satisfy (1.44).

But we have already seen how to solve (1.45) in §1B. Rewriting the differential equation as

$$(r - s - Nu)\partial_t u - MrN(n-1)\partial_s u = Mr,$$

we solve the ordinary differential equations

$$\frac{dt}{d\tau} = r - s - Nu, \qquad \frac{ds}{d\tau} = -MrN(n-1), \qquad \frac{du}{d\tau} = Mr$$

with initial conditions

$$t(0) = 0, \qquad s(0) = \sigma, \qquad u(0) = 0,$$

and find that

$$t = \tfrac{1}{2}MrN(n-2)\tau^2 + (r-\sigma)\tau, \qquad s = -MrN(n-1)\tau + \sigma, \qquad u = Mr\tau.$$

Eliminating σ and τ yields

$$u(s,t) = \frac{r - s - \sqrt{(r-s)^2 - 2MrNnt}}{Mn}.$$

(The minus sign on the square root is forced by the condition $u(s,0) = 0$.) Clearly this is analytic for s and t near 0, so the proof is complete. ∎

We conclude with a few additional remarks concerning the Cauchy-Kowalevski theorem. First, the theorem asserts the existence of a unique analytic solution in a neighborhood of one point. However, given analytic Cauchy data on an analytic hypersurface S, there is a analytic solution near any point on S, and by uniqueness any two of these solutions must agree on their common domain. Hence we can patch them together and obtain a solution on a neighborhood of S. In general, the neighborhood will depend on the particular problem. (See Exercise 1.)

Second, there remains the question of whether the Cauchy problem (1.24) might admit non-analytic solutions as well. In the linear case, the answer is negative: this is the *Holmgren uniqueness theorem*. The proof can be found in John [30], Hörmander [26], [27, vol. I], or Treves [52].

A major drawback of the Cauchy-Kowalevski theorem is that it gives little control over the dependence of the solution on the Cauchy data. Consider the following example in $\mathbb{R}^2$, due to Hadamard:

$$(1.46) \qquad \begin{aligned} &\partial_1^2 u + \partial_2^2 u = 0, \\ &u(x_1, 0) = 0, \qquad \partial_2 u(x_1, 0) = k e^{-\sqrt{k}} \sin kx_1, \end{aligned}$$

where k is a positive integer. This problem is non-characteristic since $\partial_1^2 + \partial_2^2$ is elliptic, and one easily checks that the solution is

$$u(x_1, x_2) = e^{-\sqrt{k}}(\sin kx_1)(\sinh kx_2).$$

As $k \to \infty$, the Cauchy data and their derivatives of all orders tend uniformly to zero since $e^{-\sqrt{k}}$ decays faster than polynomially. But if $x_2 \neq 0$, $\lim e^{-\sqrt{k}} \sinh kx_2 = \infty$, so as $k \to \infty$ the solution oscillates more and more rapidly with greater and greater amplitude, and in the end it blows up altogether. The solution for the limiting case $k = \infty$ is of course $u \equiv 0$. This example shows that the solution of the Cauchy problem may not depend continuously on the Cauchy data in most of the usual topologies on functions.

The proof of the Cauchy-Kowalevski theorem obviously depends in an essential way on the assumption of analyticity. In the linear case, there is a more general version of the theorem that requires analyticity only in x, not in t; see Treves [52]. However, the analyticity hypothesis cannot be discarded completely, as we shall see in the next section.

EXERCISES

1. Let S be the unit circle in the complex plane. A function ϕ on S is analytic if and only if it is the restriction to S of a holomorphic function on an annulus $A_{ab} = \{z : a < |z| < b\}$ where $a < 1 < b$; in this case, if $\sum_{-\infty}^{\infty} c_m e^{im\theta}$ is the Fourier series of ϕ, $\sum_{-\infty}^{\infty} c_m z^m$ is the Laurent series of its holomorphic extension to A.

a. Use this fact to find an explicit solution of the Cauchy problem for the Laplacian on S:

$$\Delta u = 0 \text{ on a neighborhood of } S, \ u = \phi \text{ and } \partial_r u = \psi \text{ on } S. \quad (*)$$

Here $z = re^{i\theta}$, so ∂_r is the normal derivative on S, and ϕ and ψ are analytic on S. (Hint: $z^m = r^m e^{im\theta}$ and $\overline{z^{-m}} = r^{-m} e^{im\theta}$ are both harmonic functions that agree with $e^{im\theta}$ on S.)

b. Show that if $a < 1 < b$, there exist ϕ and ψ for which the solution to $(*)$ exists on the annulus A_{ab} but cannot be extended beyond A_{ab}. (Hint: for any disc D there are holomorphic functions on D that cannot be extended holomorphically beyond D.)

2. Carry out explicitly the reduction of the Cauchy problem for Laplace's equation $\Delta u = f$ to a first order system as in Theorem (1.31), taking S to be the hyperplane $x_n = 0$.

3. To see how the ideas in the proof of the Cauchy-Kowalevski theorem work in a simpler setting, prove the following theorem about *ordinary differential equations in the complex domain*. Suppose $p(z) = \sum_0^\infty p_m z^m$ and $q(z) = \sum_0^\infty q_m z^m$ are holomorphic in the disc $|z| < R$, and consider the initial value problem

$$u'' = p(z)u' + q(z)u, \qquad u(0) = c_0, \quad u'(0) = c_1. \quad (**)$$

a. Show that if $u(z) = \sum_0^\infty c_m z^m$ satisfies $(**)$, then

$$c_{m+2} = \frac{1}{(m+2)(m+1)} \sum_0^m [(j+1)c_{j+1}p_{m-j} + c_j q_{m-j}],$$

so that the coefficients c_m are uniquely determined by $(**)$.

b. Suppose $P_m \geq |p_m|$ and $Q_m \geq |q_m|$ for all $m \geq 0$, and let $P(z) = \sum_0^\infty P_m z^m$, $Q(z) = \sum_0^\infty Q_m z^m$. Show that if $U(z) = \sum_0^\infty C_m z^m$ satisfies $U'' = P(z)U' + Q(z)U$, and $C_0 \geq |c_0|$ and $C_1 \geq |c_1|$, then $C_m \geq |c_m|$ for all m.

c. Suppose $r < R$. Show that the conditions of (b) are satisfied if we take $P_m = Kr^{-m}$ and $Q_m = K(m+1)r^{-m}$ for K sufficiently large, so that $P(z) = (1 - r^{-1}z)^{-1}$ and $Q(z) = (1 - r^{-1}z)^{-2}$. Show also that the general solution of $U'' = P(z)U' + Q(z)U$ in this case is a linear combination of $(1 - r^{-1}z)^\alpha$ and $(1 - r^{-1}z)^\beta$ for suitable exponents α and β.

d. Conclude that $(**)$ has a unique holomorphic solution u in the disc $|z| < R$.

E. Local Solvability: The Lewy Example

It was long believed that any "reasonable" partial differential equation (with no boundary conditions imposed) should have many solutions. In particular, suppose we have a linear equation $\sum_{|\alpha| \leq k} a_\alpha \partial^\alpha u = f$ where f and the coefficients a_α are C^∞. Given $x_0 \in \mathbb{R}^n$, can we find a solution u (not necessarily C^∞) of this equation in some neighborhood of x_0? If f and the a_α's are analytic and $a_\alpha(x_0) \neq 0$ for some α with $|\alpha| = k$, the Cauchy-Kowalevski theorem shows that the answer is yes. Indeed, we can choose a vector ξ which is non-characteristic for $\sum a_\alpha \partial^\alpha$ at x_0 (and hence at all x in a neighborhood of x_0) and solve the Cauchy problem with zero Cauchy data on the hyperplane through x_0 orthogonal to ξ.

One might well expect that the assumption of analyticity can be omitted. But in 1957 Hans Lewy [33] destroyed all hopes for such a theorem with the following embarrassingly simple counterexample. Consider the differential operator L defined on $\mathbb{R}^3$ with coordinates (x, y, t) by

$$(1.47) \qquad L = \frac{\partial}{\partial x} + i\frac{\partial}{\partial y} - 2i(x + iy)\frac{\partial}{\partial t}.$$

(1.48) Theorem.
Let f be a continuous real-valued function depending only on t. If there is a C^1 function u of (x, y, t) satisfying $Lu = f$ in some neighborhood of the origin, then f is analytic at $t = 0$.

Proof: Suppose $Lu = f$ in the set where $x^2 + y^2 < R^2$ and $|t| < R$ $(R > 0)$. Set $z = x + iy$; write z in polar coordinates as $re^{i\theta}$ and set $s = r^2$. Consider the quantity V, a function of t and r (or equivalently of t and s) defined for $0 < r < R$ and $|t| < R$ by the contour integral

$$V = \int_{|z|=r} u(x, y, t)\, dz = ir\int_0^{2\pi} u(r\cos\theta,\, r\sin\theta,\, t)\, e^{i\theta}\, d\theta.$$

By Green's theorem,

$$V = i\iint_{|z| \leq r} \left[\frac{\partial u}{\partial x} + i\frac{\partial u}{\partial y}\right](x, y, t)\, dx\, dy$$

$$= i\int_0^r \int_0^{2\pi} \left[\frac{\partial u}{\partial x} + i\frac{\partial u}{\partial y}\right](\rho\cos\theta,\, \rho\sin\theta,\, t)\, \rho\, d\rho\, d\theta.$$

Hence

$$\frac{\partial V}{\partial r} = i\int_0^{2\pi} \left[\frac{\partial u}{\partial x} + i\frac{\partial u}{\partial y}\right](r\cos\theta,\, r\sin\theta,\, t)\, r\, d\theta$$

$$= \int_{|z|=r} \left[\frac{\partial u}{\partial x} + i\frac{\partial u}{\partial y}\right](x, y, t)\, r\frac{dz}{z}.$$

The equation $Lu = f$ then implies

$$\frac{\partial V}{\partial s} = \frac{1}{2r} \frac{\partial V}{\partial r} = \int_{|z|=r} \left[\frac{\partial u}{\partial x} + i \frac{\partial u}{\partial y} \right] (x, y, t) \frac{dz}{2z}$$

$$= i \int_{|z|=r} \frac{\partial u}{\partial t} (x, y, t) \, dz + \int_{|z|=r} f(t) \frac{dz}{2z} = i \frac{\partial V}{\partial t} + \pi i f(t).$$

Thus if we set $F(t) = \int_0^t f(\tau) \, d\tau$, the quantity $U(t, s) = V(t, s) + \pi F(t)$ satisfies

$$\frac{\partial U}{\partial t} + i \frac{\partial U}{\partial s} = 0.$$

This is the Cauchy-Riemann equation, so U is a holomorphic function of $w = t + is$ in the region $0 < s < R^2$, $|t| < R$, and U is continuous up to the line $s = 0$. Moreover, $V = 0$ when $s = 0$, so $U(0, t) = \pi F(t)$ is real-valued. Therefore, by the Schwarz reflection principle, the formula $U(t, -s) = \overline{U(t, s)}$ gives a holomorphic continuation of U to a full neighborhood of the origin. In particular, $U(t, 0) = \pi F(t)$ is analytic in t, hence so is $f = F'$. ∎

There is really nothing special about the origin in Theorem 1.48. In fact, a change-of-variable argument that we outline in Exercise 1 shows that for any $(x_0, y_0, t_0) \in \mathbb{R}^3$, the equation

$$Lu(x, y, t) = f(t + 2y_0 x - 2x_0 y)$$

has no C^1 solution in any neighborhood of (x_0, y_0, t_0) unless $f(\tau)$ is analytic at $\tau = t_0$.

Once this is known, it is not hard to show that there are C^∞ functions g on $\mathbb{R}^3$ such that the equation $Lu = g$ has no solution $u \in C^{1+\alpha}$ $(\alpha > 0)$ in any neighborhood of any point. The idea is as follows. Pick a C^∞ periodic function f on $\mathbb{R}$ that is not analytic at any point, and pick a countable dense set $\{(x_j, y_j, t_j)\}_1^\infty$ in $\mathbb{R}^3$. Then there is a sequence of positive constants $\{c_j\}$ tending to zero rapidly enough so that the series

$$g_a(x, y, t) = \sum_{j=1}^{\infty} a_j c_j f(t + 2y_j x - 2x_j y)$$

defines a C^∞ function on $\mathbb{R}^3$ for any bounded sequence $a = \{a_j\}$. One can then show that for "most" sequences a, in the sense of Baire category in the space l^∞ of bounded sequences, the equation $Lu = g_a$ has no $C^{1+\alpha}$

solution near any point. For the details of this argument, see Lewy [33] or John [30].

This construction also leads immediately to an example of a *homogeneous* equation with no nontrivial solution. Namely, if f is a C^∞ function on $\mathbb{R}^3$ such that the equation $Lu = f$ has no solution near any point, then the equation $Lv - fv = 0$ has no solution except $v \equiv 0$. Indeed, suppose v is a solution that is nonzero on an open set Ω. By shrinking Ω we can assume that a single-valued branch of the logarithm can be defined on $v(\Omega)$, and then $u = \log v$ is a solution of $Lu = f$ on Ω.

A couple of years after Lewy proved Theorem (1.48), Hörmander embedded it into a more general result that initiated the theory of local solvability of differential operators. We make a formal definition: A linear differential operator L with C^∞ coefficients is said to be **locally solvable** at x_0 if there is a neighborhood Ω of x_0 such that for every $f \in C_c^\infty(\Omega)$ there exists $u \in \mathcal{D}'(\Omega)$ with $Lu = f$. Hörmander's theorem is then as follows; see [26] for the proof.

(1.49) Theorem.
Let L be a linear differential operator with C^∞ coefficients on Ω, let $P(x, \xi) = \chi_L(x, \xi)$ be the characteristic form of L, and let

$$Q(x, \xi) = \sum_{1}^{n} \left[\frac{\partial P}{\partial \xi_j}(x, \xi) \frac{\partial \overline{P}}{\partial x_j}(x, \xi) - \frac{\partial P}{\partial x_j}(x, \xi) \frac{\partial \overline{P}}{\partial \xi_j}(x, \xi) \right].$$

  a. If $x_0 \in \Omega$ and there is a $\xi \in \mathbb{R}^n$ such that $P(x_0, \xi) = 0$ but $Q(x_0, \xi) \neq 0$, then L is not locally solvable at x_0.

  b. If for each $x \in \Omega$ there is a $\xi \in \mathbb{R}^n$ such that $P(x, \xi) = 0$ but $Q(x, \xi) \neq 0$, then there is an $f \in C^\infty(\Omega)$ such that the equation $Lu = f$ has no distribution solution on any open subset of Ω.

It is easy to check that the Lewy operator (1.47) satisfies the hypothesis of (b) on $\Omega = \mathbb{R}^n$. Thus we obtain a strengthening of Theorem (1.48): there exist functions $f \in C^\infty$ for which the Lewy equation has no solution in $\mathcal{D}'$, not merely in C^1.

Theorem (1.49) was the starting point for a considerable body of research into necessary and sufficient conditions for local solvability. For an account of this work, we refer the reader to the expository articles of Treves [50], [51], and to Beals and Fefferman [6]. We mention also that Greiner, Kohn, and Stein have found a necessary and sufficient condition on f for the Lewy equation $Lu = f$ to be locally solvable; see Stein [46, §XIII.4].

EXERCISES

1. Given $(x_0, y_0, t_0) \in \mathbb{R}^3$, define the transformation T of $\mathbb{R}^3$ by

$$T(x, y, t) = (x - x_0, \, y - y_0, \, t - t_0 - 2y_0 x + 2x_0 y).$$

Show that if L is the Lewy operator (1.47), then $L(u \circ T) = (Lu) \circ T$ for any function u, and conclude that solving $Lu = f(t + 2y_0 x - 2x_0 y)$ near (x_0, y_0, t_0) is equivalent to solving $L(u \circ T) = f(t - t_0)$ near the origin.

2. (Addendum to Exercise 1.) Show that the binary operation $*$ on $\mathbb{R}^3$ defined by

$$(a, b, c) * (x, y, t) = (a + x, \, b + y, \, c + t + 2bx - 2ay)$$

makes $\mathbb{R}^3$ into a group. (This group is known as the 3-dimensional *Heisenberg group*. Exercise 1 says that the Lewy operator is invariant under left translations on this group, as the transformation T is left translation by $(-x_0, -y_0, -t_0)$.)

3. The Lewy operator (1.47) arises in complex analysis because the equation $Lu = 0$ is, in a sense, the restriction of the Cauchy-Riemann equations on $\mathbb{C}^2$ to the hypersurface $\{(z_1, z_2) : \operatorname{Im} z_2 = |z_1|^2\}$. More precisely, define $\Phi : \mathbb{R}^3 \to \mathbb{C}^2$ by

$$\Phi(x, y, t) = \big(x + iy, \, t + i(x^2 + y^2)\big).$$

Suppose f is a holomorphic function on $\mathbb{C}^2$, so that it satisfies the Cauchy-Riemann equations $(\partial_{u_j} + i\partial_{v_j})f = 0$ for $j = 1, 2$, where $z_j = u_j + iv_j$. Show that $L(f \circ \Phi) = 0$.

4. Local non-solvability may occur for relatively trivial reasons when the characteristic form of an operator vanishes at a point. Show, for example, that the equation $x\partial_y u - y\partial_x u = x^2 + y^2$ has no continuous solutions in any neighborhood of $(0,0)$ in $\mathbb{R}^2$. (Hint: show that $x\partial_y - y\partial_x = \partial_\theta$ in polar coordinates.)

F. Constant-Coefficient Operators: Fundamental Solutions

A couple of years before Lewy discovered his example, Malgrange and Ehrenpreis independently proved that every linear differential operator with constant coefficients has a fundamental solution (a concept we shall define below). An immediate corollary is that every constant-coefficient operator is locally solvable, and one can deduce regularity properties of the solutions by examination of the fundamental solution. In this section we shall derive these results, following an argument of Nirenberg [38].

Let

$$L = \sum_{|\alpha| \leq k} c_\alpha \partial^\alpha$$

be a differential operator with constant coefficients. The natural tool for studying such operators is the Fourier transform; if f is any tempered distribution, we have

(1.50) $(Lu)\widehat{\ }(\xi) = P(\xi)\widehat{u}(\xi),$

where

$$P(\xi) = \sum_{|\alpha| \leq k} c_\alpha (2\pi i \xi)^\alpha.$$

P is called the **symbol** of L; this notation relating L and P will be maintained throughout this section.

We begin by considering the question of local solvability of L. In view of (1.50), if $f \in C_c^\infty$, it would seem that we should be able to solve $Lu = f$ by taking $\widehat{u} = \widehat{f}/P$, that is,

(1.51) $$u(x) = \int e^{2\pi i x \cdot \xi} \frac{\widehat{f}(\xi)}{P(\xi)} \, d\xi.$$

The trouble with this is that usually the polynomial P will have zeros, so that $\widehat{f}/P$ is not a locally integrable function and the integral (1.51) is not well-defined. However, since $f \in C_c^\infty$, $\widehat{f}$ extends to an entire holomorphic function on $\mathbb{C}^n$ by Proposition (0.30). The idea will therefore be to make sense of (1.51) by deforming the contour of integration so as to avoid the zeros of P.

To this end, we make a simplification. By a rotation of coordinates we can assume that the vector $(0, \ldots, 0, 1)$ is non-characteristic for L, which means that the coefficient of ξ_n^k in $P(\xi)$ is nonzero. After dividing everything through by it, we may — and shall — assume that this coefficient is 1, so that

$$P(\xi) = \xi_n^k + \text{terms of lower order in } \xi_n.$$

For each fixed $\xi' = (\xi_1, \ldots, \xi_{n-1}) \in \mathbb{R}^{n-1}$, we consider $P(\xi) = P(\xi', \xi_n)$ as a polynomial in the single *complex* variable ξ_n. Let $\lambda_1(\xi'), \ldots, \lambda_k(\xi')$ be its zeros, counted according to multiplicity and arranged so that for $i \leq j$, $\operatorname{Im} \lambda_i(\xi') \leq \operatorname{Im} \lambda_j(\xi')$ and $\operatorname{Re} \lambda_i(\xi') \leq \operatorname{Re} \lambda_j(\xi')$ if $\operatorname{Im} \lambda_i(\xi') = \operatorname{Im} \lambda_j(\xi')$. By Rouché's theorem, a small perturbation of ξ' produces a small perturbation of the zeros of $P(\xi', \xi_n)$, so the functions $\operatorname{Im} \lambda_j(\xi')$ are continuous functions of ξ'. Before proceeding to the main results, we need two lemmas.

(1.52) Lemma.
There is a measurable function $\phi : \mathbb{R}^{n-1} \to [-k, k]$ such that for all $\xi' \in \mathbb{R}^{n-1}$,

$$\min\{|\phi(\xi') - \operatorname{Im} \lambda_j(\xi')| : 1 \leq j \leq k\} \geq 1.$$

Proof: The idea is simple: There are at most k distinct points among $\operatorname{Im} \lambda_j(\xi')$ $(1 \leq j \leq k)$, so at least one of the $k + 1$ intervals $[2m - k - 1, 2m - k + 1)$ $(0 \leq m \leq k)$ must contain none of them, and we can take $\phi(\xi')$ to be the midpoint of that interval. That is, for $0 \leq m \leq k$, let

$$V_m = \{\xi' : \operatorname{Im} \lambda_j(\xi') \notin [2m - k - 1, 2m - k + 1) \text{ for } j = 1, \ldots, k\}.$$

Then the sets V_m cover $\mathbb{R}^{n-1}$, and they are Borel sets since $\operatorname{Im} \lambda_j$ is continuous, so we can take

$$\phi(x) = 2m - k \text{ for } x \in V_m \setminus \bigcup_0^{m-1} V_l \qquad (0 \leq m \leq k). \qquad \blacksquare$$

(1.53) Lemma.
Let $g(z)$ be a monic polynomial of degree k in the complex variable z such that $g(0) \neq 0$, and let $\lambda_1, \ldots, \lambda_k$ be its zeros. Then $|g(0)| \geq (d/2)^k$ where $d = \min|\lambda_j|$.

Proof: We have $g(z) = (z - \lambda_1) \cdots (z - \lambda_k)$, so

$$\left|\frac{g(z)}{g(0)}\right| = \prod_1^k \left|1 - \frac{z}{\lambda_k}\right| \leq 2^k \text{ for } |z| \leq d.$$

Moreover, $g^{(k)}(z) \equiv k!$, so by the Cauchy integral formula,

$$k! = |g^{(k)}(0)| = \left|\frac{k!}{2\pi i} \int_{|z|=d} \frac{g(z)}{z^{k+1}} \, dz\right| \leq \frac{k! 2^k |g(0)|}{d^k},$$

which is the desired result. $\blacksquare$

(1.54) Theorem.
If L is a differential operator with constant coefficients on $\mathbb{R}^n$ and $f \in C_c^\infty(\mathbb{R}^n)$, there exists $u \in C^\infty(\mathbb{R}^n)$ such that $Lu = f$.

Proof: We employ the notation introduced above. Let ϕ be as in Lemma (1.52), and set

$$(1.55) \qquad u(x) = \int_{\mathbb{R}^{n-1}} \int_{\mathrm{Im}\,\xi_n = \phi(\xi')} e^{2\pi i x \cdot \xi} \frac{\widehat{f}(\xi)}{P(\xi)} \, d\xi_n \, d\xi'.$$

By Lemma (1.53) (applied to $g(z) = P(\xi', \xi_n + z)$) together with Lemma (1.52), we see that $|P(\xi)| \geq 2^{-k}$ when $\mathrm{Im}\,\xi_n = \phi(\xi')$. Moreover, by Proposition (0.30), $\widehat{f}(\xi)$ is rapidly decaying as $|\mathrm{Re}\,\xi| \to \infty$ when $|\mathrm{Im}\,\xi|$ remains bounded. Hence the integrand in (1.55) is bounded and rapidly decaying at infinity, so the integral is absolutely convergent. For the same reason, we can differentiate under the integral as often as we please and conclude that u is C^∞ and that

$$Lu(x) = \int_{\mathbb{R}^{n-1}} \int_{\mathrm{Im}\,\xi_n = \phi(\xi')} e^{2\pi i x \cdot \xi} \widehat{f}(\xi) \, d\xi_n \, d\xi'.$$

But now the integrand is an entire function which is rapidly decaying as $|\mathrm{Re}\,\xi| \to \infty$, so by Cauchy's theorem we can deform the contour of integration in ξ_n back to the real axis. By the Fourier inversion theorem, then, $Lu = f$. ∎

The content of Theorem (1.54) can be usefully rephrased as follows. A **fundamental solution** for the constant-coefficient operator L is a distribution K on $\mathbb{R}^n$ such that $LK = \delta$, where δ is the point mass at the origin. On the one hand, Theorem (1.54) is an immediate corollary of the existence of a fundamental solution, for if $f \in C_c^\infty$ we can take $u = K * f$: then $Lu = LK * f = \delta * f = f$. On the other hand, the proof of Theorem (1.54) easily yields a fundamental solution.

(1.56) The Malgrange-Ehrenpreis Theorem.
Every differential operator L with constant coefficients has a fundamental solution.

Proof: With notation as above, define a linear functional K on C_c^∞ by

$$\langle K, f \rangle = \int_{\mathbb{R}^{n-1}} \int_{\mathrm{Im}\,\xi_n = \phi(\xi')} \frac{\widehat{f}(-\xi)}{P(\xi)} \, d\xi_n \, d\xi'.$$

As in the proof of Theorem (1.54), the integral is bounded by

$$C \sup_{|\operatorname{Im}\xi|\leq k} (1+|\xi|)^{-n-1}|\widehat{f}(\xi)| \leq C' \sum_{|\alpha|\leq n+1} \|\partial^\alpha f\|_\infty,$$

where C and C' depend only on the support of f, so K is a distribution. Moreover, $\langle LK, f\rangle = \langle K, L'f\rangle$ where L' is the operator with symbol $P(-\xi)$, so that $(L'f)\widehat{\ }(-\xi) = P(\xi)\widehat{f}(-\xi)$. Hence, as in the proof of Theorem (1.54),

$$\langle LK, f\rangle = \int_{\mathbb{R}^{n-1}} \int_{\operatorname{Im}\xi_n=\phi(\xi')} \widehat{f}(-\xi)\,d\xi = \int_{\mathbb{R}^n} \widehat{f}(\xi)\,d\xi = f(0) = \langle \delta, f\rangle.$$

Alternatively, one could observe that $K * f$ is the function u defined by (1.55), so that $LK * f = Lu = f$ for all f and hence $LK = \delta$.   ∎

With a fundamental solution K in hand, we can solve the equation $Lu = f$ not only when $f \in C_c^\infty$ but when f is any distribution with compact support; of course, the solution u will then be a distribution. Indeed, if $f \in \mathcal{E}'$, we have

$$(1.57) \qquad\qquad L(K * f) = K * Lf = f,$$

since $L(K * f)$ and $K * Lf$ are both equal to $LK * f = \delta * f = f$. These relations can often be extended to f's which are not compactly supported, but the class of f's for which they hold will depend on the nature of K. We shall see some specific examples in Chapters 2, 4, and 5.

Fundamental solutions are useful not only for producing solutions of differential equations but also for studying their regularity properties. In particular, we have the following important result.

A differential operator L with C^∞ coefficients is called **hypoelliptic** if any distribution u on an open set Ω such that Lu is C^∞ on Ω must itself be C^∞ on Ω, that is, if all solutions of the equation $Lu = f$ must be C^∞ wherever f is C^∞. (The origin of this term is the fact that all elliptic operators are hypoelliptic, a fact which we shall prove in Chapter 6.)

(1.58) Theorem.
If L is a differential operator with constant coefficients, the following are equivalent:
a. Some fundamental solution for L is C^∞ on $\mathbb{R}^n \setminus \{0\}$.
b. Every fundamental solution for L is C^∞ on $\mathbb{R}^n \setminus \{0\}$.
c. L is hypoelliptic.

Proof: If K is a fundamental solution for L, then $LK = \delta$ is C^∞ on $\mathbb{R}^n \setminus \{0\}$, so (c) implies (b). (b) trivially implies (a), so it remains to show that (a) implies (c). For this we need a lemma.

(1.59) Lemma.
Suppose f and g are distributions on $\mathbb{R}^n$, f is C^∞ on $\mathbb{R}^n \setminus \{0\}$, and g has compact support. Then $f * g$ is C^∞ on $\mathbb{R}^n \setminus (\text{supp } g)$.

Proof: Given $x \notin \text{supp } g$, choose $\epsilon > 0$ small enough so that $B_\epsilon(x)$ and supp g are disjoint, and choose $\phi \in C_c^\infty(B_{\epsilon/2}(0))$ such that $\phi = 1$ on $B_{\epsilon/4}(0)$. Then we can write

$$f * g = (\phi f) * g + [(1 - \phi)f] * g.$$

On the one hand, $(1 - \phi)f$ is a C^∞ function, so $[(1 - \phi)f] * g$ is C^∞ everywhere. On the other hand,

$$\text{supp}[(\phi f) * g] \subset \{x + y : x \in \text{supp } \phi, \ y \in \text{supp } g\},$$

which is disjoint from $B_{\epsilon/2}(x)$. Hence, on $B_{\epsilon/2}(x)$, $f * g = [(1 - \phi)f] * g$ is C^∞. ∎

Returning to the proof of Theorem (1.58), let K be a fundamental solution for L that is C^∞ on $\mathbb{R}^n \setminus \{0\}$. Suppose u is a distribution on an open set $\Omega \subset \mathbb{R}^n$ such that Lu is C^∞ on Ω. If $x \in \Omega$, we pick $\epsilon > 0$ so that $B_\epsilon(x) \subset \Omega$, and we shall show that u is C^∞ on $B_{\epsilon/2}(x)$.

Pick $\phi \in C_c^\infty(B_\epsilon(x))$ with $\phi = 1$ on $B_{\epsilon/2}(x)$. Then $L(\phi u) = \phi Lu + v$ where $v = 0$ on $B_{\epsilon/2}(x)$ and outside $B_\epsilon(x)$. $K * (\phi Lu)$ is C^∞ since $\phi Lu \in C_c^\infty$, and $K * v$ is C^∞ on $B_{\epsilon/2}(x)$ by Lemma (1.59). But by (1.57),

$$\phi u = K * L(\phi u) = K * \phi Lu + K * v,$$

so on $B_{\epsilon/2}(x)$, $u = \phi u$ is C^∞. ∎

EXERCISES

1. Let $L = \sum_0^k c_j (d/dx)^j$ be an ordinary differential operator with constant coefficients. Let v be the solution of $Lv = 0$ satisfying $v(0) = \cdots = v^{(k-2)}(0) = 0$, $v^{(k-1)}(0) = c_k^{-1}$. Define $K(x) = 0$ if $x \leq 0$, $K(x) = v(x)$ if $x > 0$. Show that K is a fundamental solution for L.

2. Show that the characteristic function of $\{(x,y) : x > 0, \ y > 0\}$ is a fundamental solution for $\partial_x \partial_y$ in $\mathbb{R}^2$.

3. Show that $K(x,y) = [2\pi i(x + iy)]^{-1}$ is a fundamental solution for the Cauchy-Riemann operator $L = \partial_x + i\partial_y$ on $\mathbb{R}^2$. Hint: if $\phi \in C_c^\infty$,

$$\langle LK, \phi \rangle = -\langle K, L\phi \rangle = \frac{-1}{2\pi i} \lim_{\epsilon \to 0} \iint_{x^2+y^2 > \epsilon^2} \frac{\partial_x \phi + i\partial_y \phi}{x + iy} \, dx \, dy.$$

Use Green's theorem to show that this equals

$$\lim_{\epsilon \to 0} \frac{1}{2\pi i} \int_{x^2+y^2=\epsilon^2} \phi(x,y) \frac{dx + i\,dy}{x + iy} = \lim_{\epsilon \to 0} \frac{1}{2\pi} \int_0^{2\pi} \phi(\epsilon \cos \theta, \epsilon \sin \theta) \, d\theta.$$

4. Suppose L is a constant-coefficient differential operator. Modify the proof of Theorem (1.58) to show that if there is a distribution K that is C^∞ away from the origin and satisfies $LK = \delta + f$ where f is C^∞, then L is hypoelliptic.

Chapter 2
THE LAPLACE OPERATOR

For the next four chapters we depart from general theory to investigate the great trinity of operators from mathematical physics: the Laplace, heat, and wave operators. These operators are of fundamental importance not only because of their applications but because they are archetypes of more general phenomena; indeed, much of the theory of partial differential equations as it now stands has its roots in the study of these operators.

We begin with what is perhaps the most important of all partial differential operators, the **Laplace operator** or **Laplacian** Δ on $\mathbb{R}^n$ defined by

$$\Delta = \sum_1^n \partial_j^2 = \nabla \cdot \nabla.$$

It is useful to have a physical model in mind when thinking of Δ, and perhaps the simplest (among several) comes from the theory of electrostatics. According to Maxwell's equations, an electrostatic field E in space (a vector field representing the electrostatic force on a unit positive charge) is related to the charge density in space, f, by the equation $\nabla \cdot E = f$ (provided the units of measurement are properly chosen: one usually finds $4\pi f$ in place of f) and also satisfies curl $E = 0$. (In n dimensions, curl E is the matrix $(\partial_i E_j - \partial_j E_i)$.) The second condition means that, at least locally, E is the gradient of a function $-u$, determined up to an additive constant, called the electrostatic potential. We therefore have $\Delta u = -\nabla \cdot E = -f$, so the Laplacian relates the potential to the charge density. For more details, see, e.g., Kellogg [31].

Throughout this chapter we assume implicitly that we are working in $\mathbb{R}^n$ with $n > 1$. Much of the theory is valid also for $n = 1$ but becomes more or less trivial there.

A. Symmetry Properties of the Laplacian

The Laplacian is not only important in its own right but also forms the spatial component of the heat operator $\partial_t - \Delta$ and the wave operator $\partial_t^2 - \Delta$. Why is it so ubiquitous? The answer, which we shall now prove, is that it commutes with translations and rotations and generates the ring of all differential operators with this property. Hence, the Laplacian is likely to turn up in the description of any physical process whose underlying physics is homogeneous (independent of position) and isotropic (independent of direction).

More precisely, to say that an operator L on functions on $\mathbb{R}^n$ commutes with translations (rotations) means that $L(f \circ T) = (Lf) \circ T$ for any translation (rotation) on $\mathbb{R}^n$. We also say that a function f on $\mathbb{R}^n$ is **radial** if is rotation-invariant ($f \circ T = f$ for all rotations T). Thus f is radial precisely when it is constant on every sphere about the origin, or equivalently when $f(x)$ depends only on $|x|$.

(2.1) Theorem.
Suppose L is a partial differential operator on $\mathbb{R}^n$. Then L commutes with translations and rotations if and only if L is a polynomial in Δ — that is, $L = \sum a_j \Delta^j$ for some constants a_j.

Proof: We leave it to the reader (Exercise 1) to verify that L commutes with translations if and only if L has constant coefficients, say $L = \sum c_\alpha \partial^\alpha$. In this case, we have $(Lu)\widehat{\ }(\xi) = P(\xi)\widehat{f}(\xi)$ where $P(\xi) = \sum c_\alpha (2\pi i \xi)^\alpha$. Since the Fourier transform commutes with rotations (Proposition (0.21)), it follows easily that L commutes with rotations if and only if P is radial. In particular, this is true when $L = \sum a_j \Delta^j$, i.e., $P(\xi) = \sum a_j (2\pi |\xi|)^{2j}$. On the other hand, since composition with linear maps preserves homogeneity, P is radial if and only if each homogeneous piece $P_j(\xi) = \sum_{|\alpha|=j} c_\alpha (2\pi i \xi)^\alpha$ is radial. But this means that each P_j depends only on $|\xi|$; since it is homogeneous of degree j we must have $P_j(\xi) = b_j (2\pi i |\xi|)^j$ for some constant b_j, and since P_j is a polynomial we must have $b_j = 0$ when j is odd. In short, $P(\xi) = \sum b_{2j} (2\pi i |\xi|)^{2j}$, so $L = \sum b_{2j} \Delta^j$. ∎

Since the Laplacian commutes with rotations, it preserves the class of radial functions, on which it reduces to an ordinary differential operator called the **radial part** of the Laplacian. For future reference, we compute it explicitly.

(2.2) Proposition.
If $f(x) = \phi(r)$ where $x \in \mathbb{R}^n$ and $r = |x|$, then

$$\Delta f(x) = \phi''(r) + \frac{n-1}{r}\phi'(r).$$

Proof: Since $\partial r/\partial x_j = x_j/r$, we have

$$\Delta f(x) = \sum_1^n \partial_j \left[\frac{x_j}{r}\phi'(r)\right] = \sum_1^n \left[\frac{x_j^2}{r^2}\phi''(r) + \frac{1}{r}\phi'(r) - \frac{x_j^2}{r^3}\phi'(r)\right]$$

$$= \phi''(r) + \frac{n}{r}\phi'(r) - \frac{1}{r}\phi'(r). \qquad \blacksquare$$

(2.3) Corollary.
If $f(x) = \phi(r)$ is a radial function on $\mathbb{R}^n$, then f satisfies $\Delta f = 0$ on $\mathbb{R}^n \backslash \{0\}$ if and only if $\phi(r) = a + br^{2-n}$ ($n \neq 2$) or $\phi(r) = a + b\log r$ ($n = 2$), where a and b are constants.

Proof: $\Delta f = 0$ means that $\phi''(r)/\phi'(r) = (1-n)/r$. Integration gives $\log \phi'(r) = (1-n)\log r + \log c$, or $\phi'(r) = cr^{1-n}$. One more integration yields the desired result. $\qquad \blacksquare$

EXERCISES

1. Show that a differential operator L on $\mathbb{R}^n$ commutes with translations if and only if L has constant coefficients. (Hint: consider the action of L on the monomials x^α.)

2. Show that the wave operator $\partial_t^2 - \partial_x^2$ on $\mathbb{R}^2$ commutes with the Lorentz transformations (hyperbolic rotations) $T_\theta = \begin{pmatrix} \cosh\theta & \sinh\theta \\ \sinh\theta & \cosh\theta \end{pmatrix}$ ($\theta \in \mathbb{R}$).

B. Basic Properties of Harmonic Functions

A C^2 function u on an open subset of $\mathbb{R}^n$ is said to be **harmonic** if $\Delta u = 0$. (We shall soon see, in Corollary (2.20), that the hypothesis $u \in C^2$ can be relaxed without changing anything.) We proceed to derive some of the basic properties of harmonic functions. In what follows, we will need to integrate over various hypersurfaces S that are boundaries of domains Ω in $\mathbb{R}^n$; $d\sigma$ will denote the surface measure on S and ν will denote the unit normal vector field on S pointing out of Ω.

(2.4) Green's Identities.
If Ω is a bounded domain with smooth boundary S and u, v are C^1 functions on $\overline{\Omega}$, then

(2.5)
$$\int_S v\partial_\nu u \, d\sigma = \int_\Omega (v\Delta u + \nabla v \cdot \nabla u) \, dx,$$

(2.6)
$$\int_S (v\partial_\nu u - u\partial_\nu v) \, d\sigma = \int_\Omega (v\Delta u - u\Delta v) \, dx.$$

Proof: (2.5) is just the divergence theorem (0.4) applied to the vector field $v\nabla u$. (2.6) follows from (2.5) by interchanging u and v and then subtracting. ∎

(2.7) Corollary.
If u is harmonic on Ω then $\int_S \partial_\nu u \, d\sigma = 0$.

Proof: Take $v = 1$. ∎

The following theorem states that the value of a harmonic function at a point is equal to its mean value on any sphere about that point. Here and in what follows, ω_n denotes the area of the unit sphere in $\mathbb{R}^n$ (see Proposition (0.7)):

$$\omega_n = \frac{2\pi^{n/2}}{\Gamma(n/2)}.$$

(2.8) The Mean Value Theorem.
Suppose u is harmonic on an open set Ω. If $x \in \Omega$ and $r > 0$ is small enough so that $\overline{B_r(x)} \subset \Omega$, then

$$u(x) = \frac{1}{r^{n-1}\omega_n} \int_{S_r(x)} u(y) \, d\sigma(y) = \frac{1}{\omega_n} \int_{S_1(0)} u(x + ry) \, d\sigma(y).$$

Proof: We first remark that the second equality follows from the change of variable $y \to x + ry$, and that by composing with a translation we may assume that $x = 0$. To prove the first equality, then, we use Green's identity (2.6), where we take u to be our harmonic function, $v(y) = |y|^{2-n}$ if $n \neq 2$ or $v(y) = \log|y|$ if $n = 2$, and $\Omega = B_r(0) \setminus \overline{B_\epsilon(0)}$ where $0 < \epsilon < r$. By Corollary (2.3), v is harmonic in Ω, and by (0.1), $\partial_\nu v$ is the constant $(2 - n)r^{1-n}$ on $S_r(0)$ and the constant $-(2 - n)\epsilon^{1-n}$ on $S_\epsilon(0)$. (The minus sign is there because the orientation of $S_\epsilon(0)$ is the opposite of the usual one, and the factor $(2 - n)$ should be omitted when $n = 2$.) Thus, by (2.6),

$$
\begin{aligned}
0 &= \int_{S_r(0)} (v \partial_\nu u - u \partial_\nu v)\, d\sigma - \int_{S_\epsilon(0)} (v \partial_\nu u - u \partial_\nu v)\, d\sigma \\
&= r^{2-n} \int_{S_r(0)} \partial_\nu u\, d\sigma + \epsilon^{2-n} \int_{S_\epsilon(0)} \partial_\nu u\, d\sigma \\
&\quad - (2 - n) r^{1-n} \int_{S_r(0)} u\, d\sigma + (2 - n)\epsilon^{1-n} \int_{S_\epsilon(0)} u\, d\sigma,
\end{aligned}
$$

with suitable modifications when $n = 2$. By Corollary (2.7), the first two terms in the last sum vanish, so

$$
\frac{1}{r^{n-1}\omega_n} \int_{S_r(0)} u\, d\sigma = \frac{1}{\epsilon^{n-1}\omega_n} \int_{S_\epsilon(0)} u\, d\sigma.
$$

But u is continuous, so the right hand side, being the mean value of u on $S_\epsilon(0)$, converges to $u(0)$ as $\epsilon \to 0$. ∎

(2.9) Corollary.
If u, Ω, and r are as above,

$$
u(x) = \frac{n}{r^n \omega_n} \int_{B_r(x)} u(y)\, dy = \frac{n}{\omega_n} \int_{B_1(0)} u(x + ry)\, dy.
$$

Proof: Multiply both sides of the equation

$$
u(x) = \frac{1}{\omega_n} \int_{S_1(0)} u(x + \rho ry)\, d\sigma(y)
$$

by $\rho^{n-1}\, d\rho$ and integrate from 0 to 1. ∎

(2.10) The Converse of the Mean Value Theorem.
Suppose that u is continuous on an open set Ω and that whenever $x \in \Omega$ and $\overline{B_r(x)} \subset \Omega$ we have

$$
u(x) = \frac{1}{\omega_n} \int_{S_1(0)} u(x + ry)\, d\sigma(y).
$$

Then $u \in C^\infty(\Omega)$ and u is harmonic on Ω.

Proof: Choose $\phi \in C_c^\infty(B_1(0))$ such that $\int \phi = 1$ and $\phi(x) = \psi(|x|)$ for some $\psi \in C_c^\infty(\mathbb{R})$. Given $\epsilon > 0$, set $\phi_\epsilon(x) = \epsilon^{-n}\phi(\epsilon^{-1}x)$ and $\Omega_\epsilon = \{x : \overline{B_\epsilon(x)} \subset \Omega\}$. Then if $x \in \Omega_\epsilon$, the function $y \to \phi_\epsilon(x - y)$ is supported in Ω, and we have

$$\int u(y)\phi_\epsilon(x - y)\, dy = \int u(x - y)\phi_\epsilon(y)\, dy = \int_{B_\epsilon(0)} u(x - y)\phi(\epsilon^{-1}y)\epsilon^{-n}\, dy$$

$$= \int_{B_1(0)} u(x - \epsilon y)\phi(y)\, dy = \int_0^1 \int_{S_1(0)} u(x - r\epsilon y)\psi(r)r^{n-1}\, d\sigma(y)\, dr$$

$$= \omega_n u(x) \int_0^1 \psi(r)r^{n-1}\, dr = u(x) \int_0^1 \int_{S_1(0)} \phi(ry)r^{n-1}\, d\sigma(y)\, dr$$

$$= u(x) \int \phi(y)\, dy = u(x).$$

In the first member of this string of equalities we can clearly differentiate under the integral as often as we please, since $\phi_\epsilon \in C_c^\infty$. Conclusion: $u \in C^\infty(\Omega_\epsilon)$, and since ϵ is arbitrary, $u \in C^\infty(\Omega)$. Finally, if $x \in \Omega_\epsilon$, the mean value of u on $S_r(x)$ is independent of r for $r < \epsilon$, so by the substitution $z = ry$ and Green's identity (2.5) (with $v = 1$),

$$0 = \frac{d}{dr} \int_{S_1(0)} u(x + ry)\, d\sigma(y) = \int_{S_1(0)} y \cdot \nabla u(x + ry)\, d\sigma(y)$$

$$= \int_{S_r(0)} (r^{-1}z) \cdot \nabla u(x + z)\, r^{1-n}\, d\sigma(z)$$

$$= r^{1-n} \int_{S_r(0)} \partial_\nu u\, d\sigma = r^{1-n} \int_{B_r(0)} \Delta u.$$

Thus the integral of Δu over any ball in Ω vanishes, so $\Delta u = 0$ in Ω.   ∎

(2.11) Corollary.
If u is harmonic on Ω then $u \in C^\infty(\Omega)$.

Proof: Apply Theorems (2.8) and (2.10) in succession.   ∎

(2.12) Corollary.
If $\{u_k\}$ is a sequence of harmonic functions on Ω which converges uniformly on compact subsets of Ω to a limit u, then u is harmonic in Ω.

Proof: Since each u_k satisfies the hypotheses of (2.10), so does u.   ∎

(2.13) The Maximum Principle.

Suppose Ω is a connected open set. If u is harmonic and real-valued on Ω and $\sup_{x \in \Omega} u(x) = A < \infty$, then either $u(x) < A$ for all $x \in \Omega$ or $u(x) = A$ for all $x \in \Omega$.

Proof: Clearly $\{x \in \Omega : u(x) = A\}$ is relatively closed in Ω. But by the mean value theorem, if $u(x) = A$ then $u(y) = A$ for all y in a ball about x (otherwise the mean value on spheres about x would be less than A), so this set is also open. Hence it is either Ω or $\varnothing$. ∎

(2.14) Corollary.

Suppose $\overline{\Omega}$ is compact. If u is harmonic and real-valued on Ω and continuous on $\overline{\Omega}$, then the maximum value of u on $\overline{\Omega}$ is achieved on $\partial\Omega$.

Proof: The maximum is achieved somewhere; if at an interior point, u is constant on the connected component containing that point, so the maximum is also achieved on the boundary. ∎

(2.15) The Uniqueness Theorem.

Suppose $\overline{\Omega}$ is compact. If u_1 and u_2 are harmonic functions on Ω which are continuous on $\overline{\Omega}$ and $u_1 = u_2$ on $\partial\Omega$, then $u_1 = u_2$ on Ω.

Proof: The real and imaginary parts of $u_1 - u_2$ and $u_2 - u_1$ are harmonic on Ω, hence must achieve their maxima on $\partial\Omega$; these maxima are therefore zero, so $u_1 = u_2$. ∎

The mean value theorem pertains only to harmonic functions, but the maximum principle and its corollaries are valid for solutions of much more general partial differential equations. See Protter and Weinberger [40], Miranda [37], and Exercises 3 and 4.

(2.16) Liouville's Theorem.

If u is bounded and harmonic on $\mathbb{R}^n$, then u is constant.

Proof: For any $x \in \mathbb{R}^n$ and $R > |x|$, by Corollary (2.9) we have

$$|u(x) - u(0)| = \frac{n}{R^n \omega_n} \left| \int_{B_R(x)} u(y)\, dy - \int_{B_R(0)} u(y)\, dy \right| \leq \frac{n}{R^n \omega_n} \|u\|_\infty \int_D dy,$$

where D is the symmetric difference of the balls $B_R(x)$ and $B_R(0)$. D is contained in the set where $R - |x| < |y| < R + |x|$, so

$$|u(x) - u(0)| \le \frac{n}{R^n \omega_n} \|u\|_\infty \int_{R-|x|<|y|<R+|x|} dy = \frac{n}{R^n} \|u\|_\infty \int_{R-|x|}^{R+|x|} r^{n-1} dr$$

$$= \|u\|_\infty \frac{(R + |x|)^n - (R - |x|)^n}{R^n},$$

which vanishes as $R \to \infty$. Hence $u(x) = u(0)$.   ∎

EXERCISES

1. Prove the maximum modulus principle for complex harmonic functions: If u is harmonic on Ω and continuous on $\overline{\Omega}$, the maximum value of $|u|$ on $\overline{\Omega}$ is achieved on $\partial\Omega$. (Hint: If $|u|$ achieves its maximum M at x_0 then $u(x_0) = e^{i\theta} M$; consider $v = \text{Re}(e^{-i\theta} u)$.)

2. Suppose $u \in C^2(\Omega)$ and $x \in \Omega$. Show that

$$\Delta u(x) = \lim_{r \to 0} \frac{2n}{r^2} \left[\frac{1}{\omega_n} \int_{S_1(0)} u(x + ry) \, d\sigma(y) - u(x) \right].$$

This gives another proof of the converse of the mean value theorem for C^2 functions. (Hint: Consider the second-order Taylor polynomial of u about x. By symmetry considerations, $\int_{S_1(0)} x_j = \int_{S_1(0)} x_j x_k = 0$ for $j \ne k$ and $\int_{S_1(0)} x_k^2 = n^{-1} \int_{S_1(0)} \sum_1^n x_j^2 = n^{-1} \int_{S_1(0)} 1$.)

3. Here is another proof of Corollary (2.14) that works for more general operators. Let Ω be a bounded domain in $\mathbb{R}^n$, and let

$$L = \sum a_{jk}(x) \partial_j \partial_k + \sum b_j(x) \partial_j,$$

where a_{jk} and b_j are continuous functions on $\overline{\Omega}$ and the matrix (a_{jk}) is positive definite on $\overline{\Omega}$.

a. Show that if $v \in C^2(\Omega)$ is real-valued and $Lv > 0$ in Ω, then v cannot have a local maximum in Ω. (Hint: Given $x_0 \in \Omega$, by a rotation of coordinates one can assume that the matrix $(a_{jk}(x_0))$ is diagonal [cf. the discussion of coordinate changes in §1A].)

b. Show that if $x_0 \notin \overline{\Omega}$ and $M > 0$ is sufficiently large, then $w(x) = \exp[-M|x - x_0|^2]$ satisfies $Lw > 0$ in Ω.

c. Suppose $u \in C^2(\Omega) \cap C(\overline{\Omega})$ is real-valued and $Lu = 0$ in Ω. Show that $\max_{\overline{\Omega}} u = \max_{\partial\Omega} u$. (Hint: Show that this conclusion holds for $v = u + \epsilon w$ where w is as in (b) and $\epsilon > 0$.)

4. Let L be as in Exercise 3, and let $Mu = Lu + c(x)u$ where c is continuous and nonpositive on $\overline{\Omega}$.

   a. Assume $u \in C^2(\Omega) \cap C(\overline{\Omega})$. By modifying the argument of Exercise 3, show that if $u \ge 0$ and $Mu = 0$ in Ω then $\max_{\overline{\Omega}} u = \max_{\partial\Omega} u$.

   b. Show that the uniqueness theorem (2.15) holds for M: if u_1 and u_2 are solutions of $Mu = 0$ on Ω and $u_1 = u_2$ on $\partial\Omega$ then $u_1 = u_2$ in Ω. (Hint: Consider $\Omega' = \{x \in \Omega : u_1 - u_2 > 0\}$.)

   c. Show that this conclusion may fail if $c > 0$. (Make life simple: take $L = (d/dx)^2$ on $\mathbb{R}$.)

5. The only distributions whose support is $\{0\}$ are the linear combinations of the point mass at 0 and its derivatives (Folland [14], Rudin [41]). Use this fact to prove a generalization of Liouville's theorem: If u is harmonic on $\mathbb{R}^n$ and $|u(x)| \le C(1 + |x|)^N$ for some $C, N > 0$, then u is a polynomial. (Hint: The estimate on u implies that u is tempered and so has a Fourier transform.)

6. Suppose u is a harmonic function on a disc $D \subset \mathbb{R}^2$. Show that there is a harmonic function v on D, uniquely determined up to an additive constant, such that $\partial_x v = -\partial_y u$ and $\partial_y v = \partial_x u$. Show also that $w = u + iv$ is holomorphic on D, i.e, satisfies the Cauchy-Riemann equation $(\partial_x + i\partial_y)w = 0$. (Hint: One way to define v is via line integrals of the differential form $(\partial_x u)\, dy - (\partial_y u)\, dx$.)

C. The Fundamental Solution

In this section we compute a fundamental solution for the Laplacian and give some applications.

One way to find a fundamental solution is by Fourier analysis. Since $(\Delta u)\widehat{}(\xi) = -4\pi^2 |\xi|^2 \hat{u}(\xi)$, on a formal level the inverse Fourier transform of $[-4\pi^2 |\xi|^2]^{-1}$ should be a fundamental solution. When $n > 2$, this is exactly correct: $|\xi|^{-2}$ is integrable near the origin by (0.5), so it defines a tempered distribution whose inverse Fourier transform is a fundamental solution. A similar result holds for $n = 1$ and $n = 2$ provided one "renormalizes" $|\xi|^{-2}$ so as to make it a tempered distribution. We shall show how this works, and more generally how to compute the Fourier transforms of distributions of the form $|\xi|^{-\alpha}$, in §4B.

However, a more elementary way to obtain a fundamental solution is as follows. Since the Laplacian commutes with rotations, it should have a radial fundamental solution, which must be a function of $|x|$ that is harmonic on $\mathbb{R}^n \setminus \{0\}$. By Corollary (2.3), such a function must be of the form $a + b|x|^{2-n}$ if $n \neq 2$ or $a + b \log|x|$ if $n = 2$. (Note that these functions are locally integrable by (0.5), so they define distributions.) Since the constant function a is harmonic even at 0, it contributes nothing and can be omitted. It remains to show that the constant b can be chosen so as to obtain a fundamental solution, and here is the result.

(2.17) Theorem.
Let

$$(2.18) \quad N(x) = \frac{|x|^{2-n}}{(2-n)\omega_n} \quad (n > 2); \qquad N(x) = \frac{1}{2\pi} \log|x| \quad (n = 2).$$

Then N is a fundamental solution for Δ.

Proof: The standard way to prove this result is via Green's identities, and we invite the reader to perform this calculation (Exercise 1). Here we shall adopt a different method whose computations will be used again later. Namely, for $\epsilon > 0$ we consider a smoothed-out version N^ϵ of N,

$$(2.19) \quad \begin{aligned} N^\epsilon(x) &= \frac{(|x|^2 + \epsilon^2)^{(2-n)/2}}{(2-n)\omega_n} \quad (n > 2); \\[1em] N^\epsilon(x) &= \frac{\log(|x|^2 + \epsilon^2)}{\pi} \quad (n = 2). \end{aligned}$$

$N^\epsilon \to N$ pointwise as $\epsilon \to 0$, and N^ϵ and N are dominated by a fixed locally integrable function for $\epsilon \leq 1$ (namely, $|N|$ when $n > 2$, or $|\log|x|| + 1$ when $n = 2$), so by the dominated convergence theorem, $N^\epsilon \to N$ in the topology of distributions when $\epsilon \to 0$. Hence we need to show that $\Delta N^\epsilon \to \delta$ as $\epsilon \to 0$, i.e., that $\langle \Delta N^\epsilon, \phi \rangle \to \phi(0)$ for any $\phi \in C_c^\infty$.

A simple calculation using (2.2) shows that

$$\Delta N^\epsilon(x) = n\omega_n^{-1}\epsilon^2(|x|^2 + \epsilon^2)^{-(n+2)/2} = \epsilon^{-n}\psi(\epsilon^{-1}x),$$

where

$$\psi(x) = \Delta N^1(x) = n\omega_n^{-1}(|x|^2 + 1)^{-(n+2)/2}.$$

Also, $\Delta N^\epsilon(-x) = \Delta N^\epsilon(x)$. Hence, by Theorem (0.13),

$$\langle \Delta N^\epsilon, \phi \rangle = \int \Delta N^\epsilon(-x)\phi(x)\,dx = \phi * \Delta N^\epsilon(0) \to a\phi(0),$$

where $a = \int \psi(x)\,dx$. But by integration in polar coordinates and the substitution $s = r^2/(r^2 + 1)$, $ds = 2r\,dr/(r^2 + 1)^2$,

$$\int \psi(x)\,dx = n\int_0^\infty (r^2 + 1)^{-(n+2)/2}r^{n-1}\,dr = \frac{n}{2}\int_0^1 s^{(n-2)/2}\,ds = 1,$$

and the proof is complete. ∎

(2.20) Corollary.
Δ *is hypoelliptic: that is, if u is a distribution such that $\Delta u \in C^\infty(\Omega)$, then $u \in C^\infty(\Omega)$. In particular, every distribution solution of $\Delta u = 0$ is a harmonic function.*

   Proof: Apply Theorem (1.58).   ∎

   Our name N for the fundamental solution is in honor of Newton, since for $n = 3$ N is the Newtonian potential, i.e., the gravitational potential generated by a unit mass at the origin. In terms of electrostatics, N is the Coulomb potential, i.e., the electrostatic potential generated by a unit negative charge at the origin.
   We can now solve the inhomogeneous Laplace equation $\Delta u = f$ for any distribution f of compact support — namely, $u = f * N$. In fact, this formula also works for functions f without compact support provided they satisfy conditions to ensure convergence of the appropriate integrals. Here is a representative result along these lines:

(2.21) Theorem.
Suppose that $f \in L^1(\mathbb{R}^n)$, and that $\int |f(y)| \log|y| \, dy < \infty$ in case $n = 2$. Then $f*N$ is well-defined as a locally integrable function, and $\Delta(f*N) = f$.

   Proof: We assume $n > 2$ and leave the case $n = 2$ to the reader. Let χ_r be the characteristic function of $B_r(0)$. Then by (0.5), $\chi_1 N \in L^1$ and $(1 - \chi_1)N \in L^\infty$ (in fact, $\chi_1 N \in L^p$ for $p < n/(n-2)$ and $(1-\chi_1)N \in L^p$ for $p > n/(n-2)$), so $f * (\chi_1 N) \in L^1$ and $f * [(1-\chi_1)N] \in L^\infty$. Moreover, $\chi_r f \rightarrow f$ in L^1 as $r \rightarrow \infty$, so $(\chi_r f) * (\chi_1 N) \rightarrow f * (\chi_1 N)$ in L^1 and $(\chi_r f)*[(1-\chi_1)N] \rightarrow f*[(1-\chi_1)N]$ in L^∞. In particular, $\chi_r f$ and $(\chi_r f)*N$ converge respectively to f and $f*N$ in the topology of distributions. Thus, since $\chi_r f$ has compact support,

$$\Delta[f * N] = \lim \Delta[(\chi_r f) * N] = \lim \chi_r f = f. \qquad \blacksquare$$

   Another interesting application of the fundamental solution N is the following representation of a harmonic function on a domain Ω in terms of its Cauchy data on $\partial\Omega$. For this purpose it is convenient to regard N as a function of two variables $x, y \in \mathbb{R}^n$ by the formula

(2.22) $$N(x, y) = N(x - y).$$

When we differentiate $N(x, y)$, we shall indicate whether the differentiation is with respect to x or y by affixing the subscript x or y to the derivative; e.g., $\partial_y^\alpha N(x, y)$.

(2.23) Theorem.
Let Ω be a bounded domain with C^1 boundary S. If $u \in C^1(\overline{\Omega})$ is harmonic in Ω, then

$$(2.24) \quad u(x) = \int_S [u(y)\partial_{\nu_y} N(x, y) - \partial_\nu u(y) N(x, y)] \, d\sigma(y) \qquad (x \in \Omega).$$

Proof: Let $N^\epsilon(x, y) = N^\epsilon(x - y)$, the N^ϵ on the right being defined by (2.19). Since $\Delta u = 0$ in Ω, by Green's identity (2.6) we have

$$\int_\Omega u(y) \Delta_y N^\epsilon(x, y) \, dy = \int_S [u(y)\partial_{\nu_y} N^\epsilon(x, y) - \partial_\nu u(y) N^\epsilon(x, y)] \, d\sigma(y).$$

As $\epsilon \to 0$, the right side of this equation tends to the right side of (2.24) for each $x \in \Omega$. (Since $x \neq y$ for $x \in \Omega$ and $y \in S$, the singularities of N do not appear here.) On the other hand, the left side is just $u * (\Delta N^\epsilon)(x)$ if we set $u = 0$ outside Ω, so the proof of Theorem (2.17), together with Theorem (0.13), shows that $u * (\Delta N^\epsilon)(x) \to u(x)$ as $\epsilon \to 0$ for $x \in \Omega$. ∎

The formula (2.24) suggests that we might try to solve the Cauchy problem

$$(2.25) \qquad \Delta u = 0 \text{ on } \Omega, \qquad u = f \text{ and } \partial_\nu u = g \text{ on } S,$$

by the formula

$$(2.26) \qquad u(x) = \int_S [f(y)\partial_{\nu_y} N(x, y) - g(y) N(x, y)] \, d\sigma(y).$$

This won't work in general, for we know by the uniqueness theorem (2.15) that the solution of (2.25) (if it exists) is completely determined by f alone. The function u defined by (2.26) will be harmonic in Ω, since $N(x, y)$ and $\partial_{\nu_y} N(x, y)$ are harmonic functions of $x \in \Omega$ for $y \in S$, but it will not have the right boundary values unless f and g are related by a certain pseudodifferential equation on S. (See §4B, where a special case is worked out.)

As a consequence of Theorem (2.23), we obtain the analytic version of Corollary (2.20).

(2.27) Theorem.
If f is analytic on an open set $\Omega \subset \mathbb{R}^n$ and u is a distribution on Ω such that $\Delta u = f$, then u is analytic on Ω.

Proof: We already know that $u \in C^\infty(\Omega)$, and it suffices to show that for each $x_0 \in \Omega$, u is analytic in a neighborhood of x_0. By the Cauchy-Kowalevski theorem we can find an analytic function u' defined on some ball $B_r(x_0) \subset \Omega$ and satisfying $\Delta u' = f$ there. Let $B = B_{r/2}(x_0)$, $S = S_{r/2}(x_0)$, and $v = u - u'$. Then $v \in C^\infty(\overline{B})$ and $\Delta v = 0$ on B, so by Theorem (2.23),

$$v(x) = \int_S \left[v(y)\partial_{\nu_y} N(x,y) - \partial_\nu v(y) N(x,y) \right] \, d\sigma(y) \qquad (x \in B).$$

But for $y \in S$ the functions $N(x,y)$ and $\partial_{\nu_y} N(x,y)$ extend to holomorphic functions of the *complex* variable x in the region where $|\operatorname{Re} x - x_0| < r/4$ and $|\operatorname{Im} x| < r/4$. (In the definition (2.18) of N, simply interpret $|x|$ as $[\sum x_j^2]^{1/2}$.) It follows that $v(x)$ extends holomorphically to the same region, since one can pass complex derivatives under the integral sign. In particular, v is analytic on $B_{r/4}(x_0)$, and hence so is $u = v + u'$. ∎

After Corollary (2.20) and Theorem (2.27), it is natural to ask what one can say about the smoothness of solutions of $\Delta u = f$ when f has only a finite amount of differentiability. The simplest guess would be that if $f \in C^k(\Omega)$ then $u \in C^{k+2}(\Omega)$, but this turns out to be false (except in the one-dimensional case, where it is trivially true). However, analogous results are valid if one replaces $C^k(\Omega)$ by slightly more sophisticated function spaces. One option is to replace continuous derivatives by L^2 derivatives, and we shall explore this in §6B. Another is to consider the Hölder spaces $C^{k+\alpha}(\Omega)$. We have the following result, due to Hölder himself:

(2.28) Theorem.
Suppose $k \geq 0$, $0 < \alpha < 1$, and Ω is an open set in $\mathbb{R}^n$. If $f \in C^{k+\alpha}(\Omega)$ and u is a distribution solution of $\Delta u = f$ on Ω, then $u \in C^{k+2+\alpha}(\Omega)$.

Proof: It suffices to establish the case $k = 0$, since $\Delta(\partial^\beta u) = \partial^\beta f$. It then suffices to prove that if $f \in C^\alpha(\Omega)$ then $u \in C^{2+\alpha}(\Omega')$ for any open Ω' with compact closure in Ω. Given such an Ω', pick $\phi \in C_c^\infty(\Omega)$ such that $\phi = 1$ on Ω', and let $g = \phi f$. Then $\Delta(g * N) = \phi f = f$ on Ω', so $u - (g * N)$ is harmonic and hence C^∞ on Ω'. It is therefore enough to prove that if g is a C^α function with compact support, then $g * N \in C^{2+\alpha}$. This we now do.

To this end, we consider the regularized kernel N^ϵ defined by (2.19) and its derivatives

$$N_j^\epsilon(x) = \partial_j N^\epsilon(x) = \omega_n^{-1} x_j (|x|^2 + \epsilon^2)^{-n/2},$$

$$N_{ij}^\epsilon(x) = \partial_i \partial_j N^\epsilon(x) = \begin{cases} -n\omega_n^{-1} x_i x_j (|x|^2 + \epsilon^2)^{-(n+2)/2} & (i \neq j), \\ \omega_n^{-1}(|x|^2 + \epsilon^2 - nx_j^2)(|x|^2 + \epsilon^2)^{-(n+2)/2} & (i = j). \end{cases}$$

$g * N^\epsilon \in C^\infty$ since $N^\epsilon \in C^\infty$, and we have $\partial_j(g * N^\epsilon) = g * N_j^\epsilon$ and $\partial_i \partial_j (g * N^\epsilon) = g * N_{ij}^\epsilon$. We also need the pointwise limits of N_j^ϵ and N_{ij}^ϵ as $\epsilon \to 0$:

$$N_j(x) = \omega_n^{-1} x_j |x|^{-n},$$

$$N_{ij}(x) = \begin{cases} -n\omega_n^{-1} x_i x_j |x|^{-n-2} & (i \neq j), \\ \omega_n^{-1}(|x|^2 - nx_j^2)|x|^{-n-2} & (i = j). \end{cases}$$

Let χ_1 be the characteristic function of $B_1(0)$ as in the proof of Theorem (2.21). Then $\chi_1 N^\epsilon \to \chi_1 N$ in L^1 and $(1 - \chi_1)N^\epsilon \to (1 - \chi_1)N$ uniformly (uniformly on compact sets in case $n = 2$). Since g is bounded with compact support, it follows easily that $g * N^\epsilon \to g * N$ uniformly (on compact sets), so $g * N$ is continuous. Likewise, $\chi_1 N_j^\epsilon \to \chi_1 N_j$ in L^1 and $(1 - \chi_1)N_j^\epsilon \to (1 - \chi_1)N_j$ uniformly, so $g * N_j^\epsilon \to g * N_j$ uniformly. This also shows that $N_j^\epsilon \to N_j$ in the topology of distributions, so the locally integrable function N_j is the distribution derivative $\partial_j N$, and hence $\partial_j(g * N) = g * N_j$ is continuous.

This simple argument does not work for the second derivatives, because the functions N_{ij} are not locally integrable at the origin; this follows from (0.5) since they are all homogeneous of degree $-n$. We must take more care to see what happens to $g * N_{ij}^\epsilon$ as $\epsilon \to 0$.

Let us consider first the case $i \neq j$. The functions N_{ij}^ϵ and their limit N_{ij} are odd functions of x_i (and x_j), and it follows that their integrals over any annulus $a < |x| < b$ vanish. For $\epsilon > 0$ we can even take $a = 0$, and we have, for any $b > 0$,

$$g * N_{ij}^\epsilon(x) = \int g(x - y)N_{ij}^\epsilon(y)\, dy - g(x) \int_{|y|<b} N_{ij}^\epsilon(y)\, dy$$

$$= \int_{|y|<b} [g(x - y) - g(x)]N_{ij}^\epsilon(y)\, dy + \int_{|y|>b} g(x - y)N_{ij}^\epsilon(y)\, dy.$$

Now we can let $\epsilon \to 0$ to obtain

$$\lim_{\epsilon \to 0} g * N_{ij}^\epsilon(x) = \int_{|y|<b} [g(x - y) - g(x)]N_{ij}(y)\, dy + \int_{|y|>b} g(x - y)N_{ij}(y)\, dy.$$

This works because

$$\left| [g(x - y) - g(x)]N_{ij}^\epsilon(y) \right| \leq \left| [g(x - y) - g(x)]N_{ij}(y) \right| \leq C|y|^{\alpha - n},$$

which is integrable on $|y| < b$, so the dominated convergence theorem can be applied and the limiting integrals are absolutely convergent. In fact, the convergence is uniform in x since these estimates are, so that $\partial_i \partial_j (g * N)$ is continuous. Moreover, since b is arbitrary, we can let $b \to \infty$ to obtain

$$\partial_i \partial_j (g * N)(x) = \lim_{b \to \infty} \int_{|y| < b} [g(x - y) - g(x)] N_{ij}(y) \, dy.$$

A similar result holds for $i = j$. We have

$$N_{jj}^\epsilon(x) = \frac{1}{n\epsilon^n} \psi(\epsilon^{-1}x) + \tilde{N}_{jj}^\epsilon(x),$$

where ψ is as in the proof of Theorem (2.17) and

$$\tilde{N}_{jj}^\epsilon(x) = \omega_n^{-1}(|x|^2 - nx_j^2)(|x|^2 + \epsilon^2)^{-(n+2)/2}.$$

Now, the integral I_j of $\tilde{N}_{jj}^\epsilon(y)$ over an annulus $a < |y| < b$, like that of N_{ij}^ϵ with $i \neq j$, vanishes. The reason is that I_j is independent of j by symmetry in the coordinates, so nI_j is the integral of $\sum_1^n \tilde{N}_{jj}^\epsilon$; but $\sum_1^n \tilde{N}_{jj}^\epsilon \equiv 0$. Hence, the preceding argument, together with the proof of Theorem (2.17), shows that $\partial_j^2 (g * N)$ is continuous and that

$$\partial_j^2 (g * N)(x) = \frac{1}{n} g(x) + \lim_{b \to \infty} \int_{|y| < b} [g(x - y) - g(x)] N_{jj}(y) \, dy.$$

At this point we have shown that $g * N \in C^2$, and the proof will be completed by establishing the following general result and applying it to the kernels $K = N_{ij}$.

(2.29) Theorem.
Let K be a C^1 function on $\mathbb{R}^n \setminus \{0\}$ that is homogeneous of degree $-n$ ($K(rx) = r^{-n} K(x)$ for $r > 0$) and satisfies $\int_{a < |y| < b} K(y) \, dy = 0$ for all $a, b > 0$. If g is a C^α function with compact support $(0 < \alpha < 1)$, then the function

$$h(x) = \lim_{b \to \infty} \int_{|z| < b} [g(x - z) - g(x)] K(z) \, dz$$

belongs to C^α.

Proof: h is well-defined by the argument given above for the case $K = N_{ij}$. Given $y \in \mathbb{R}^n$, we wish to estimate $h(x + y) - h(x)$. Let us write

$h = h_1 + h_2$ where

$$h_1(x) = \int_{|z|<3|y|} [g(x-z) - g(x)]K(z)\,dz,$$

$$h_2(x) = \lim_{b\to\infty} \int_{3|y|<|z|<b} [g(x-z) - g(x)]K(z)\,dz.$$

We have

$$|h_1(x)| \le C_1 \int_{|z|<3|y|} |z|^{\alpha-n}\,dz = C_2 \int_0^{3|y|} r^{\alpha-1}\,dr = C_3|y|^\alpha$$

for all x, and hence

$$|h_1(x+y) - h_1(x)| \le |h_1(x+y)| + |h_1(x)| \le 2C_3|y|^\alpha.$$

On the other hand,

$$h_2(x+y) = \lim_{b\to\infty} \int_{3|y|<|z|<b} [g(x+y-z) - g(x)]K(z)\,dz$$

$$= \lim_{b\to\infty} \int_{3|y|<|z+y|<b} [g(x-z) - g(x)]K(z+y)\,dz,$$

so

$$h_2(x+y) - h_2(x)$$

$$(2.30) \qquad = \lim_{b\to\infty} \int_{3|y|<|z|<b} [g(x-z) - g(x)][K(z+y) - K(z)]\,dz$$

$$+ \lim_{b\to\infty} E_1(b) + E_2,$$

where $E_1(b)$ and E_2 are errors coming from the disparity in the regions of integration. $E_1(b)$ is the error coming from the symmetric difference between the regions $|z| < b$ and $|z+y| < b$, which is contained in the annulus $b - |y| < |z| < b + |y|$. Assuming $b \gg |y|$, in this annulus we have $|z| \approx |z+y| \approx b$, so $|K(z)| \le Cb^{-n}$ and $|K(z+y)| \le Cb^{-n}$. Hence $E_1(b)$ is dominated by

$$\int_{b-|y|<|z|<b+|y|} \|g\|_\infty b^{-n}\,dz = C_5\|g\|_\infty b^{-n}[(b+|y|)^n - (b-|y|)^n],$$

which vanishes as $b \to \infty$. E_2 is the error coming from the symmetric difference of the regions $|z| > 3|y|$ and $|z+y| > 3|y|$, which is contained in the annulus $2|y| < |z| < 4|y|$. In this annulus we have $|z| \approx |z+y| \approx |y|$, so

$$|E_2| \le C_4 \int_{2|y|<|z|<4|y|} |y|^{-n+\alpha}\,dz = C_5|y|^\alpha.$$

Finally, to estimate the main term in (2.30), we observe that for $|z| > 3|y|$,

$$|K(z+y) - K(z)| = \left| \int_0^1 y \cdot \nabla K(x+ty)\, dt \right| \le |y| \sup_{0 \le t \le 1} |\nabla K(x+ty)|.$$

Since K is homogeneous of degree $-n$, ∇K is homogeneous of degree $-n-1$, so

$$|K(z+y) - K(z)| \le C_6 |y| \sup_{0 \le t \le 1} |z+ty|^{-n-1} \le C_7 |y||z|^{-n-1} \qquad (|z| > 3|y|).$$

Hence the main term in (2.30) is bounded by

$$C_8 \int_{|z| > 3|y|} |z|^\alpha |y||z|^{-n-1}\, dz = C_9 |y| \int_{3|y|}^\infty r^{\alpha-2}\, dr = C_{10}|y|^\alpha.$$

(Note that the condition $\alpha < 1$ is needed here.)   Combining all these estimates, we have $|h(x+y) - h(x)| \le C|y|^\alpha$ as desired.   ∎

Theorem (2.28) remains valid if Δ is replaced by an arbitrary elliptic operator L with smooth coefficients. (If L is of order m, the theorem is that if $Lu \in C^{k+\alpha}$ then $u \in C^{k+m+\alpha}$.) The proof may be found in Stein [46, §VI.5] or Taylor [48, §XI.2]. However, the essential ideas for this general result are all contained in the arguments above.

EXERCISES

1. Theorem (2.17) is equivalent to the assertion that $\langle N, \Delta\phi \rangle = \phi(0)$ for any $\phi \in C_c^\infty$. Prove this by applying Green's identity (2.6) with $u = N$, $v = \phi$, and $\Omega = B_r(0) \setminus B_\epsilon(0)$, where r is large enough so that supp $\phi \subset B_r(0)$.

2. Show that the formula (2.18) for N in the case $n > 2$ also yields a fundamental solution for $\Delta = (d/dx)^2$ in the case $n = 1$. (The proof of Theorem (2.17) works for $n = 1$, but a simpler argument is available.)

3. Work out the proof of Theorem (2.21) for the case $n = 2$.

4. Generalize Theorem (2.21) for $n > 2$ to include f in other L^p spaces.

5. Show that the following function is a fundamental solution for Δ^2 on $\mathbb{R}^n$:

$$\frac{|x|^{4-n}}{2(4-n)(2-n)\omega_n} \quad (n \ne 2, 4);$$

$$\frac{-\log|x|}{4\omega_4} \quad (n = 4); \qquad \frac{|x|^2 \log|x|}{8\pi} \quad (n = 2).$$

Can you generalize to find fundamental solutions for higher powers of Δ?

6. Show that $(4\pi|x|)^{-1}e^{-c|x|}$ is a fundamental solution for $-\Delta + c^2$ ($c \in \mathbb{C}$) on $\mathbb{R}^3$.

7. Show that Theorem (2.23) remains valid if the hypothesis that u is harmonic is replaced by the hypothesis that $u \in C^2(\overline{\Omega})$ and the term $\int_\Omega N(x, y)\Delta u(y)\, dy$ is added to the right side of (2.24). Show also that this result remains valid if N is replaced by $N - c$, for any constant c.

8. Suppose u is a C^2 function on an open set Ω. Apply the result of Exercise 7, with Ω replaced by $B_r(x)$ and with $c = r^{2-n}/(2-n)\omega_n$, to show that if $\overline{B_r(x)} \subset \Omega$,

$$u(x) = \int_{B_r(x)} \frac{[|x-y|^{2-n} - r^{2-n}]}{(2-n)\omega_n} \Delta u(y)\, dy + \frac{1}{\omega_n r^{n-1}} \int_{S_r(x)} u(y)\, d\sigma(y).$$

(Here we assume $n > 2$; a similar formula holds for $n = 2$.) Combining this with Exercise 1 in §2B, conclude that if $u \in C^2(\Omega)$ is real-valued, then u has the "sub-mean-value property"

$$u(x) \le \frac{1}{\omega_n r^{n-1}} \int_{S_r(x)} u(y)\, d\sigma(y) \text{ whenever } \overline{B_r(x)} \subset \Omega$$

if and only if $\Delta u \ge 0$ in Ω. Functions with this sub-mean-value property are called *subharmonic*.

D. The Dirichlet and Neumann Problems

In this section we begin a study of boundary value problems for the Laplacian. The two most important problems, to which we shall devote most of our attention, are the so-called Dirichlet and Neumann problems. Throughout this discussion, Ω will be a domain in $\mathbb{R}^n$ with smooth boundary S.

The Dirichlet Problem: *Given functions f on Ω and g on S, find a function u on $\overline{\Omega}$ satisfying*

$$(2.31) \qquad\qquad \Delta u = f \text{ on } \Omega, \qquad u = g \text{ on } S.$$

The Neumann Problem: *Given functions f on Ω and g on S, find a function u on $\overline{\Omega}$ satisfying*

$$(2.32) \qquad\qquad \Delta u = f \text{ on } \Omega, \qquad \partial_\nu u = g \text{ on } S.$$

Of course, we should be more precise about the smoothness assumptions on f, g, and u, and if Ω is unbounded we shall want to impose conditions on their behavior at infinity. However, for the time being we shall work only on the formal level and assume that Ω is bounded. We shall not, however, assume that Ω is connected. (This added generality is only rarely useful, but it makes the theory in Chapter 3 turn out more neatly.)

The uniqueness theorem (2.15) shows that the solution to the Dirichlet problem (if it exists) will be unique, at least if we require $u \in C(\overline{\Omega})$. For the Neumann problem uniqueness does not hold: we can add to u any function that is constant on each connected component of Ω. Moreover, there is an obvious necessary condition for solvability of the Neumann problem: if u satisfies (2.31) and Ω' is a connected component of Ω, by Green's identity (2.5) (with $v = 1$) we have

$$\int_{\Omega'} f = \int_{\Omega'} \Delta u = \int_{\partial\Omega'} \partial_\nu u = \int_{\partial\Omega'} g,$$

which imposes a restriction on f and g.

The Dirichlet problem is easily reduced to the cases where either $f = 0$ or $g = 0$. Indeed, if we can find functions v and w satisfying

(2.33) $\Delta v = f$ on Ω, $v = 0$ on S,

(2.34) $\Delta w = 0$ on Ω, $w = g$ on S,

then $u = v + w$ will satisfy (2.31). Moreover, the problems (2.33) and (2.34) are more or less equivalent. Indeed, suppose we can solve (2.33) and wish to solve (2.34). Assume that g has an extension $\widetilde{g}$ to $\overline{\Omega}$ which is C^2; then we can find v satisfying

$$\Delta v = \Delta\widetilde{g} \text{ on } \Omega, \qquad v = 0 \text{ on } S,$$

and take $u = \widetilde{g} - v$. On the other hand, suppose we can solve (2.34) and wish to solve (2.33). Extend f to be zero outside Ω and set $v' = f * N$, so that $\Delta v' = f$. We then solve

$$\Delta w = 0 \text{ on } \Omega, \qquad w = v' \text{ on } S,$$

and take $v = v' - w$. Henceforth when we consider the Dirichlet problem we shall usually assume either that $f = 0$ or that $g = 0$.

Similar remarks apply to the Neumann problem: it splits into the cases $f = 0$ and $g = 0$, and these are roughly equivalent. To derive the analogue of (2.34) from that of (2.33), we assume that there exists $\widetilde{g} \in C^2(\overline{\Omega})$ such that $\partial_\nu \widetilde{g} = g$ on S and solve

$$\Delta v = \Delta \widetilde{g} \text{ on } \Omega, \qquad \partial_\nu v = 0 \text{ on } S.$$

To go the other way, we set $v' = f * N$ and solve

$$\Delta w = 0 \text{ on } \Omega, \qquad \partial_\nu w = \partial_\nu v' \text{ on } S.$$

There are many approaches to the Dirichlet and Neumann problems, and we shall investigate several of them. This is instructive because the various methods yield somewhat different results, and also because the techniques involved are applicable to other problems. In fact, we shall solve the Dirichlet problem by Dirichlet's principle (§2F), layer potentials (Chapter 3), and L^2 estimates (Chapter 7), and the last two methods will also solve the Neumann problem. In addition, we shall obtain explicit solutions on a half-space (§2G) and a ball (§2H). At this point, we sketch yet another approach — still on the formal level — using the notion of Green's function.

E. The Green's Function

The **Green's function**[*] for the bounded domain Ω with smooth boundary S is the function $G(x, y)$ on $\Omega \times \overline{\Omega}$ determined by the following properties:
  i. $G(x, \cdot) - N(x, \cdot)$ is harmonic on Ω and continuous on $\overline{\Omega}$, where N is defined by (2.22) and (2.18), and
  ii. $G(x, y) = 0$ for each $x \in \Omega$ and $y \in S$.
Clearly G is unique: for each $x \in \Omega$, $G(x, \cdot) - N(x, \cdot)$ is the unique solution of the Dirichlet problem (2.34) with $g(y) = -N(x, y)$. Thus if we can solve the Dirichlet problem, obtaining a continuous solution from continuous boundary data, we can find the Green's function.

(Green himself gave a simple physical "proof" of the existence of G. Let S be a perfectly conducting shell enclosing a vacuum in Ω, and let S be grounded so the potential on S is zero. Let a unit negative charge be placed at $x \in \Omega$. This will induce a distribution of positive charge on S to keep the potential zero, and $G(x, y)$ is the potential at y induced by the

---

[*] The ubiquitous use of "Green's function" rather than the more grammatical "Green function" is an example of what Fowler [19] called "cast-iron idiom."

charges at x and on S. Unfortunately, to impart mathematical substance to this argument is a decidedly nontrivial task.)

On the other hand, if we can find the Green's function, we obtain simple formulas for the solution of the Dirichlet problem. To see how this works, we shall have to make some assertions which we are not yet able to prove.

(2.35) Claim.
Let Ω be a bounded domain with C^∞ boundary S. The Green's function G for Ω exists, and for each $x \in \Omega$, G is C^∞ on $\overline{\Omega} \setminus \{x\}$.

Granting this claim, we have:

(2.36) Lemma.
$G(x, y) = G(y, x)$ for all $x, y \in \Omega$.

Proof: Given x and y, set $u(z) = G(x, z)$ and $v(z) = G(y, z)$. Then $\Delta u(z) = \delta(x - z)$ and $\Delta v(z) = \delta(y - z)$ where δ is the Dirac distribution, so a formal application of Green's identity (2.6) yields

$$G(x, y) - G(y, x) = \int_\Omega \left[G(x, z)\delta(y - z) - G(y, z)\delta(x - z) \right] dz$$

$$= \int_S \left[G(x, z)\partial_{\nu_z} G(y, z) - G(y, z)\partial_{\nu_z} G(x, z) \right] d\sigma(z) = 0,$$

since $G(x, z) = G(y, z) = 0$ for $z \in S$. This argument may be made rigorous by replacing G by $G - N + N^\epsilon$ and letting $\epsilon \to 0$ as in the proof of Theorem (2.23), or alternatively by excising small balls about x and y from Ω and letting their radii shrink to zero as in the proof of the mean value theorem. Details are left to the reader.   ∎

Because of this symmetry, G may be extended naturally to $\overline{\Omega} \times \overline{\Omega}$ by setting $G(x, y) = 0$ for $x \in S$. Also, $G(\cdot, y) - N(\cdot, y)$ is a harmonic function on Ω for each y.

Now, to solve the inhomogeneous equation with homogeneous boundary conditions (2.33), we set $f = 0$ outside Ω and define

$$v(x) = \int_\Omega G(x, y)f(y)\, dy = f * N(x) + \int_\Omega [G(x, y) - N(x, y)]f(y)\, dy.$$

The Laplacian of the first term on the right is f, and the second term is harmonic in x. Also, $v(x) = 0$ for $x \in S$ since the same is true of $G(x, \cdot)$.

Next, consider the homogeneous equation with inhomogeneous boundary conditions (2.34). We assume that g is continuous on S, and we wish to find a solution w which is continuous on $\overline{\Omega}$. We can reason as follows: suppose the solution w is known, and suppose that $w \in C^1(\overline{\Omega})$. Applying Green's identity (2.6) (together with some limiting process as in the proof of Lemma (2.36)), we obtain

$$w(x) = \int_\Omega w(y)\delta(x,y)\,dy = \int_\Omega \left[w(y)\Delta_y G(x,y) - \Delta w(y)G(x,y)\right] dy$$
$$= \int_S w(y)\partial_{\nu_y} G(x,y)\,d\sigma(y)$$

for $x \in \Omega$, since $G(x,y) = 0$ for $y \in S$. This formula represents w on Ω in terms of its boundary values on S.

Therefore, the obvious candidate for the solution of (2.34) is

$$(2.37) \qquad\qquad w(x) = \int_S g(y)\partial_{\nu_y} G(x,y)\,d\sigma(y).$$

Since $\partial_{\nu_y} G(x,y)$ is harmonic in x and continuous in y for $x \in \Omega$ and $y \in S$, it is clear that w is harmonic in Ω.

(2.38) Claim.
If $g \in C(S)$ and w is defined by (2.37) on Ω, then w extends continuously to $\overline{\Omega}$ and $w = g$ on S.

The function $\partial_{\nu_y} G(x,y)$ on $\Omega \times S$ is calleed the **Poisson kernel** for Ω, and (2.37) is called the **Poisson integral formula** for the solution of the Dirichlet problem.

As mentioned above, we shall force the Dirichlet problem into submission by other methods, and afterwards, in §7H, we shall return to this discussion and prove Claims (2.35) and (2.38). (We shall also verify them directly for the unit ball in §2H.)

EXERCISES

1. Complete the proof of Lemma (2.36).

2. Show that the Green's function for $(d/dx)^2$ on $(0,1)$ is $G(x,y) = x(y-1)$ for $x < y$, $G(x,y) = y(x-1)$ for $x > y$.

F. Dirichlet's Principle

Given a bounded domain Ω with smooth boundary S, we define the Hermitian form D on $C^1(\overline{\Omega})$ by

$$D(u, v) = \int_\Omega \nabla u \cdot \overline{\nabla v}.$$

$D(u, u) = \int_\Omega |\nabla u|^2$ is the so-called **Dirichlet integral** of u; physically it represents the potential energy in Ω of the electrostatic field $-\nabla u$.

We note that $u \to D(u, u)^{1/2}$ is a seminorm on $C^1(\overline{\Omega})$, and $D(u, u) = 0$ if and only if u is constant on each connected component of Ω. Let $H_1(\Omega)$ be the completion of $C^1(\overline{\Omega})$ with respect to the norm

$$\|u\|_{(1)} = \left[D(u, u) + \int_\Omega |u|^2 \right]^{1/2}.$$

$H_1(\Omega)$ can be regarded as a subspace of $L^2(\Omega)$, consisting of functions $u \in L^2(\Omega)$ whose distribution derivatives $\partial_j u$ are also in $L^2(\Omega)$, and it is a Hilbert space with inner product $\langle u \mid v \rangle_{(1)} = D(u, v) + \int_\Omega u\overline{v}$. We shall study it in more detail in §6E.

(2.39) Proposition.
There is a constant $C > 0$ such that $\int_S |u|^2 \leq C\|u\|_{(1)}^2$ for all $u \in C^1(\overline{\Omega})$.

Proof: Extend the normal vector field ν on S in some smooth fashion to be a vector field on $\overline{\Omega}$. (For example, extend it to a neighborhood of S by making it constant on each normal line to S, then multiply it by a smooth cutoff function.) By the divergence theorem (0.4),

$$\int_S |u|^2 = \int_S (|u|^2 \nu) \cdot \nu = \sum_1^n \int_\Omega \partial_j(|u|^2 \nu_j)$$

$$\leq \sum_1^n \int_\Omega \left[|u(\partial_j \overline{u})\nu_j| + |(\partial_j u)\overline{u}\nu_j| + |u|^2 |\partial_j \nu_j| \right].$$

Thus, letting $C' = \sup_{\Omega} \sum_1^n (|\nu_j| + |\partial_j \nu_j|)$,

$$\int_S |u|^2 \leq C' \sum_1^n \int_\Omega (|u \partial_j \overline{u}| + |\overline{u} \partial_j u| + |u|^2)$$

$$\leq C' \left(2 \sum_1^n \left[\int_\Omega |u|^2 \right]^{1/2} \left[\int_\Omega |\partial_j u|^2 \right]^{1/2} + n \int_\Omega |u|^2 \right)$$

$$\leq C' \left(2n \int_\Omega |u|^2 + \sum_1^n \int_\Omega |\partial_j u|^2 \right)$$

$$= 2nC' \int_\Omega |u|^2 + C' D(u, u),$$

where we have used the Schwarz inequality and the fact that $2ab \leq a^2 + b^2$ for all positive numbers a, b. Thus we can take $C = 2nC'$. ∎

(2.40) Corollary.
The restriction map $u \to u|S$ from $C^1(\overline{\Omega})$ to $C^1(S)$ extends continuously to a map from $H_1(\Omega)$ to $L^2(S)$.

It follows that elements of $H_1(\Omega)$ have boundary values on S which are well-defined as elements of $L^2(S)$; we denote the boundary values of $u \in H_1(\Omega)$ by $u|S$. However, not every L^2 function on S — indeed, not every continuous function on S — is the restriction to S of an element of $H_1(\Omega)$. Roughly speaking, the restriction of a function in $H_1(\Omega)$ must possess "L^2 derivatives of order $\frac{1}{2}$" on S. See Exercise 3 and Theorem (6.47).)

Let $H_1^0(\Omega)$ be the closure of $C_c^\infty(\Omega)$ in $H_1(\Omega)$. Clearly, if $f \in H_1^0(\Omega)$ then $f|S = 0$. (The converse is also true. We shall not prove this, but see Proposition (6.50) and the remarks preceding it.)

We propose to solve the following version of the Dirichlet problem (2.34). We assume that the boundary function g is the restriction to S of some $f \in H_1(\Omega)$, and we take the statement "$w = g$ on S" to mean that $w - f \in H_1^0(\Omega)$. Thus, given $f \in H_1(\Omega)$, the problem is to find a harmonic function $w \in H_1(\Omega)$ such that $w - f \in H_1^0(\Omega)$.

(2.41) Theorem.
Suppose $w \in H_1(\Omega)$. Then w is harmonic in Ω if and only if w is orthogonal to $H_1^0(\Omega)$ with respect to D, that is, $D(w, v) = 0$ for all $v \in H_1^0(\Omega)$.

Proof: By Green's identity, if $w \in C^1(\overline{\Omega})$ and $v \in C_c^\infty(\Omega)$ then $\int_\Omega w \overline{\Delta v} = -D(w, v)$, there being no boundary term since v vanishes near the boundary. Passing to limits, we see that this identity remains true for any $w \in H_1(\Omega)$. Hence, w is harmonic in $\Omega \iff w$ satisfies $\Delta w = 0$ in Ω in the sense of distributions (Corollary (2.20)) $\iff \int w \overline{\Delta v} = 0$ for all $v \in C_c^\infty(\Omega) \iff D(w, v) = 0$ for all $v \in C_c^\infty(\Omega) \iff D(w, v) = 0$ for all $v \in H_1^0(\Omega)$.　∎

Since the functions $u \in H_1(\Omega)$ with $D(u, u) = 0$ are locally constant, hence harmonic, and no such function except 0 belongs to $H_1^0(\Omega)$, it follows from elementary Hilbert space theory that each $f \in H_1(\Omega)$ can be written uniquely as $f = w + v$ where w is harmonic and $v \in H_1^0(\Omega)$. (The Hilbert space in question is $H_1(\Omega)$ modulo locally constant functions, with inner product D.) Thus w, the orthogonal projection of f onto the harmonic space, is the solution of the Dirichlet problem posed above.

From the norm-minimizing properties of orthogonal projections in a Hilbert space, it also follows that solving this Dirichlet problem is equivalent to minimizing the Dirichlet integral in a certain class of functions. This approach to solving the Dirichlet problem via the calculus of variations is the classical Dirichlet principle:

Dirichlet's Principle.
If f and w are in $H_1(\Omega)$, the following three conditions are equivalent:
a. *w is harmonic in Ω and $w - f \in H_1^0(\Omega)$.*
b. *$D(w, w) \le D(u, u)$ for all $u \in H_1(\Omega)$ such that $u - f \in H_1^0(\Omega)$.*
c. *$D(w - f, w - f) \le D(u, u)$ for all $u \in H_1(\Omega)$ such that $u - f$ is harmonic in Ω.*

The reader who is acquainted with the checkered history of Dirichlet's principle — it was stated by Dirichlet, used by Riemann, discredited by Weierstrass, and rehabilitated much later by Hilbert — may be surprised at the simplicity of the above arguments. Several points should be kept in mind. In the first place, 19th-century mathematicians did not have Hilbert spaces, or the theory of Lebesgue integration with which to construct Hilbert spaces of functions, at their disposal. In an incomplete inner product space like $C^1(\overline{\Omega})$ there is no guarantee that orthogonal projections will exist. Neither did they have the notion of distribution, much less a proof that distribution solutions of $\Delta u = 0$ are genuinely harmonic, a fact which was essential for the proof of Theorem (2.41). On the other hand, we have solved the Dirichlet problem in a weaker sense than the old mathematicians would have wished: we had to assume that the boundary function is the restriction of a function in $H_1(\Omega)$, and we only showed that the solution assumes its boundary values in the sense of Corollary (2.37). The first restriction is unavoidable in the context of Dirichlet's principle, but we would like to know, for example, that if the boundary data are continuous on S then the solution is continuous on $\overline{\Omega}$. This can be proved in the setting of Dirichlet's principle (see John [30]), but we shall derive it by different methods in Chapter 3.

EXERCISES

In the following exercises, Ω is the unit disc in $\mathbb{R}^2$ and S is the unit circle. For $m \geq 0$ we set $e_m(x, y) = (x + iy)^m$, and for $m < 0$ we set $e_m(x, y) = (x - iy)^{|m|}$.

1. Show that $\{e_m\}_{-\infty}^{\infty}$ is an orthogonal set with respect to the inner product on $L^2(\Omega)$ and also with respect to the Dirichlet form D on Ω, and hence with respect to the inner product $(\cdot \,|\, \cdot)_{(1)}$ on $H_1(\Omega)$. Compute $\|e_m\|_{(1)}$ for all m, and conclude that a series $\sum c_m e_m$ converges in $H_1(\Omega)$ if and only if $\sum |m|\,|c_m|^2 < \infty$. (Hint: polar coordinates.)

2. Show that $\{e_m\}_{-\infty}^{\infty}$ is an orthogonal basis for the space of harmonic functions in $H_1(\Omega)$. (It is trivial to verify that each e_m is harmonic. To see that they span all harmonic functions, you can use Exercise 6 in §2B together with the fact that every holomorphic function in $H_1(\Omega)$ has an expansion $\sum_0^{\infty} c_m e_m$ [a Taylor series!]).

3. If $f = \sum c_m e_m \in H_1(\Omega)$, the restriction $f|S$ is the function in $L^2(S)$ whose Fourier series is $\sum c_m e^{im\theta}$. Conclude that a function $\sum c_m e^{im\theta}$ is the restriction of a function in $H_1(\Omega)$ (in fact, of a harmonic function in $H_1(\Omega)$) if and only if $\sum |m|\,|c_m|^2 < \infty$. Find a sequence $\{c_m\}$ such that $\sum |c_m| < \infty$ but $\sum |m|\,|c_m|^2 = \infty$, and hence exhibit a continuous function on S that is not the restriction of a function in $H_1(\Omega)$.

G. The Dirichlet Problem in a Half-Space

In this section we shall solve the Dirichlet problem in the half-space $\{x \in \mathbb{R}^n : x_n > 0\}$ by computing the Green's function and the Poisson kernel explicitly. Actually, we shall change notation slightly: we replace n by $n+1$ and denote x_{n+1} by t. The domain in question is then

$$\mathbb{R}_+^{n+1} = \{(x, t) \in \mathbb{R}^n \times \mathbb{R} : t > 0\},$$

and the Laplacian on it is

$$\Delta_x + \partial_t^2 = \sum_1^n \frac{\partial^2}{\partial x_j^2} + \frac{\partial^2}{\partial t^2}.$$

Finding the Green's function for $\mathbb{R}_+^{n+1}$ is just a matter of making the following simple observation, In physical terms, if a charge is placed at (x, t) and an equal but opposite charge is placed at $(x, -t)$, the induced potential at the points equidistant from these two charges — namely, the points $(x, 0)$ — will always be zero. Thus, the Green's function for $\mathbb{R}_+^{n+1}$ is

$$G((x, t), (y, s)) = N(x - y, t - s) - N(x - y, -t - s),$$

where N is the fundamental solution given by (2.18) with n replaced by $n + 1$. This clearly enjoys the defining properties for a Green's function: it satisfies $\Delta_{(y,s)} G((x, t), (y, s)) = \delta(x - y, t - s)$ for $t, s > 0$, and it vanishes when $t = 0$ or $s = 0$.

From this we immediately have the solution of the Dirichlet problem

$$\Delta_x u + \partial_t^2 u = f \text{ on } \mathbb{R}_+^{n+1}, \qquad u(x, 0) = 0,$$

namely,

$$u(x, t) = \int_0^\infty \int_{\mathbb{R}^n} G((x, t), (y, s)) f(y, s) \, dy \, ds.$$

More precisely, if f is, say, bounded with compact support in $\overline{\mathbb{R}_+^{n+1}}$ (these conditions can be relaxed), u will be continuous on $\overline{\mathbb{R}_+^{n+1}}$ with $u(x, 0) = 0$, and u will satisfy $(\Delta_x + \partial_t^2) u = f$ (in the sense of distributions) on $\mathbb{R}_+^{n+1}$.

To solve the dual Dirichlet problem,

$$(2.42) \qquad \Delta_x u + \partial_t^2 u = 0 \text{ on } \mathbb{R}_+^{n+1}, \qquad u(x, 0) = g(x),$$

we compute the Poisson kernel. Since the outward normal derivative on $\partial \mathbb{R}_+^{n+1}$ is $-\partial/\partial t$, the Poisson kernel is

$$-\frac{\partial}{\partial s} G((x, t), (y, s)) \Big|_{s=0} = -\frac{\partial}{\partial s} [N(x - y, t - s) - N(x - y, -t - s)]_{s=0}$$

$$= \frac{2t}{\omega_{n+1}(|x - y|^2 + t^2)^{(n+1)/2}}.$$

According to (2.37), the candidate for a solution to (2.42) is then

$$u(x, t) = \int_{\mathbb{R}^n} \frac{2t}{\omega_{n+1}(|x - y|^2 + t^2)^{(n+1)/2}} g(y) \, dy.$$

In other words, if we set

$$(2.43) \qquad P_t(x) = \frac{2t}{\omega_{n+1}(|x|^2 + t^2)^{(n+1)/2}} \qquad (x \in \mathbb{R}^n, \ t > 0),$$

which is what is usually called the **Poisson kernel** for $\mathbb{R}^{n+1}_+$, the proposed solution is

$$u(x,t) = g * P_t(x) \quad \text{(convolution on } \mathbb{R}^n).$$

We now verify that this works. The key observations are that

$$P_t(x) = t^{-n} P_1(t^{-1}x)$$

and that

$$\int P_1(x)\,dx = \frac{2\omega_n}{\omega_{n+1}} \int_0^\infty \frac{r^{n-1}\,dr}{(r^2+1)^{(n+1)/2}}$$

$$= \frac{2\Gamma(\tfrac{1}{2}(n+1))}{\pi^{1/2}\Gamma(\tfrac{1}{2}n)} \int_0^\infty \frac{r^{n-1}\,dr}{(r^2+1)^{(n+1)/2}},$$

which by the substitution $s = r^2/(r^2+1)$ (so $r^2 = s/(1-s)$) equals

$$\frac{\Gamma(\tfrac{1}{2}(n+1))}{\Gamma(\tfrac{1}{2}n)\Gamma(\tfrac{1}{2})} \int_0^1 s^{(n/2)-1}(1-s)^{-1/2}\,ds = 1,$$

by the well-known formula for the beta integral. (See Folland [17, Appx. 3] or Hochstadt [25, §3.4].)

(2.44) Theorem.
Suppose $g \in L^p(\mathbb{R}^n)$, where $1 \le p \le \infty$. Then $u(x,t) = g * P_t(x)$ is well-defined on $\mathbb{R}^{n+1}_+$ and is harmonic there. If g is bounded and continuous, then u is continuous on $\overline{\mathbb{R}^{n+1}_+}$ and $u(x,0) = g(x)$. If $g \in L^p$ where $p < \infty$, then $u(\cdot,t) \to g$ in the L^p norm as $t \to 0$.

Proof: We note that $P_t \in L^1 \cap L^\infty$, so $P_t \in L^q$ for all $q \in [1,\infty]$; hence the integral $g * P_t(x)$ is absolutely convergent for all x and t, and the same is true if P_t is replaced by its derivatives $\Delta_x P_t$ or $\partial_t^2 P_t$. Since $G((x,t),(y,s))$ is harmonic as a function of (x,t) for $(x,t) \ne (y,s)$, $P_t(x)$ is harmonic on $\mathbb{R}^{n+1}_+$, and hence so is u:

$$\Delta_x u + \partial_t^2 u = g * (\Delta_x + \partial_t^2) P_t = 0.$$

The remaining assertions follow from the calculations preceding the theorem together with Theorem (0.13).   ∎

The solution to the problem (2.42) is not unique; for example, if $u(x,t)$ is a solution then so is $u(x,t) + ct$ for any $c \in \mathbb{C}$. However, if g is bounded and continuous on $\mathbb{R}^n$, then $u(x,t) = g * P_t(x)$ is the unique *bounded* solution; see Exercise 1 in §2I. Here we shall prove the analogous result with boundedness replaced by vanishing at infinity.

(2.45) Theorem.
If g is continuous and vanishes at infinity on $\mathbb{R}^n$, then $u(x,t) = g * P_t(x)$ vanishes at infinity on $\overline{\mathbb{R}^{n+1}_+}$, and it is the unique solution of (2.42) with this property.

   Proof: Assume for the moment that g has compact support, say $g(x) = 0$ for $|x| > a$. Then $g \in L^1$, and $\|g * P_t\|_\infty \leq \|g\|_1 \|P_t\|_\infty \leq Ct^{-n}$, so $u(x,t) \to 0$ as $t \to \infty$ uniformly in x. On the other hand, if $0 \leq t \leq R$,

$$|u(x,t)| \leq \|g\|_1 \sup_{|y|<a} |P_t(x-y)| \leq CR|x|^{-n-1}$$

for $|x| > 2a$, so $u(x,t) \to 0$ as $x \to \infty$ uniformly for $t \in [0, R]$. This proves that u vanishes at infinity when g has compact support. For general g, choose a sequence $\{g_n\}$ of compactly supported functions that converge uniformly to g, and let $u_n(x,t) = g_n * P_t(x)$. Then u_n vanishes at infinity, and $u_n \to u$ uniformly on $\overline{\mathbb{R}^{n+1}_+}$ since

$$\|u_n - u\|_\infty = \sup_t \|(g_n - g) * P_t\|_\infty \leq \sup_t \|g_n - g\|_\infty \|P_t\|_1 = \|g_n - g\|_\infty \to 0.$$

Hence u vanishes at infinity.
   Now suppose v is another solution, and let $w = v - u$. Then w vanishes at infinity and also on the hyperplane $t = 0$. Thus, given $\epsilon > 0$, if R is sufficiently large we have $|w| < \epsilon$ on the boundary of the region $|x| < R$, $0 < t < R$. By the maximum principle (cf. Exercise 1 in §2B) it follows that $|w| < \epsilon$ on this region. Letting $\epsilon \to 0$ and $R \to \infty$, we conclude that $w \equiv 0$. ∎

   If $t, s > 0$, the function $u(x,t) = P_{s+t}(x)$ vanishes at infinity on $\overline{\mathbb{R}^{n+1}_+}$ and satisfies (2.42) with $g(x) = P_s(x)$, so it follows from Theorem (2.45) that

$$P_{s+t} = P_s * P_t.$$

That is, the functions P_t form a semigroup under convolution, and the corresponding operators $g \to g * P_t$ form a semigroup under composition, called the **Poisson semigroup**. This is a contraction semigroup on L^p for $1 \leq p \leq \infty$, since

$$\|g * P_t\|_p \leq \|g\|_p \|P_t\|_1 = \|g\|_p,$$

and it is strongly continuous on L^p for $p < \infty$ by Theorem (2.44).

Some of the results of this section can also be obtained by using the Fourier transform on $\mathbb{R}^n$. When $n = 1$ this works out rather simply; see Exercise 1. We shall consider the case $n > 1$ in §4B.

EXERCISES

1. Assume that $n = 1$. Show that $\widehat{P}_t(\xi) = e^{-2\pi t |\xi|}$, either directly or via the Fourier inversion theorem. Use this result to give simple proofs that $u(x,t) = g * P_t(x)$ satisfies (2.42) if $g, \widehat{g} \in L^1$, and that $P_s * P_t = P_{s+t}$.

2. The formula (2.43) for P_t makes sense for all $t \in \mathbb{R}$, not just $t > 0$. Show that if $f \in L^1(\mathbb{R}^n)$ and $u(x,t) = f * P_t(x)$ for all $(x,t) \in \mathbb{R}^{n+1}$, then $\Delta_x u + \partial_t^2 u = 2f(x)\delta'(t)$.

H. The Dirichlet Problem in a Ball

We now solve the Dirichlet problem for the unit ball in $\mathbb{R}^n$, first by computing the Green's function and Poisson kernel explicitly, and then by expansion in spherical harmonics. We remark that these results are easily extended to arbitrary balls by translating and dilating the coordinates. Throughout this section, we employ the notation

$$B = B_1(0), \qquad S = \partial B = S_1(0).$$

The Green's function for B may be found by an idea similar to the one we used for the half-space in §2G. Namely, the potential generated by a unit charge at $x \in B$ can be cancelled on S by placing a charge of opposite sign at the point $x/|x|^2$ obtained by "reflecting" x in S. As it turns out, the magnitude of the second charge should be not 1 but $|x|^{2-n}$, and a slight modification must be made when $n = 2$. (We shall obtain more insight into this when we consider the Kelvin transform in §2I.) To see that this works, we use the following lemma.

(2.46) Lemma.
If $x, y \in \mathbb{R}^n$, $x \neq 0$, and $|y| = 1$, then

$$|x - y| = \big| \, |x|^{-1}x - |x|y \, \big|.$$

Proof: We have

$$|x - y|^2 = |x|^2 - 2x \cdot y + 1 = \big| |x|y \big|^2 - 2(|x|^{-1}x) \cdot (|x|y) + \big| |x|^{-1}x \big|^2$$
$$= \big| |x|^{-1}x - |x|y \big|^2. \qquad \blacksquare$$

Now, assuming $n > 2$, let us define

$$G(x, y) = N(x - y) - |x|^{2-n} N(|x|^{-2}x - y)$$
$$= \frac{1}{(2 - n)\omega_n} \left[|x - y|^{2-n} - \big| |x|^{-1}x - |x|y \big|^{2-n} \right].$$

From the first equation it is clear that $G(x, y) - N(x, y)$ is harmonic in y for $y \neq |x|^{-2}x$, and in particular for $y \in B$ when $x \in B$. The second equation, together with Lemma (2.46), shows that $G(x, y) = 0$ for $y \in S$. It also makes clear how to define G at $x = 0$:

$$G(0, y) = \frac{1}{(2 - n)\omega_n} \left[|y|^{2-n} - 1 \right].$$

When $n = 2$, the analogous formula is

$$G(x, y) = \frac{1}{2\pi} \left[\log |x - y| - \log \big| |x|^{-1}x - |x|y \big| \right] \qquad (x \neq 0),$$
$$G(0, y) = \frac{1}{2\pi} \log |y|.$$

Again it is clear that G enjoys the defining properties of a Green's function.

Clearly G satisfies Claim (2.35). The symmetry property $G(x, y) = G(y, x)$ is not obvious from the formula for G, but it is not hard to verify directly by a calculation like the proof of Lemma (2.46).

Now that we know the Green's function, we can compute the Poisson kernel

$$P(x, y) = \partial_{\nu_y} G(x, y) \qquad (x \in B, \ y \in S).$$

Indeed, by (0.1) we have $\partial_{\nu_y} = y \cdot \nabla_y$ on S, so for all $n \geq 2$,

$$P(x, y) = \frac{-1}{\omega_n} \left[\frac{y \cdot (x - y)}{|x - y|^n} - \frac{|x|y \cdot (|x|^{-1}x - |x|y)}{\big| |x|^{-1}x - |x|y \big|^n} \right].$$

Since $|y| = 1$, Lemma (2.46) then implies that

(2.47) $$P(x, y) = \frac{1 - |x|^2}{\omega_n |x - y|^n}.$$

It is a fairly simple matter to prove Claim (2.38) for $\Omega = B$; in fact, we shall obtain the analogue of Theorem (2.44). A bit of notation: if u is a continuous function on B and $0 < r < 1$, we define the function u_r on S by

$$u_r(y) = u(ry).$$

(2.48) Theorem.
If $f \in L^1(S)$ and P is given by (2.47), set

$$u(x) = \int_S P(x,y)f(y)\,d\sigma(y) \qquad (x \in B).$$

Then u is harmonic on B. If f is continuous, u extends continuously to $\overline{B}$ and $u = f$ on S. If $f \in L^p(S)$ $(1 \le p < \infty)$, then $u_r \to f$ in the L^p norm as $r \to 1$.

Proof: For each $x \in B$, $P(x,y)$ is a bounded function of $y \in S$, so $u(x)$ is well-defined for $x \in B$. It is harmonic since P is harmonic in x (since $G(x,y)$ is, or by a direct calculation).
    Next, we claim that
i. $\int_S P(x,y)\,d\sigma(y) = 1$ for all $x \in B$.
ii. For any $y_0 \in S$ and any neighborhood V of y_0 in S,

$$\lim_{r \to 1} \int_{S \backslash V} P(ry_0, y)\,d\sigma(y) = 0.$$

(ii) is obvious, since $|ry_0 - y|^{-n}$ is bounded uniformly for $0 < r < 1$ and $y \in S \backslash V$, and $1 - |ry_0|^2 = 1 - r^2 \to 0$ as $r \to 1$. To prove (i), we note that since P is harmonic in x, for $0 < r < 1$ and $y \in S$ the mean value theorem implies that

$$1 = \omega_{n-1}P(0,y) = \int_S P(ry', y)\,d\sigma(y').$$

But another application of Lemma (2.44) shows that $P(ry', y) = P(ry, y')$, so with $x = ry$, we have $1 = \int_S P(x,y')\,d\sigma(y')$.
    Now, suppose f is continuous, and hence uniformly continuous since S is compact. Given $\epsilon > 0$, choose $\delta > 0$ so small that $|f(x) - f(y)| < \epsilon$ whenever $|x - y| < \delta$, and set $V_x = \{y \in S : |x - y| < \delta\}$. Then, for any $x \in S$ and $r < 1$, by (i) we have

$$|f(x) - u(rx)| = \left| \int_S [f(x) - f(y)]P(rx, y)\,d\sigma(y) \right|$$

$$\le \epsilon \int_{V_x} P(rx, y)\,d\sigma(y) + 2\|f\|_\infty \int_{S \backslash V_x} P(rx, y)\,d\sigma(y).$$

By (i), the first term on the right is less than ϵ, and by (ii), the second term is also less than ϵ if $1 - r$ is small enough. Hence $u_r \to f$ uniformly as $r \to 1$, and it follows that u extends continuously to $\overline{B}$ with $u = f$ on S.

Finally, suppose $f \in L^p$, $1 \leq p < \infty$. Given $\epsilon > 0$, choose $g \in C(S)$ with $\|g - f\|_p < \epsilon/3$. Setting $v(x) = \int_S P(x, y)g(y)\, d\sigma(y)$, we have

$$\|f - u_r\|_p \leq \|f - g\|_p + \|g - v_r\|_p + \|v_r - u_r\|_p.$$

The first term on the right is $< \epsilon/3$, and if $1 - r$ is small enough the second term is $< \epsilon/3$ since uniform convergence implies L^p convergence on S. We claim that the linear mapping $f \to u_r$ is bounded on $L^p(S)$ with norm ≤ 1 for all r, so the third term is also $< \epsilon/3$. This assertion follows from the generalized Young's inequality (0.10), since $|P| = P$ and

$$\int P(rx, y)\, d\sigma(y) = 1 = \omega_n P(0, y) = \int P(rx, y)\, d\sigma(x)$$

by (i) and the mean value theorem. ∎

We now turn to the theory of spherical harmonics.

Let $\mathcal{P}_k$ be the space of homogeneous polynomials of degree k on $\mathbb{R}^n$, and let

$$\mathcal{H}_k = \{P \in \mathcal{P}_k : \Delta P = 0\},$$
$$H_k = \{P|S : P \in \mathcal{H}_k\}.$$

That is, $\mathcal{H}_k$ is the space of homogeneous harmonic polynomials of degree k and H_k is the space of their restrictions to the unit sphere. The elements of H_k are called **spherical harmonics** of degree k. The restriction map from $\mathcal{H}_k$ to H_k is an isomorphism; its inverse is the map $Y \to P$ where $P(x) = |x|^k Y(|x|^{-1} x)$.

We denote by r^2 the function $x \to \sum_1^n x_j^2$ on $\mathbb{R}^n$, an element of $\mathcal{P}_2$.

(2.49) Proposition.
$\mathcal{P}_k = \mathcal{H}_k \oplus r^2 \mathcal{P}_{k-2}$, where $r^2 \mathcal{P}_{k-2} = \{r^2 P : P \in \mathcal{P}_{k-2}\}$.

Proof: For $P, Q \in \mathcal{P}_k$, let $\{P, Q\} = P(\partial)\overline{Q}$; that is, if $P(x) = \sum a_\alpha x^\alpha$ then $\{P, Q\} = \sum a_\alpha \partial^\alpha \overline{Q}$. The form $\{P, Q\}$ is linear in P and conjugate-linear in Q, and it is scalar-valued: applying a derivative of degree k to a polynomial of degree k yields a number. Moreover, one readily checks that $\{x^\alpha, x^\beta\} = \alpha!$ if $\beta = \alpha$ and $\{x^\alpha, x^\beta\} = 0$ otherwise. Hence, in general,

$$\left\{\sum a_\alpha x^\alpha, \sum b_\beta x^\beta\right\} = \sum \alpha! a_\alpha \overline{b}_\alpha,$$

so the form $\{\cdot, \cdot\}$ is a scalar product on $\mathcal{P}_k$.

Now, notice that for any $P \in \mathcal{P}_{k-2}$ and $Q \in \mathcal{P}_k$,

$$\{r^2 P, Q\} = P(\partial)r^2(\partial)\overline{Q} = P(\partial)\overline{\Delta Q} = \{P, \Delta Q\}.$$

This immediately implies that $\mathcal{H}_k$ is the orthogonal complement of $r^2 \mathcal{P}_{k-2}$ with respect to $\{\cdot, \cdot\}$, which completes the proof. ∎

(2.50) Corollary.
$\mathcal{P}_k = \mathcal{H}_k \oplus r^2\mathcal{H}_{k-2} \oplus r^4\mathcal{H}_{k-4} \oplus \cdots$.

Proof: Induction on k.   ∎

(2.51) Corollary.
The restriction to the unit sphere of any element of $\mathcal{P}_k$ is a sum of spherical harmonics of degree at most k.

Proof: $r^2 = 1$ on S.   ∎

(2.52) Euler's Lemma.
If $Q \in \mathcal{P}_k$ then $\sum x_j \partial_j Q(x) = kQ(x)$.

Proof: Exercise.   ∎

(2.53) Theorem.
$L^2(S) = \bigoplus_0^\infty H_k$, *the expression on the right being an orthogonal direct sum with respect to the scalar product on $L^2(S)$.*

Proof: By Corollary (2.50) and the Weierstrass approximation theorem (which, by the way, we shall prove in §4A), the linear span of the H_k's is dense in $L^2(S)$. We must show that $H_j \perp H_k$ if $j \neq k$. Given $Y_j \in H_j$ and $Y_k \in H_k$, let P_j and P_k be their harmonic extensions in $\mathcal{H}_j$ and $\mathcal{H}_k$. By Green's identity (2.6), Euler's lemma (2.52), and (0.1),

$$0 = \int_B (P_j\Delta\overline{P}_k - P_k\Delta\overline{P}_j) = \int_S (P_j\partial_\nu\overline{P}_k - \overline{P}_k\partial_\nu P_j) = (k-j)\int_S P_j\overline{P}_k$$
$$= (k-j)\langle Y_j \mid Y_k\rangle. \quad ∎$$

Remark: Since Δ commutes with rotations (Theorem (2.1)), the spaces H_k are invariant under rotations, and one can show that they have no nontrivial invariant subspaces; see Exercise 8. Theorem (2.53) thus provides the decomposition of $L^2(S)$ into irreducible subspaces under the action of the rotation group.

Let
$$d_k = \dim H_k = \dim \mathcal{H}_k.$$

For future purposes we shall need to compute d_k.

(2.54) Proposition.
$$\dim \mathcal{P}_k = \frac{(k+n-1)!}{k!(n-1)!}.$$

Proof: Since the monomials x^α with $|\alpha| = k$ are a basis for $\mathcal{P}_k$, $\dim \mathcal{P}_k$ is the number of ways we can choose an ordered n-tuple $(\alpha_1, \ldots, \alpha_n)$ of non-negative integers whose sum is k. Think of it this way: we line up k black balls in a row and wish to divide them into n groups of consecutive balls with cardinalities $\alpha_1, \ldots, \alpha_n$. To mark the division between two adjacent groups we interpose a white ball between two black balls; for this purpose we need $n-1$ white balls. The number of ways we can do this is the number of ways we can take $k+n-1$ black balls and choose $n-1$ of them to be painted white, which is $(k+n-1)!/k!(n-1)!$. ∎

(2.55) Corollary.
$$d_k = (2k+n-2)\frac{(k+n-3)!}{k!(n-2)!}.$$

Proof: By Proposition (2.49), $d_k = \dim \mathcal{P}_k - \dim \mathcal{P}_{k-2}$. ∎

(2.56) Corollary.
$$d_k = O(k^{n-2}) \text{ as } k \to \infty.$$

Proof: d_k is a polynomial of degree $n-2$ in k. ∎

Some remarks on the low-dimensional cases are in order. If $n = 1$, S consists of two points, and there are only two independent harmonic polynomials, 1 and x. 1 spans $\mathcal{H}_0$ and x spans $\mathcal{H}_1$, and $\mathcal{H}_k = \{0\}$ for $k > 1$. If $n = 2$, $\mathcal{H}_0$ is spanned by 1 and, for $k > 0$, $\mathcal{H}_k$ is spanned by $(x_1 + ix_2)^k$ and $(x_1 - ix_2)^k$. (These polynomials clearly belong to $\mathcal{H}_k$, and $\dim \mathcal{H}_k = 2$.) Thus if we ser $x_1 + ix_2 = re^{i\theta}$ and take θ as coordinate on the unit circle S, we see that H_k is spanned by $e^{ik\theta}$ and $e^{-ik\theta}$. The decomposition of $L^2(S)$ into spherical harmonics is therefore just the usual Fourier series expansion.

Back to the general case: For each $x \in S$, consider the linear functional $Y \to Y(x)$ on H_k. Since $H_k \subset L^2(S)$ is a finite-dimensional Hilbert space, there is a unique $Z_k^x \in H_k$ such that

$$Y(x) = \langle Y \mid Z_k^x \rangle \qquad (Y \in H_k).$$

Z_k^x is called the **zonal harmonic of degree k with pole at x.** We list some of the amusing proerties of Z_k^x in the following theorem.

(2.57) Theorem.

Suppose $x, y \in S$.

  a. *For any orthonormal basis $Y_1, \ldots, Y_{d_k}$ of H_k, $Z_k^x(y) = \sum_1^{d_k} \overline{Y_j(x)} Y_j(y)$.*

  b. *Z_k^x is real-valued, and $Z_k^x(y) = Z_k^y(x)$.*

  c. *$Z_k^{Tx}(Ty) = Z_k^x(y)$ for any rotation T.*

  d. *$Z_k^x(x) = d_k/\omega_n$.*

  e. *$\|Z_k^x\|_2^2 = d_k/\omega_n$.*

  f. *$\|Z_k^x\|_\infty = d_k/\omega_n$.*

Proof: (a) holds since

$$Z_k^x(y) = \sum \langle Z_k^x \mid Y_j \rangle Y_j(y) = \sum \overline{\langle Y_j \mid Z_k^x \rangle} Y_j(y) = \sum \overline{Y_j(x)} Y_j(y).$$

Moreover, we can take the basis $\{Y_j\}$ to consist of real-valued functions, so (b) follows from (a). Next, if $Y \in H_k$, since $d\sigma$ is rotation-invariant we have

$$\langle Y \mid Z_k^{Tx} \circ T \rangle = \int_S Y(y) \overline{Z_k^{Tx}(Ty)} \, d\sigma(y) = \int_S Y(T^{-1}y) Z_k^{Tx}(y) \, d\sigma(y)$$
$$= Y \circ T^{-1}(Tx) = Y(x).$$

But Z_k^x is uniquely determined by the property $\langle Y \mid Z_k^x \rangle = Y(x)$, so (c) follows.

By taking $x = y$ in (c) and noting that rotations act transitively on S, we see that $Z_k^x(x)$ is independent of x. But by (a), $Z_k^x(x) = \sum |Y_j(x)|^2$, so

$$d_k = \sum \|Y_j\|_2^2 = \sum \int_S |Y_j(x)|^2 \, d\sigma(x) = \int_S Z_k^x(x) \, d\sigma(x) = \omega_n Z_k^x(x),$$

which proves (d). (e) follows from (a), (d), and the Parseval equation:

$$\|Z_k^x\|_2^2 = \sum \left| \langle Z_k^x \mid Y_j \rangle \right|^2 = \sum |Y_j(x)|^2 = Z_k^x(x) = \frac{d_k}{\omega_n}.$$

Finally, (f) follows from (d), (e) and the Schwarz inequality: for all $y \in S$,

$$|Z_k^x(y)| = \left| \langle Z_k^x \mid Z_k^y \rangle \right| \leq \|Z_k^x\|_2 \|Z_k^y\|_2 = \frac{d_k}{\omega_n}. \qquad \blacksquare$$

(2.58) Theorem.

$Z_k^x(y) = F_k(x \cdot y)$, where F_k is an explicitly computable polynomial of degree k in one variable.

Proof: Let $e = (1, 0, \ldots, 0) \in S$, and let $P \in \mathcal{H}_k$ be the harmonic extension of Z_k^e; thus $P(rx) = r^k Z_k^e(x)$ for $x \in S$ and $r > 0$. From Theorem (2.55c) we see that $P(Ty) = P(y)$ $(y \in \mathbb{R}^n)$ for any rotation T that leaves e fixed. Thus, if we set $y' = (y_2, \ldots, y_n)$ and write $P(y) = \sum_0^k y_1^{k-j} P_j(y')$, the polynomials P_j are homogeneous of degree j and rotation-invariant on $\mathbb{R}^{n-1}$. But the only functions which are homogeneous and rotation-invariant are constants times powers of $|y'|$, and these are polynomials only if the power is a nonnegative even integer. Thus $P_j(y') = C_j |y'|^j$ where $C_j = 0$ if j is odd, so

$$P(y) = \sum_0^{[k/2]} C_{2j} |y'|^{2j} y_1^{k-2j}.$$

Now, P is harmonic, so by Proposition (2.2) and the identity $\Delta(fg) = (\Delta f)g + 2\nabla f \cdot \nabla g + f(\Delta g)$,

$$0 = \Delta P(y)$$
$$= \sum_0^{[k/2]} C_{2j} \left[2j(2j+n-3)|y'|^{2j-2} y_1^{k-2j} + (k-2j)(k-2j-1)|y'|^{2j} y_1^{k-2j-2} \right]$$
$$= \sum_0^{[k/2]-1} \left[C_{2j}(k-2j)(k-2j-1) + C_{2j+2}(2j+2)(2j+n-1) \right] |y'|^{2j} y_1^{k-2j-2}.$$

Thus

$$C_{2j+2} = -\frac{(k-2j)(k-2j-1)}{(2j+2)(2j+n-1)} C_{2j},$$

so the C_{2j}'s are determined by recursion once we know C_0, and $C_0 = P(e) = Z_k^e(e) = d_k/\omega_n$ by Theorem (2.55d).

If we restrict y to the unit sphere we have $|y'|^2 = 1 - y_1^2$, so

$$Z_k^e(y) = \sum_0^{[k/2]} C_{2j}(1 - y_1^2)^{2j} y_1^{k-2j} = \sum_0^k b_j y_1^j = \sum_0^k b_j (e \cdot y)^j,$$

where the b_j's can be computed from the C_{2j}'s and the binomial theorem. Finally, given $x \in S$, let T be a rotation such that $Te = x$. By Theorem (2.55c),

$$Z_k^x(y) = Z_k^e(T^{-1}y) = \sum b_j (e \cdot T^{-1}y)^j = \sum b_j (Te \cdot y)^j = \sum b_j (x \cdot y)^j,$$

which completes the proof. ∎

The polynomials $F_k(t) = \sum b_j t^j$ are well known special functions. When $n = 2$, F_k is a constant times the **Chebyshev polynomial** of degree k; see Exercise 4. When $n \geq 3$, F_k is a constant times the **Gegenbauer** (or **ultraspherical**) **polynomial** $C_k^{(n-2)/2}$ of degree k associated to the parameter $(n-2)/2$; see Exercise 6. For $n = 3$, $C_k^{1/2}$ is known as the **Legendre polynomial** of degree k. For more about these polynomials, see the Bateman Manuscript Project [4] or Hochstadt [25].

Let π_k denote the orthogonal projection of $L^2(S)$ onto H_k. The theory of zonal harmonics furnishes us with a simple formula for π_k:

(2.59) Proposition.
If $f \in L^2(S)$,

$$\pi_k f(x) = \int_S F_k(x \cdot y) f(y) \, d\sigma(y).$$

Proof: We have $f = \pi_k f + g$ where $g \perp H_k$. Since $Z_k^x \in H_k$,

$$\int_S F_k(x \cdot y) f(y) \, d\sigma(y) = \langle f \mid Z_k^x \rangle = \langle \pi_k f \mid Z_k^x \rangle = \pi_k f(x). \qquad \blacksquare$$

It is clear how to solve the Dirichlet problem

$$\Delta u = 0 \text{ on } B, \qquad u = f \text{ on } S$$

by spherical harmonics. If f is itself a spherical harmonic, u is just its extension as a harmonic polynomial. But every f is a sum of spherical harmonics, so we can extend by linearity.

(2.60) Theorem.
a. If $f \in L^2(S)$, the harmonic function u on B with $u = f$ on S is given by

$$u(rx) = \sum_0^\infty r^k \pi_k f(x) \qquad (x \in S, \ 0 \leq r \leq 1).$$

The series converges in $L^2(S)$ for each fixed $r \leq 1$, and absolutely and uniformly in $y = rx$ for $x \in S$ and $r \leq r_0$ for any $r_0 < 1$.
b. The Poisson kernel (2.47) has the expansion

$$P(rx, y) = \sum_0^\infty r^k F_k(x \cdot y) \qquad (x, y \in S, \ 0 \leq r < 1).$$

The series converges uniformly for $x, y \in S$ and $r \leq r_0$ for any $r_0 < 1$.

Proof: (a) Clearly the series $\sum r^k \pi_k f$ converges in the $L^2(S)$ norm for $r \leq 1$ and tends to f in this norm as $r \to 1$. Moreover, since

$$\|Z_k^x\|_2 \leq (d_k/\omega_n)^{1/2} \leq C(1+k)^{(n-2)/2}$$

by (2.57e) and (2.56), for $r \leq r_0 < 1$ we have

$$\sum |r^k \pi_k f(x)| \leq \sum r_0^k |\langle f \mid Z_k^x \rangle| \leq \|f\|_2 \sum r_0^k \|Z_k^x\|_2$$
$$\leq C\|f\|_2 \sum r_0^k (1+k)^{(n-2)/2} < \infty,$$

where we have used (2.59) and the Schwarz inequality. Thus the series converges absolutely and uniformly. Since $\pi_k f \in H_k$, we have $r^k \pi_k f \in \mathcal{H}_k$, so $\sum r^k \pi_k f$ is harmonic by Corollary (2.12).

(b) If $r \leq r_0$, by (2.57f) and (2.52) we have

$$\sum |r^k F_k(x \cdot y)| \leq \sum r^k |Z_k^x(y)| \leq \sum r_0^k \frac{d_k}{\omega_n} \leq C \sum r_0^k (1+k)^{n-2} < \infty,$$

so the series $\sum r^k F_k(x \cdot y)$ converges absolutely and uniformly to a continuous function of $y \in S$ for each $x \in S$ and $r < 1$. Thus to complete the proof it suffices to show that for each $x \in S$, $r < 1$, and $f \in C(S)$,

$$\int_S P(rx, y) f(y) \, d\sigma(y) = \int_S \sum r^k F_k(x \cdot y) f(y) \, d\sigma(y).$$

But this follows by interchanging the summation and integration and using (2.58) and part (a). ∎

Spherical harmonics also lead to a solution of the dual Dirichlet problem

(2.61) $\Delta u = f$ on B, $u = 0$ on S.

The idea here is to find an orthonormal basis for $L^2(B)$ consisting of eigenfunctions for Δ that vanish on S.

We adopt the following convention: if $Y \in H_k$, we regard Y as a function on $\mathbb{R}^n \setminus \{0\}$ by extending it to be homogeneous of degree 0:

$$Y(x) = Y(|x|^{-1}x).$$

(2.62) Lemma.
If $Y \in H_k$ then $\Delta Y = -k(k + n - 2)r^{-2}Y$.

Proof: Since $P(x) = |x|^k Y(|x|^{-1}x) = r^k Y(x)$ is harmonic, by (2.2) we have

$$0 = \Delta(r^k Y) = (\Delta r^k)Y + 2\nabla(r^k) \cdot \nabla Y + r^k \Delta Y$$

$$= \left[k(k-1)r^{k-2} + \frac{n-1}{r} kr^{k-1} \right] Y + 2kr^{k-2}\sum x_j \partial_j Y + r^k \Delta Y$$

$$= k(k+n-2)r^{k-2}Y + r^k \Delta Y,$$

since $\sum x_j \partial_j Y = r\partial Y/\partial r = 0$ by (0.1).   ∎

(2.63) Lemma.
If $Y \in H_k$ and $F(x) = f(|x|) = f(r)$, then

$$\Delta(FY)(x) = \left[f''(r) + \frac{n-1}{r} - \frac{k(k+n-2)}{r^2} f(r) \right] Y(x).$$

Proof: As in the proof of Lemma (2.62), we have

$$\nabla F \cdot \nabla Y = \frac{f'(r)}{r}\sum x_j \partial_j Y(x) = f'(r)\frac{\partial Y}{\partial r} = 0,$$

so by (2.2) and (2.62),

$$\Delta(FY) = (\Delta F)Y + 2\nabla F \cdot \nabla Y + F\Delta Y$$

$$= \left[f''(r) + \frac{n-1}{r} - \frac{k(k+n-2)}{r^2} f(r) \right] Y(x).   ∎$$

We propose to solve the eigenvalue problem

$$\Delta u = -\lambda^2 u \text{ on } B, \qquad u = 0 \text{ on } S.$$

(We call the eigenvalue $-\lambda^2$ because it will turn out to be negative, but for the moment, λ is just a complex number.) If we assume that u has the form $u(x) = f(|x|)Y(x)$ where $Y \in H_k$, by Lemma (2.63) we are led to the ordinary differential equation

$$(2.64) \qquad f''(r) + \frac{n-1}{r}f'(r) + \left[\lambda^2 - \frac{k(k+n-2)}{r^2} \right] f(r) = 0.$$

For boundary conditions, we have $f(1) = 0$, and we shall also require that $u(x) = [r^{-k}f(r)][r^k Y(x)]$ be smooth at $x = 0$, i.e., that $r^{-k}f(r)$ be smooth at $r = 0$. (The justification for this is that $\Delta + \lambda^2$ is hypoelliptic, as we shall show in §6B.)

(2.64) is essentially a Bessel equation. Indeed, if we set $\rho = \lambda r$ and $g(r) = f(r/\lambda)$, so that $g(\rho) = f(r)$ and $g'(\rho) = \lambda^{-1}f'(r)$, then (2.64) becomes

$$g''(\rho) + \frac{n-1}{\rho}g'(\rho) + \left[1 - \frac{k(k+n-2)}{\rho^2}\right]g(\rho) = 0.$$

Now set $h(\rho) = \rho^{(n-2)/2}g(\rho)$. Substituting $\rho^{(2-n)/2}h(\rho)$ for $g(\rho)$ and simplifying, we obtain

$$(2.65) \qquad h''(\rho) + \frac{1}{\rho}h'(\rho) + \left[1 - \frac{[k+(n-2)/2]^2}{\rho^2}\right]h(\rho) = 0.$$

This is Bessel's equation of order $k + (n-2)/2$. We shall assume the following facts about solutions of this equation; see Folland [17] or Hochstadt [25] for more details.

In the first place, the only solutions that are bounded at $\rho = 0$ are multiples of $J_{k+(n-2)/2}(\rho)$, where

$$J_\alpha(t) = \sum_0^\infty \frac{(-1)^j}{j!\Gamma(\alpha+1+j)}\left(\frac{t}{2}\right)^{\alpha+2j}$$

is the Bessel function of the first kind of order α. Undoing our changes of variables, we see that

$$f(r) = Cr^{(2-n)/2}J_{k+(n-2)/2}(\lambda r) = C'r^k \sum_0^\infty \frac{(-1)^j}{j!\Gamma((n/2)+k+j)}\left(\frac{\lambda r}{2}\right)^{2j}.$$

Thus $r^{-k}f(r)$ is analytic at $r = 0$, as desired, and $f(1) = 0$ if and only if $J_{k+(n-2)/2}(\lambda) = 0$.

Moreover, the zeros of $J_{k+(n-2)/2}$ are all real, and since replacement of λ by $-\lambda$ leaves $f(r)$ unchanged, it suffices to consider the positive zeros. Let $\lambda_k^1, \lambda_k^2, \ldots$ be the positive zeros of $J_{k+(n-2)/2}$, in increasing order, and let

$$C_k^l = 2^{1/2}|J_{k+(n/2)}(\lambda_k^l)|^{-1}.$$

Then

$$\{C_k^l J_{k+(n-2)/2}(\lambda_k^l r) : l = 1, 2, 3, \ldots\}$$

is an orthonormal basis for $L^2((0,1), r\,dr)$.

Now, let

$$f_k^l(r) = C_k^l r^{(2-n)/2}J_{k+(n-2)/2}(\lambda_k^l r).$$

Then $\{f_k^l\}_{l=1}^\infty$ is an orthonormal basis for $L^2((0,1), r^{n-1}\,dr)$. Moreover, let $Y_k^1, \ldots, Y_k^{d_k}$ be an orthonormal basis for H_k (as a subspace of $L^2(S)$), and set

$$F_k^{lm}(x) = f_k^l(|x|)Y_k^m(x).$$

(2.66) Theorem.
$\{F_k^{lm} : k \geq 0, \ l \geq 1, \ 1 \leq m \leq d_k\}$ *is an orthonormal basis for* $L^2(B)$.
Moreover, $\Delta F_k^{lm} = -(\lambda_k^l)^2 F_k^{lm}$ *and* $F_k^{lm}(x) = 0$ *for* $x \in S$.

Proof: The second assertion is obvious from the construction of F_k^{lm}. As for orthonormality, by integrating in polar coordinates we have

$$\int_B F_k^{lm} \overline{F_{k'}^{l'm'}} = \int_0^1 f_k^l(r)\overline{f_{k'}^{l'}(r)}r^{n-1}\,dr \int_S Y_k^m(x)\overline{Y_{k'}^{m'}(x)}\,d\sigma(x).$$

The integral over S equals $\delta_{mm'}\delta_{kk'}$, and in particular is zero unless $k = k'$, in which case the integral in r equals $\delta_{ll'}$.

To show completeness, suppose $g \in L^2(B)$ is orthogonal to all F_k^{lm}:

$$\int_0^1 \int_S f_k^l(r)Y_k^m(x)\overline{g(rx)}r^{n-1}\,d\sigma(x)\,dr = 0 \text{ for all } k, l, m.$$

Integrating first over S, by completeness of spherical harmonics we obtain

$$\int_0^1 f_k^l(r)\overline{g(rx)}r^{n-1}\,dr = 0$$

for almost every $x \in S$. But then for each such x we have $g(rx) = 0$ for almost every r. Thus $g = 0$ a.e., and the proof is complete. ∎

We can now solve the Dirichlet problem (2.61) for $f \in L^2(B)$. Namely:

$$\text{if } f = \sum a_k^{lm} F_k^{lm}, \text{ take } u = -\sum(\lambda_k^l)^{-2}a_k^{lm} F_k^{lm}.$$

Since $\lambda^{(2-n)/2}J_{k+(n-2)/2}(\lambda)$ is an entire analytic function, its zeros λ_k^l are bounded away from 0, so the series for u converges in $L^2(B)$, and it follows easily that $\Delta u = f$ in the sense of distributions. It will follow from our work in Chapter 7 that the series for u actually converges in the norm of $H_1(B)$ as defined in §2F (indeed, in an even stronger norm), so that $u = 0$ on S in the sense of Corollary (2.40). Moreover, we will show in Chapter 7 that if $f \in C^k(\overline{B})$ where $k > \frac{1}{2}n$, then $u \in C^2(\overline{B})$, so u is a classical solution of the Dirichlet problem (2.61). (Actually, it would suffice to have $f \in C^\alpha(\overline{B})$ for some $\alpha > 0$; see the remarks in §7G.)

We conclude by remarking that the same method leads to a solution of the Neumann problem

$$(2.67) \qquad\qquad \Delta u = f \text{ on } B, \qquad \partial_\nu u = 0 \text{ on } S$$

by expansion in spherical harmonics and Bessel functions. We simply replace the boundary condition $f(1) = 0$ for (2.64) by $f'(1) = 0$; the corresponding boundary condition for the Bessel equation (2.65) is

$$\tfrac{1}{2}(2 - n)J_{k+(n-2)/2}(\lambda) + \lambda J'_{k+(n-2)/2}(\lambda) = 0.$$

If $\mu_k^1, \mu_k^2, \ldots$ are the positive numbers satisfying this equation, and

$$A_k^l = \frac{\sqrt{2}\,\mu_k^l}{[(\mu_k^l)^2 - k^2 - k(n - 2)]^{1/2}|J_{k+(n-2)/2}(\mu_k^l)|},$$

then $\{A_k^l J_{k+(n-2)/2}(\mu_k^l r)\}_{l=1}^{\infty}$ is an orthonormal basis for $L^2\big((0,1),\, r\,dr\big)$ when $k > 0$. If $k = 0$, we must also include the function $n^{1/2}r^{(2-n)/2}$ corresponding to the eigenvalue $\lambda = 0$. (See Folland [17].) The rest of the discussion goes through as before: we obtain an orthonormal basis for $L^2(B)$ consisting of eigenfunctions of Δ satisfying the Neumann condition $\partial_\nu u = 0$. By expanding $f \in L^2(B)$ in terms of this basis, we obtain a solution of (2.67) provided that the component of f corresponding to the eigenvalue $\lambda = 0$ vanishes, that is, provided $\int_B f = 0$. (As we remarked in §2C, this condition is necessary for the solvability of (2.67) in any case.)

EXERCISES

1. Show that the Poisson kernel for the ball $B_R(x_0)$ is

$$P(x, y) = \frac{R^2 - |x - x_0|^2}{\omega_n R|x - y|^n}.$$

2. Suppose Ω is an open set in $\mathbb{R}^n$. Show that for any compact sets $K, K' \subset \Omega$ with K contained in the interior of K' and any multi-index α there is a constant $C_{K,\alpha}$ such that

$$\sup_{x \in K} |\partial^\alpha u(x)| \le C_{K,\alpha} \sup_{x \in K'} |u(x)|$$

for every harmonic function u on Ω. (Hint: First do the case $K = \overline{B_r(x_0)}$, $K' = \overline{B_R(x_0)}$, where $r < R$, by applying the analogue of Theorem (2.48) for the ball $B_R(x_0)$; see Exercise 1.) Conclude that if $\{u_j\}$ is a sequence of harmonic functions on Ω that converges to u uniformly on compact subsets of Ω, then $\partial^\alpha u_j \to \partial^\alpha u$ uniformly on compact subsets of Ω for every α.

3. (Harnack's inequality) Suppose u is continuous on $\overline{B_R(x_0)}$, harmonic on $B_R(x_0)$, and $u \geq 0$. Show that for $|x - x_0| = r < R$,

$$\frac{1 - (r/R)}{[1 + (r/R)]^{n-1}} u(x_0) \leq u(x) \leq \frac{1 + (r/R)}{[1 - (r/R)]^{n-1}} u(x_0).$$

(Hint: Estimate the Poisson kernel in Exercise 1 by a constant.)

4. Show that if we identify $\mathbb{R}^2$ with $\mathbb{C}$, the 2-dimensional zonal harmonics are given by

$$Z_0^{e^{i\theta}}(e^{i\phi}) = (2\pi)^{-1}, \qquad Z_k^{e^{i\theta}}(e^{i\phi}) = \pi^{-1} \cos k(\phi - \theta) \text{ for } k > 0.$$

Verify Theorem (2.57) from these formulas, and show that the polynomial F_k of Theorem (2.58) is $\pi^{-1}T_k$ for $k > 0$, where T_k is the *Chebyshev polynomial* defined by $T_k(\cos \theta) = \cos k\theta$.

5. If we identify $\mathbb{R}^2$ with $\mathbb{C}$, (2.47) and Theorem (2.60b) together with Exercise 4 show that for $n = 2$,

$$\frac{1 - r^2}{1 - 2r \cos(\theta - \phi) + r^2} = 2\pi P(re^{i\theta}, e^{i\phi}) = 1 + 2\sum_1^\infty r^k \cos k(\theta - \phi).$$

Verify this directly by summing the series. (Hint: $\cos k\theta = \operatorname{Re} e^{i\theta}$.)

6. The *Gegenbauer polynomials* C_k^λ associated to the parameter $\lambda > 0$ are defined by the generating relation

$$\sum_0^\infty C_k^\lambda(t)r^k = (1 - 2rt + r^2)^{-\lambda}.$$

By applying the operator $1 + \lambda^{-1}r(d/dr)$ to both sides of this equation and using Theorem (2.60b), show that if F_k is as in Theorem (2.58) and $n \geq 3$,

$$F_k(t) = \frac{2k + n - 2}{(n - 2)\omega_n} C_k^{(n-2)/2}(t).$$

7. Solve the Neumann problem

$$\Delta u = 0 \text{ on } B, \qquad \partial_\nu u = g \text{ on } S$$

by expanding g in spherical harmonics. How is the necessary condition $\int_S g = 0$ used?

8. Suppose V is a nonzero vector subspace of H_k that is invariant under rotations. Show that $V = H_k$. (Hint: Consider the "zonal harmonic" Z_V^x, i.e., the unique element of V such that $Y(x) = \langle Y \mid Z_V^x \rangle$ for all $Y \in V$. Z_V^x has properties analogous to Theorem (2.57), and the proof of Theorem (2.58) shows that a function with these properties must be a constant multiple of Z_k^x.)

9. Show that $J_{-1/2}(t) = \sqrt{2/\pi t}\,\cos t$ and $J_{1/2}(t) = \sqrt{2/\pi t}\,\sin t$, and then show that the orthonormal basis for $L^2(-1, 1)$ given by Theorem (2.66) in the case $n = 1$ is

$$\left\{ \cos \tfrac{1}{2} j\pi x : j = 1, 3, 5, \ldots \right\} \cup \left\{ \sin \tfrac{1}{2} j\pi x : j = 2, 4, 6, \ldots \right\}.$$

Note that if we make the change of variable $t = \tfrac{1}{2}(x + 1)$, this basis turns into the familiar Fourier sine basis $\{\sin j\pi t : j = 1, 2, 3, \ldots\}$ for $L^2(0, 1)$.

I. More about Harmonic Functions

Now that we have solved the Dirichlet problem for the ball, we can derive some more interesting facts about harmonic functions.

(2.68) The Reflection Principle.
Let Ω be an open set in $\mathbb{R}^{n+1}$ (with coordinates $x \in \mathbb{R}^n$, $t \in \mathbb{R}$) with the property that $(x, -t) \in \Omega$ if $(x, t) \in \Omega$. Let $\Omega_+ = \{(x, t) \in \Omega : t > 0\}$ and $\Omega_0 = \{(x, t) \in \Omega : t = 0\}$. If u is continuous on $\Omega_+ \cup \Omega_0$, harmonic on Ω_+, and zero on Ω_0, then u can be extended to be harmonic on Ω by setting $u(x, -t) = -u(x, t)$.

Proof: It is clear that this extension of u is continuous on Ω and harmonic on $\Omega \setminus \Omega_0$. Given $(x_0, 0) \in \Omega_0$, we shall show that u is harmonic near x_0. Let B be a ball centered at x_0 whose closure is contained in Ω. By translating and dilating the coordinates (which preserves harmonicity), we may assume that $x_0 = 0$ and B is the unit ball. Since u is continuous on $\overline{B}$, we can solve the Dirichlet problem

$$\Delta v = 0 \text{ on } B, \qquad v = u \text{ on } \partial B,$$

with $v \in C(\overline{B})$, by Theorem (2.48). By the explicit formula given there for v, the fact that $u(x, -t) = -u(x, t)$ implies that $v(x, -t) = -v(x, t)$. In particular, $v(x, 0) = 0$. Thus v agrees with u on the boundaries of the upper and lower halves of B. By the uniqueness theorem (2.15), $v = u$ on each half, so $v = u$ on B. In particular, u is harmonic on B. ∎

Suppose Ω is a neighborhood of $x_0 \in \mathbb{R}^n$. If u is a harmonic function on $\Omega \setminus \{x_0\}$, u is said to have a **removable singularity** at x_0 if u can be defined at x_0 so as to be harmonic on Ω. The following theorem says that any singularity which is weaker than that of the fundamental solution is removable.

(2.69) Theorem.
Suppose u is harmonic on $\Omega \setminus \{x_0\}$. If

$$|u(x)| = o(|x - x_0|^{2-n}) \qquad (n > 2)$$

or

$$|u(x)| = o(\log|x - x_0|^{-1}) \qquad (n = 2)$$

as $x \to x_0$, then u has a removable singularity at x_0.

Proof: By translating and dilating the coordinates, we may assume that $x_0 = 0$ and that Ω contains the closed unit ball $\overline{B}_1$. (We shall write B_r for $B_r(0)$.) We may also assume that u is real. Since u is continuous on ∂B_1, by Theorem (2.48) there exists $v \in C(\overline{B}_1)$ satisfying

$$\Delta v = 0 \text{ on } B_1, \qquad v = u \text{ on } \partial B_1.$$

We claim that $u = v$ on $B_1 \setminus \{0\}$, so we can remove the singularity by setting $u(0) = v(0)$. Given $\epsilon > 0$ and $\delta \in (0, 1)$, consider the function

$$\begin{aligned}
u(x) - v(x) - \epsilon(|x|^{2-n} - 1) & \qquad (n > 2), \\
u(x) - v(x) + \epsilon \log|x| & \qquad (n = 2)
\end{aligned}$$

on $\overline{B}_1 \setminus B_\delta$. This function is real and harmonic on $B_1 \setminus \overline{B}_\delta$ (Corollary (2.3)), continuous on the closure, zero on ∂B_1, and — by the assumption on u — negative on ∂B_δ for all sufficiently small δ. By the maximum principle, it is negative on $B_1 \setminus \{0\}$. Letting $\epsilon \to 0$, we see that $u - v \leq 0$ on $B_1 \setminus \{0\}$. By the same argument, $v - u \leq 0$ on $B_1 \setminus \{0\}$, so that $u = v$ on $B_1 \setminus \{0\}$. ∎

Our final results concern the behavior of harmonic functions at infinity. To obtain these, we first need a formula for Δ in general curvilinear coordinates, which is of interest in its own right.

Let T be a C^∞ bijection from an open set $\Omega \subset \mathbb{R}^n$ to an open set $\Omega' \subset \mathbb{R}^n$ with C^∞ inverse. Let $y = T(x)$, and let $J_T = (\partial y_i / \partial x_j)$ and

$J_{T^{-1}} = (\partial x_i / \partial y_j)$ be the Jacobian matrices of T and T^{-1}. Define the matrix (g_{ij}) on Ω' by

$$(g_{ij}(y)) = (J_{T^{-1}})^t(y)J_{T^{-1}}(y) = \left(\sum_k \frac{\partial x_k}{\partial y_i}\frac{\partial x_k}{\partial y_j}\right).$$

The inverse matrix (g^{ij}) of (g_{ij}) is then

$$(g^{ij}(y)) = (JJ^t)(T^{-1}(y)) = \left(\sum_k \frac{\partial y_i}{\partial x_k}\frac{\partial y_j}{\partial x_k}\bigg|_{T^{-1}(y)}\right).$$

Moreover, let us set

$$g = \det(g_{ij}) = (\det J_{T^{-1}})^2.$$

Then the volume elements on Ω and Ω' are related by $dx = |\det J_{T^{-1}}| \, dy = \sqrt{g} \, dy$.

(2.70) Theorem.
If u is a C^2 function on Ω and $U = u \circ T^{-1}$, then

$$(2.71) \qquad \Delta u \circ T^{-1} = \frac{1}{\sqrt{g}} \sum_{i,j=1}^{n} \frac{\partial}{\partial y_j}\left(g^{ij}\sqrt{g}\frac{\partial U}{\partial y_i}\right).$$

Proof: Given $w \in C_c^\infty(\Omega)$, let $W = w \circ T^{-1}$. Then

$$\int (\Delta u)w \, dx = \sum_k \int \frac{\partial^2 u}{\partial x_j^2}w \, dx = -\sum_k \int \frac{\partial u}{\partial x_k}\frac{\partial w}{\partial x_k} \, dx$$

$$= -\sum_{i,j,k} \int \frac{\partial U}{\partial y_i}\frac{\partial y_i}{\partial x_k}\frac{\partial W}{\partial y_j}\frac{\partial y_j}{\partial x_k}\sqrt{g} \, dy$$

$$= -\sum_{i,j} \int \frac{\partial U}{\partial y_i}\frac{\partial W}{\partial y_j}g^{ij}\sqrt{g} \, dy = \sum_{i,j} \int \frac{\partial}{\partial y_j}\left(g^{ij}\sqrt{g}\frac{\partial U}{\partial y_i}\right)W \, dy$$

$$= \sum_{i,j} \int \left(\left[\frac{1}{\sqrt{g}}\frac{\partial}{\partial y_j}\left(g^{ij}\sqrt{g}\frac{\partial U}{\partial y_i}\right)\right] \circ T\right)w \, dx.$$

This being true for all w, the result follows.   ∎

Remark: Rather than regarding $y = T(x)$ as a transformation from Ω to Ω', we can regard $y_1, \ldots, y_n$ as corvilinear coordinates on Ω, and the expression on the right of (2.71) gives the formula for the Laplacian of a function U in these coordinates. More generally, this is the expression for the Laplace-Beltrami operator on a Riemannian manifold with metric tensor (g_{ij}) in the local coordinates $y_1, \ldots, y_n$.

We are particularly interested in the transformation

$$y = T(x) = |x|^{-2}x$$

on $\mathbb{R}^n \setminus \{0\}$ obtained by inversion in the unit sphere. Since $x = |y|^{-2}y$, we see that

$$\frac{\partial x_k}{\partial y_i} = \frac{-2y_i y_k}{|y|^4} + \frac{\delta_{ik}}{|y|^2},$$

$$g_{ij} = \sum_k \left[\frac{-2y_i y_k}{|y|^4} + \frac{\delta_{ik}}{|y|^2}\right]\left[\frac{-2y_j y_k}{|y|^4} + \frac{\delta_{jk}}{|y|^2}\right] = \frac{\delta_{ij}}{|y|^4},$$

so $g^{ij} = |y|^4 \delta_{ij}$ and $g = |y|^{-4n}$. Thus, if u is a C^2 function on $\mathbb{R}^n \setminus \{0\}$ and $U(y) = u(|y|^{-2}y)$,

$$\Delta u(|y|^{-2}y) = |y|^{2n} \sum \frac{\partial}{\partial y_j}\left[|y|^{4-2n}\frac{\partial U}{\partial y_j}\right]$$

$$= |y|^{2n} \sum \left[|y|^{4-2n}\frac{\partial^2 U}{\partial y_j^2} + (4 - 2n)|y|^{2-2n}y_j\frac{\partial U}{\partial y_j}\right]$$

$$= |y|^{n+2} \sum \left[|y|^{2-n}\frac{\partial^2 u}{\partial y_j^2} + 2\frac{\partial |y|^{2-n}}{\partial y_j}\frac{\partial U}{\partial y_j}\right].$$

We can add the term $U(\partial^2 |y|^{2-n}/\partial y_j^2)$ to the last expression in square brackets without changing anything, since $\sum \partial^2 |y|^{2-n}/\partial y_j^2 = 0$ for $y \neq 0$. We therefore have

$$(2.72) \quad \Delta u(|y|^{-2}y) = |y|^{n+2} \sum \frac{\partial^2}{\partial y_j^2}(|y|^{2-n}U), \qquad U(y) = u(|y|^{-2}y).$$

If $\Omega \subset \mathbb{R}^n \setminus \{0\}$, we set

$$\tilde{\Omega} = \{|x|^{-2}x : x \in \Omega\},$$

and if u is a function on Ω, we define its **Kelvin transform** $\tilde{u}$, a function on $\tilde{\Omega}$, by

$$\tilde{u}(x) = |x|^{2-n}u(|x|^{-2}x).$$

(We have already encountered this in the construction of the Green's function for the ball in §2H.) With the notation of (2.72), we have $\tilde{U}(y) = |y|^{n-2}u(y)$, so if we replace u by $\tilde{u}$ in (2.72) we obtain

$$\Delta\tilde{u}(|y|^{-2}y) = |y|^{n+2}\Delta u(y).$$

In particular, we have proved:

(2.73) Theorem.
If u is harmonic on $\Omega \subset \mathbb{R}^n \setminus \{0\}$, its Kelvin transform $\tilde{u}$ is harmonic on $\tilde{\Omega}$.

Now suppose u is harmonic outside some bounded set; then its Kelvin transform $\tilde{u}$ is harmonic in a punctured neighborhood of the origin. We say that u is **harmonic at infinity** if $\tilde{u}$ has a removable singularity at 0. As an immediate corollary of Theorem (2.69), we have:

(2.74) Proposition.
If u is harmonic on the complement of a bounded set in $\mathbb{R}^n$, the following are equivalent:
a. u *is harmonic at infinity.*
b. $u(x) \to 0$ *as* $x \to \infty$ *if* $n > 2$, *or* $|u(x)| = o(\log|x|)$ *as* $x \to \infty$ *if* $n = 2$.
c. $|u(x)| = O(|x|^{2-n})$ *as* $x \to \infty$.

For future reference we give one more result concerning the behavior of harmonic functions at infinity. We denote by ∂_r the radial derivative, that is, the normal derivative on spheres about the origin:

$$\partial_r u(ry) = \frac{d}{dr} u(ry) = \sum y_j \partial_j u(ry) \qquad (r > 0, \ |y| = 1).$$

(2.75) Proposition.
Suppose u is harmonic at infinity. Then $|\partial_r u(x)| = O(|x|^{1-n})$ as $x \to \infty$; moreover, in case $n = 2$, $|\partial_r u(x)| = O(|x|^{-2})$ as $x \to \infty$.

Proof: By dilating the coordinates, we may assume that u is harmonic outside $B_R(0)$ for some $R < 1$. The Kelvin transform $\tilde{u}$ is then harmonic on the unit ball $B_1(0)$ (once we have removed the singularity at 0) and continuous on its closure. We can therefore expand $\tilde{u}$ in spherical harmonics according to Theorem (2.60):

$$\tilde{u}(x) = \sum_0^\infty |x|^k Y_k(|x|^{-1}x) \qquad (Y_k \in H_k).$$

But then

$$u(x) = |x|^{2-n}\tilde{u}(|x|^{-2}x) = \sum_0^\infty |x|^{2-n-k} Y_k(|x|^{-1}x).$$

Thus if we set $x = ry$ with $r > 0$ and $|y| = 1$,

$$u(x) = \sum_0^\infty r^{2-n-k} Y_k(y),$$

so that

$$(2.76) \quad \partial_r u(x) = \sum_0^\infty (2-n-k)r^{1-n-k}Y_k(y) = r^{1-n}\sum_0^\infty (2-n-k)r^{-k}Y_k(y).$$

If $r > 3$, say, then certainly

$$\left|\sum(2-n-k)r^{-k}Y_k(y)\right| \le \sum 2^{-k}|Y_k(y)|,$$

which is bounded independent of r and y since the series $\sum 2^{-k}Y_k(y) = \tilde{u}(\frac{1}{2}y)$ is absolutely and uniformly convergent in y. Thus $|\partial_r u(ry)| = O(r^{1-n})$. Moreover, if $n = 2$, the term with $k = 0$ in the series (2.76) vanishes, so the same argument yields $|\partial_r u(ry)| = O(r^{-2})$. ∎

EXERCISES

1. Suppose u and v are bounded and harmonic on the half-space $\mathbb{R}^{n+1}_+$ and continuous on its closure, and that $u(x,0) = v(x,0)$ for $x \in \mathbb{R}^n$. Show that $u = v$. (Hint: Consider $w = u - v$.)

2. Use Theorem (2.70) to calculate the Laplacian in polar coordinates on $\mathbb{R}^2$ and in spherical coordinates on $\mathbb{R}^3$.

3. Let F be a one-to-one holomorphic function on an open set $\Omega \subset \mathbb{C}$. Regarding $T = F^{-1}$ as a map from $F(\Omega) \subset \mathbb{R}^2$ to $\Omega \subset \mathbb{R}^2$ as in Theorem (2.70), show that the associated matrix (g_{ij}) is given by $g_{ij}(z) = |F'(z)|^2\delta_{ij}$, and conclude that if u is a C^2 function on $F(\Omega)$, $\Delta(u \circ F) = |F'|^2(\Delta u) \circ F$.

Chapter 3
LAYER POTENTIALS

This chapter is devoted to the solution of the Dirichlet and Neumann problems for the Laplacian by the method of layer potentials. This method reduces the problems to solving certain integral equations, for which one can use the theory of compact operators.

A. The Setup

In this chapter Ω will be a fixed bounded domain in $\mathbb{R}^n$ with C^2 boundary S, and we set $\Omega' = \mathbb{R}^n \setminus \overline{\Omega}$. Ω and Ω' will both be allowed to be disconnected; however, since S is differentiable there can only be finitely many components. (In practice one usually wants to consider connected domains, but it is of interest to allow them to have holes, i.e., to allow their complements to be disconnected; and as we shall see, the theory is quite symmetric with respect to Ω and Ω'.) We denote the connected components of Ω by $\Omega_1, \ldots, \Omega_m$ and those of Ω' by $\Omega'_0, \Omega'_1 \ldots, \Omega'_{m'}$, where Ω'_0 is the unbounded component.

To deal with the Neumann problem we need to be careful about the meaning of the normal derivative, since we don't want to clutter up the theory with extraneous smoothness assumptions. Recall that we have defined the normal derivative ∂_ν on a neighborhood of S by formula (0.3). We define $C_\nu(\Omega)$ to be the space of functions $u \in C^1(\Omega) \cap C(\overline{\Omega})$ such that the limit

$$\partial_{\nu-} u(x) = \lim_{t<0,\ t\to 0} \nu(x) \cdot \nabla u(x + t\nu(x))$$

exists for each $x \in S$, the convergence being uniform on S. (Thus $\partial_{\nu-} u$ is a continuous function on S.) Similarly, we define $C_\nu(\Omega')$ to be the space of functions $u \in C^1(\Omega') \cap C(\overline{\Omega'})$ such that the limit

$$\partial_{\nu+}u(x) = \lim_{t>0,\ t\to 0} \nu(x) \cdot \nabla u(x + t\nu(x))$$

exists for each $x \in S$, the convergence being uniform on S. The operators $\partial_{\nu-}$ and $\partial_{\nu+}$ are called the **interior** and **exterior normal derivatives** on S.

We can now state precisely the problems we propose to solve:

The Interior Dirichlet Problem: *Given $f \in C(S)$, find $u \in C(\overline{\Omega})$ such that u is harmonic on Ω and $u = f$ on S.*

The Exterior Dirichlet Problem: *Given $f \in C(S)$, find $u \in C(\overline{\Omega'})$ such that u is harmonic on $\Omega' \cup \{\infty\}$ and $u = f$ on S.*

The Interior Neumann Problem: *Given $f \in C(S)$, find $u \in C_\nu(\Omega)$ such that u is harmonic on Ω and $\partial_{\nu-}u = f$ on S.*

The Exterior Neumann Problem: *Given $f \in C(S)$, find $u \in C_\nu(\Omega')$ such that u is harmonic on $\Omega' \cup \{\infty\}$ and $\partial_{\nu+}u = f$ on S.*

Note that for the exterior problems we require the solution to be harmonic at infinity as discussed in §2I; the reason for this is to obtain uniqueness results. Note also that the derivative $\partial_{\nu+}$ for the exterior Neumann problem is taken along the inward-pointing normal to Ω'; this amounts to replacing f by $-f$ if we want the outward normal.

These four problems are intimately connected with each other, and we shall obtain the solutions to all of them simultaneously. To begin with, we prove the uniqueness theorems for all four problems.

(3.1) Proposition.
If u solves the interior Dirichlet problem with $f = 0$, then $u = 0$.

Proof: This is just the uniqueness theorem (2.15).   ∎

(3.2) Proposition.
If u solves the exterior Dirichlet problem with $f = 0$, then $u = 0$.

Proof: We may assume that $0 \notin \overline{\Omega}$. By Theorem (2.73), the Kelvin transform $\tilde{u}$ of u solves the interior Dirichlet problem with $f = 0$ for the bounded domain $\tilde{\Omega'} = \{x : |x|^{-2}x \in \Omega'\}$. Hence $\tilde{u} = 0$, so $u = 0$.   ∎

(3.3) Proposition.
*If u solves the interior Neumann problem with $f = 0$, then u is constant
on each component of Ω.*

Proof: By Green's identity (2.5),

$$\int_\Omega |\nabla u|^2 = -\int_\Omega u(\Delta u) + \int_S u\partial_{\nu-}u = 0.$$

Thus $\nabla u = 0$ on Ω, so u is locally constant on Ω.   ∎

Remark: In this proof, as well as the following ones, the use of Green's
identity is not quite obvious since u is not assumed to be in $C^1(\overline{\Omega})$. However, it is easily justified by replacing Ω by the domain Ω_t whose boundary
is

$$S_t = \{x + t\nu(x) : x \in S\}$$

and passing to the limit as $t \to 0$ from below or above, as appropriate. The
definitions of $\partial_{\nu-}$ and $\partial_{\nu+}$ are designed precisely to make this argument
work.

(3.4) Proposition.
*If u solves the exterior Neumann problem with $f = 0$, then u is constant on
each component of Ω', and $u = 0$ on the unbounded component Ω_0' when
$n > 2$.*

Proof: Let $r > 0$ be large enough so that $\overline{\Omega} \subset B_r = B_r(0)$. By
Green's identity (2.5),

$$\int_{B_r\backslash\Omega} |\nabla u|^2 = -\int_{B_r\backslash\Omega} u(\Delta u) - \int_S u\partial_{\nu+}u + \int_{\partial B_r} u\partial_r u$$

$$= \int_{\partial B_r} u\partial_r u,$$

where ∂_r denotes the radial derivative. Since $|u(x)| = O(|x|^{2-n})$ and
$|\partial_r u(x)| = O(|x|^{1-n})$ by Propositions (2.74) and (2.75), we have

$$\left|\int_{\partial B_r} u\partial_r u\right| \le Cr^{2-n}r^{1-n}\int_{\partial B_r} 1 \le C'r^{2-n}.$$

When $n > 2$, by letting $r \to \infty$ we obtain $\int_{\Omega'} |\nabla u|^2 = 0$. Thus $\nabla u = 0$ on
Ω', so u is locally constant on Ω', and $u = 0$ on Ω_0' since $|u(x)| = O(|x|^{2-n})$.
If $n = 2$, Proposition (2.75) gives $|\partial_r u(x)| = O(r^{-2})$, so the same argument
shows that $|\int_{\partial B_r} u\partial_r u| = O(r^{-1})$ and hence that u is locally constant on
Ω'.   ∎

We shall see that the interior and exterior Dirichlet problems are always solvable. For the Neumann problems, however, there are some necessary conditions.

(3.5) Proposition.
If the interior Neumann problem has a solution, then $\int_{\partial\Omega_j} f = 0$ for $j = 1, \ldots, m$.

Proof: This follows immediately from Corollary (2.7).   ∎

(3.6) Proposition.
If the exterior Neumann problem has a solution, then $\int_{\partial\Omega'_j} f = 0$ for $j = 1, \ldots, m'$, and also for $j = 0$ in case $n = 2$.

Proof: That $\int_{\partial\Omega'_j} f = 0$ for $j \geq 1$ follows from Corollary (2.7). If $n = 2$, let r be large enough so that $\overline{\Omega} \subset B_r = B_r(0)$. Then Corollary (2.7) gives

$$\int_{\partial B_r} \partial_r u - \int_{\partial\Omega'_0} \partial_{\nu_+} u = 0.$$

But $|\partial_r u(x)| = O(|x|^{-2})$ by Proposition (2.75), so the first term on the right vanishes as $r \to \infty$; since $\partial_{\nu_+} u = f$, we are done.   ∎

We now turn to the problem of finding solutions. To begin with, consider the interior Dirichlet problem. Our inspiration comes from the formulas (2.24) and (2.37) that represent a harmonic function in terms of its boundary values. Suppose we neglect the second term in (2.24) or the difference between G and N in (2.37) and try to solve the interior Dirichlet problem by setting

$$u(x) = \int_S \partial_{\nu_y} N(x, y) f(y) \, d\sigma(y),$$

where N is the fundamental solution for Δ defined by (2.18) and (2.22). u will be harmonic in Ω, but of course it will not have the right boundary values in general. However, in a sense it is not far wrong: we shall see that $u|S = \frac{1}{2}f + Tf$ where T is a compact operator on $L^2(S)$. Thus what we really want is to take

$$(3.7) \qquad u(x) = \int_S \partial_{\nu_y} N(x, y) \phi(y) \, d\sigma(y)$$

where $\frac{1}{2}\phi + T\phi = f$, and we can use the theory of compact operators to handle the latter equation.

The function u defined by (3.7) is called the **double layer potential with moment** ϕ, and its physical interpretation is as follows. If we think of the normal derivative $\partial_{\nu_y} N(x, y)$ as a limit of difference quotients, we see that (3.7) is the limit as $t \to 0$ of the potential induced by a charge distribution with density $t^{-1}\phi(y)$ on S together with a charge distribution with density $-t^{-1}\phi(y)$ on the parallel surface $S_t = \{y + t\nu(y) : y \in S\}$. In other words, (3.7) is the potential induced by a distribution of dipoles on S with density $\phi(y)$, the axes of the dipoles being normal to S. (See Exercise 2 of §2G for the analogous result for a half-space. There, the Poisson kernel is $2\partial_{\nu_y} N(x, y)$ and the operator T is zero.)

Similarly, we shall try to find a solution to the Neumann problem in the form

$$u(x) = \int_S N(x, y)\phi(y) \, d\sigma(y).$$

This is the **single layer potential with moment** ϕ. It is simply the potential induced by a charge distribution on S with density $-\phi(y)$.

B. Integral Operators

Before studying double and single layer potentials, we need to collect some facts about certain kinds of integral operators on the boundary S of our domain $\Omega \subset \mathbb{R}^n$. (These results also hold if S is the closure of a bounded domain in $\mathbb{R}^{n-1}$.)

Let K be a measurable function on $S \times S$, and suppose $0 < \alpha < n - 1$. We shall call K a **kernel of order** α if

$$(3.8) \qquad\qquad K(x, y) = A(x, y)|x - y|^{-\alpha}$$

where A is a bounded function on $S \times S$. We shall call K a **kernel of order zero** if

$$(3.9) \qquad\qquad K(x, y) = A(x, y) \log |x - y| + B(x, y)$$

where A and B are bounded functions on $S \times S$. We note that it is immaterial whether we measure $|x - y|$ in the ambient space or in local coordinates on S, since the two quantities have the same order of magnitude as $x - y \to 0$. Finally, we shall call K a **continuous kernel of order**

α $(0 \le \alpha < n - 1)$ if K is a kernel of order α and K is continuous on $\{(x, y) \in S \times S : x \ne y\}$.

If K is a kernel of order α, $0 \le a < n - 1$, we define the operator T_K formally by

$$T_K f(x) = \int_S K(x, y) f(y) \, d\sigma(y).$$

(3.10) Proposition.

If K is a kernel of order α, $0 \le \alpha < n - 1$, then T_K is bounded on $L^p(S)$ for $1 \le p \le \infty$. Moreover, there is a constant $C > 0$ depending only on α such that if K is supported in $\{(x, y) : |x - y| < \epsilon\}$, then

$$\|T_K f\|_p \le C \epsilon^{n-1-\alpha} \|A\|_\infty \|f\|_p \qquad (\alpha > 0),$$
$$\|T_K f\|_p \le C \epsilon^{n-1} (\|A\|_\infty (1 + |\log \epsilon|) + \|B\|_\infty) \|f\|_\infty \qquad (\alpha = 0),$$

where A and B are as in (3.8) and (3.9).

Proof: It suffices to prove the second statement, since we can always take ϵ to be the diameter of S. Using polar coordinates on S centered at $x \in S$, we see that for $\alpha > 0$,

$$\int |K(x, y)| \, d\sigma(y) \le \|A\|_\infty \int_{|x-y|<\epsilon} |x - y|^{-\alpha} \, dy$$
$$\le C_1 \|A\|_\infty \int_0^\epsilon r^{n-2-\alpha} \, dr = C_2 \|A\|_\infty \epsilon^{n-1-\alpha}.$$

Similarly, $\int |K(x, y)| \, d\sigma(x) \le C_2 \|A\|_\infty \epsilon^{n-1-\alpha}$. The same calculation shows that for $\alpha = 0$, $\int |K(x, y)| \, d\sigma(y)$ and $\int |K(x, y)| \, d\sigma(x)$ are dominated by $\epsilon^{n-1}(\|A\|_\infty (1 + |\log \epsilon|) + \|B\|_\infty)$. The proposition therefore follows from the generalized Young's inequality (0.10). ∎

(3.11) Proposition.

If K is a kernel of order α, $0 \le \alpha < n - 1$, then T_K is compact on $L^2(S)$.

Proof: Given $\epsilon > 0$, set $K_\epsilon(x, y) = K(x, y)$ if $|x - y| > \epsilon$ and $K_\epsilon(x, y) = 0$ otherwise, and set $K'_\epsilon = K - K_\epsilon$. Then K_ϵ is bounded on $S \times S$, hence is a Hilbert-Schmidt kernel, so T_{K_ϵ} is bounded on $L^2(S)$ by Theorem (0.45). On the other hand, by Proposition (3.11) the operator norm of $T_K - T_{K_\epsilon} = T_{K'_\epsilon}$ tends to zero as $\epsilon \to 0$, so T_K is compact by Theorem (0.34). ∎

(3.12) Proposition.
If K is a continuous kernel of order α, $0 \le \alpha < n - 1$, then T_K transforms bounded functions into continuous functions.

Proof: We may assume $\alpha > 0$, since a continuous kernel of order zero is also a continuous kernel of order α for any $\alpha > 0$. Thus we may write K as in (3.8). Given $x \in S$ and $\delta > 0$, set $B_\delta = \{y \in S : |x - y| < \delta\}$. If $y \in B_\delta$, we have

$$|T_K f(x) - T_K f(y)| = \left| \int_S [K(x, z) - K(y, z)] f(z) \, d\sigma(z) \right|$$
$$\le \int_{B_{2\delta}} [|K(x, z)| + |K(y, z)|] |f(z)| \, d\sigma(z)$$
$$+ \int_{S \setminus B_{2\delta}} |K(x, z) - K(y, z)| |f(z)| \, d\sigma(z).$$

The integral over $B_{2\delta}$ is bounded by

$$\|A\|_\infty \|f\|_\infty \int_{B_{2\delta}} [|x - z|^{-a} + |y - z|^{-a}] \, d\sigma(z),$$

and an integration in polar coordinates shows that this is $O(\delta^{n-1-\alpha})$. Given $\epsilon > 0$, then, we can make this term less than $\frac{1}{2}\epsilon$ by choosing δ sufficiently small. On the other hand for $y \in B_\delta$ and $z \in S \setminus B_{2\delta}$ we have $|x - z| \ge 2\delta$ and $|y - z| \ge \delta$, so the continuity of K off the diagonal implies that $K(x, z) - K(y, z)$ converges to 0 uniformly in $z \in S \setminus B_{2\delta}$ as $y \to x$. Hence the integral over $S \setminus B_{2\delta}$ will be $< \frac{1}{2}\epsilon$ if y is sufficiently close to x. ∎

It is convenient to consider T_K as an operator on $L^2(S)$ because $L^2(S)$ is a Hilbert space. However, we really want to deal with continuous functions. The following proposition assures us that when we solve the Fredholm equation $u + T_K u = f$, continuous data give us continuous solutions.

(3.13) Proposition.
Suppose K is a continuous kernel of order α, $0 \le \alpha < n - 1$. If $u \in L^2(S)$ and $u + T_K u \in C(S)$, then $u \in C(S)$.

Proof: Given $\epsilon > 0$, choose $\phi \in C(S \times S)$ such that $0 \le \phi \le 1$, $\phi(x, y) = 1$ for $|x - y| < \frac{1}{2}\epsilon$, and $\phi(x, y) = 0$ for $|x - y| > \epsilon$. Set $K_0 = \phi K$ and $K_1 = (1 - \phi) K$. Then by the Schwarz inequality,

$$|T_{K_1} u(x) - T_{K_1} u(y)| \le \|u\|_2 \left[\int |K_1(x, z) - K_1(y, z)|^2 \, d\sigma(z) \right]^{1/2}.$$

Since K_1 is continuous, the integral on the right tends to zero as $y \to x$; thus $T_{K_1} u$ is continuous.

Now if we set $g = (u + T_K u) - T_{K_1} u$, we see that g is continuous and that $u + T_{K_0} u = g$. By Proposition (3.11), if ϵ is sufficiently small the operator norm of T_{K_0} on both L^2 and L^∞ will be less than 1. Then $I + T_{K_0}$ is invertible, and u is expressed in terms of g by the geometric series $u = \sum_0^\infty (-T_{K_0})^j g$. By Proposition (3.12), each term in the series is continuous, and the series converges in the norm of L^∞, i.e., uniformly, so u is continuous.   ∎

EXERCISES

1. Suppose $K \in C(S \times S)$. Show that T_K maps $L^p(S)$ into $C(S)$ for $1 \le p \le \infty$. Show also that if $\{f_j\}$ is a bounded sequence in $L^p(S)$ then $\{T_K f_j\}$ has a uniformly convergent subsequence; in particular, T_K is a compact operator on $L^p(S)$. (Use the Arzela-Ascoli theorem.)

2. Show that if K is a continuous kernel of order α on S, $0 \le \alpha < n - 1$, then T_K is a compact operator on $L^p(S)$ for $1 \le p \le \infty$. (Use Exercise 1.)

3. Extend Proposition (3.12) to show that if K is a continuous kernel of order α then T_K maps $L^p(S)$ into $C(S)$ for $p > (n-1)/(n-1-\alpha)$.

C. Double Layer Potentials

Let ϕ be a continuous function on S. In the section we study the properties of the double layer potential with moment ϕ,

$$(3.14) \qquad u(x) = \int_S \partial_{\nu_y} N(x,y) \phi(y) \, d\sigma(y) \qquad (x \in \mathbb{R}^n \setminus S).$$

To begin with, we note that for $x \in \mathbb{R}^n \setminus S$ and $y \in S$,

$$\partial_{\nu_y} N(x,y) = -\frac{(x - y) \cdot \nu(y)}{\omega_n |x - y|^n}$$

is continuous in y and harmonic in x (since x-derivatives commute with y-derivatives), and it is $O(|x|^{1-n})$ as $x \to \infty$. Therefore, u is harmonic on $\mathbb{R}^n \setminus S$ and $|u(x)| = O(|x|^{1-n})$ as $x \to \infty$, so that u is also harmonic at infinity. More interesting and more subtle is the behavior of u near S. Before investigating this, we state a technical lemma.

(3.15) Lemma.
There is a constant $c > 0$ such that for all $x, y \in S$,
$$|(x - y) \cdot \nu(y)| \leq c|x - y|^2.$$

Proof: Since $|(x - y) \cdot \nu(y)| \leq |x - y|$ for all x, y, it suffices to assume that $|x - y| \leq 1$. Given $y \in S$, by a translation and rotation of coordinates we may assume that $y = 0$ and $\nu(y) = (0, \ldots, 0, 1)$. Hence $(x - y) \cdot \nu(y) = x_n$, and near y, S is the graph of an equation $x_n = f(x_1, \ldots, x_{n-1})$ where $f \in C^2$, $f(0) = 0$, and $\nabla f(0) = 0$. By Taylor's theorem, then,
$$|(x - y) \cdot \nu(y)| = |f(x_1, \ldots, x_{n-1})| \leq c|(x_1, \ldots, x_{n-1})|^2 \leq c|x|^2 = c|x - y|^2$$
for $|x| \leq 1$, where c depends only on a bound for the second partial derivatives of f. Since S is compact and of class C^2, there is such a bound that holds for all $y \in S$, and we are done. ∎

We shall give a special name to the kernel $\partial_{\nu_y} N(x, y)$ when x and y are both on S; namely, we set

(3.16) $K(x, y) = \partial_{\nu_y} N(x, y)$ $(x \in S, \ y \in S, \ x \neq y)$.

The reason for this bit of notation is twofold. First, the kernel K is going to play a special role in our theory. Second, it is a little dangerous to regard K as really the normal derivative of N for $x \in S$; there are some delta-functions lurking in the shadows, as we shall see at the end of this section.

(3.17) Proposition.
K is a continuous kernel of order $n - 2$ on S.

Proof: We have
$$K(x, y) = \frac{A(x, y)}{|x - y|^{n-2}}, \quad \text{where } A(x, y) = -\frac{(x - y) \cdot \nu(y)}{\omega_n |x - y|^2}.$$
$K(x, y)$ is clearly continuous for $x \neq y$, so the result follows from Lemma (3.15). ∎

It is therefore reasonable to extend the potential u to S by setting

(3.18) $u(x) = \displaystyle\int_S K(x, y)\phi(y) \, d\sigma(y) = T_K \phi(x)$ $(x \in S)$.

By Proposition (3.12), the restriction of u to S is continuous on S. However, u is not continuous on $\mathbb{R}^n$: there is a jump when we approach points on S by points in $\mathbb{R}^n \setminus S$. This phenomenon shows up most clearly in the simplest case, where $\phi \equiv 1$.

(3.19) Proposition.

$$\int_S \partial_{\nu_y} N(x, y)\, d\sigma(y) = \begin{cases} 1 & \text{if } x \in \Omega, \\ 0 & \text{if } x \in \Omega'; \end{cases}$$

$$\int_S K(x, y)\, d\sigma(y) = \tfrac{1}{2} \quad \text{if } x \in S.$$

Proof: The result for $x \in \Omega'$ follows immediately from Corollary (2.7), since $N(x, y)$ is C^∞ on $\overline{\Omega}$ and harmonic on Ω as a function of y when $x \in \Omega'$. On the other hand, if $x \in \Omega$, let $\epsilon > 0$ be small enough so that $\overline{B}_\epsilon = \overline{B_\epsilon(x)} \subset \Omega$. We can then apply Corollary (2.7) to $N(x, \cdot)$ on the domain $\Omega \setminus \overline{B}_\epsilon$ as in the proof of the mean value theorem:

$$0 = \int_S \partial_{\nu_y} N(x, y)\, d\sigma(y) - \frac{\epsilon^{1-n}}{\omega_n} \int_{\partial B_\epsilon} d\sigma(y)$$

$$= \int_S \partial_{\nu_y} N(x, y)\, d\sigma(y) - 1.$$

Now suppose $x \in S$, and again let $B_\epsilon = B_\epsilon(x)$. Set

$$S_\epsilon = S \setminus (S \cap B_\epsilon), \quad \partial B'_\epsilon = \partial B_\epsilon \cap \Omega, \quad \partial B''_\epsilon = \{y \in \partial B_\epsilon : \nu(x) \cdot y < 0\}.$$

(Thus $\partial B''_\epsilon$ is the hemisphere of ∂B_ϵ lying on the same side of the tangent plane to S at x as Ω.) On the one hand, clearly

$$\int_S K(x, y)\, d\sigma(y) = \lim_{\epsilon \to 0} \int_{S_\epsilon} K(x, y)\, d\sigma(y).$$

On the other hand, since $N(x, \cdot)$ is harmonic on $\Omega \setminus \overline{B}_\epsilon$ and smooth up to the boundary $S_\epsilon \cup \partial B'_\epsilon$, by Corollary (2.7) we have

$$0 = \int_{S_\epsilon} K(x, y)\, d\sigma(y) + \int_{\partial B'_\epsilon} \partial_{\nu_y} N(x, y)\, d\sigma(y).$$

Thus, taking into account the proper orientation on $\partial B'_\epsilon$,

$$\int_S K(x, y)\, d\sigma(y) = -\lim_{\epsilon \to 0} \int_{\partial B'_\epsilon} \partial_{\nu_y} N(x, y)\, d\sigma(y) = \lim_{\epsilon \to 0} \frac{\epsilon^{1-n}}{\omega_n} \int_{\partial B'_\epsilon} d\sigma(y).$$

But since S is C^2, the symmetric difference between $\partial B'_\epsilon$ and $\partial B''_\epsilon$ is contained in an "equatorial strip"

$$\{y \in \partial B_\epsilon : |y \cdot \nu(x)| \le c(\epsilon)\}, \qquad c(\epsilon) = O(\epsilon^2),$$

whose area is $O(\epsilon^n)$. Hence

$$\int_{\partial B'_\epsilon} d\sigma(y) = \int_{\partial B''_\epsilon} d\sigma(y) + O(\epsilon^n) = \tfrac{1}{2}\epsilon^{n-1}\omega_n + O(\epsilon^n),$$

and the result follows.   ∎

To extend this result to general densities ϕ, we need two preliminary lemmas.

(3.20) Lemma.
There is a constant $C < \infty$ such that for all $x \in \mathbb{R}^n \setminus S$,

$$\int_S |\partial_{\nu_y} N(x, y)| \, d\sigma(y) \leq C.$$

Proof: Let $\text{dist}(x, S)$ denote the distance from x to the nearest point of S. Fix $\delta > 0$ with the following two properties: (i) $\delta < 1/(2c)$ where c is the constant in Lemma (3.15); (ii) the set of x such that $\text{dist}(x, S) < \frac{1}{2}\delta$ is a tubular neighborhood of S as in Proposition (0.2). Thus, if $\text{dist}(x, S) < \frac{1}{2}\delta$, there is a unique $x_0 \in S$ and $t \in (-\frac{1}{2}\delta, \frac{1}{2}\delta)$ such that $x = x_0 + t\nu(x_0)$; x_0 is the closest point in S to x by the Lagrange multiplier principle.

Case I: $\text{dist}(x, S) \geq \frac{1}{2}\delta$. Then $|\partial_{\nu_y} N(x, y)| \leq C_1 \delta^{1-n}$ for all $y \in S$, so

$$\int_S |\partial_{\nu_y} N(x, y)| \, d\sigma(y) \leq C_1 \delta^{1-n} \int_S d\sigma = C_2.$$

Case II: $\text{dist}(x, S) < \frac{1}{2}\delta$. Let x_0 be the unique point of S such that $x = x_0 + t\nu(x_0)$ with $|t| < \frac{1}{2}\delta$, and let $B_\delta = \{y \in S : |x_0 - y| < \delta\}$. We estimate the integral of $|\partial_{\nu_y} K|$ over $S \setminus B_\delta$ and over B_δ separately. If $y \in S \setminus B_\delta$, then

$$|x - y| \geq |x_0 - y| - |x - x_0| \geq \delta - \tfrac{1}{2}\delta = \tfrac{1}{2}\delta.$$

Hence $|\partial_{\nu_y} N(x, y)| \leq C_1 \delta^{1-n}$, so the integral over $S \setminus B_\delta$ is bounded by C_2 as above. To estimate the integral over B_δ, we note that

$$\omega_n |\partial_{\nu_y} N(x, y)| = \frac{|(x - y) \cdot \nu(y)|}{|x - y|^n} \leq \frac{|(x - x_0) \cdot \nu(y)| + |(x_0 - y) \cdot \nu(y)|}{|x - y|^n}$$
$$\leq \frac{|x - x_0| + c|x_0 - y|^2}{|x - y|^n},$$

by Lemma (3.15). Moreover,

$$|x - y|^2 = |x - x_0|^2 + |x_0 - y|^2 + 2(x - x_0) \cdot (x_0 - y),$$

and by Lemma (3.15) again,

$$|2(x - x_0) \cdot (x_0 - y)| = 2|x - x_0||\nu(x_0) \cdot (x_0 - y)| \leq 2c|x - x_0||x_0 - y|^2.$$

In particular, since $|x_0 - y| < \delta < 1/(2c)$, this last quantity is less than $|x - x_0||x_0 - y|$, whence

$$|x - y|^2 \geq \tfrac{1}{2}(|x - x_0|^2 + |x_0 - y|^2).$$

Therefore,

$$|\partial_{\nu_y} N(x, y)| \leq C_3 \frac{|x - x_0| + c|x_0 - y|^2}{(|x - x_0|^2 + |x_0 - y|^2)^{n/2}}$$

$$\leq \frac{C_3|x - x_0|}{(|x - x_0|^2 + |x_0 - y|^2)^{n/2}} + \frac{C_3 c}{|x_0 - y|^{n-2}},$$

so if we set $r = |x_0 - y|$ and $a = |x - x_0|$, integrate in polar coordinates, and make the substitution $r = as$, we obtain

$$\int_{B_\delta} |\partial_{\nu_y} N(x, y)| \leq C_4 \int_0^\delta \left[\frac{a}{(a^2 + r^2)^{n/2}} + \frac{1}{r^{n-2}} \right] r^{n-2} \, dr$$

$$\leq C_4 \int_0^\infty \frac{s^{n-2}}{(1 + s^2)^{n/2}} \, ds + C_4 \delta,$$

This last integral converges since the integrand is $O(s^{-2})$ as $s \to \infty$, so we are done. ∎

(3.21) Lemma.
Suppose $\phi \in C(S)$ and $\phi(x_0) = 0$ for some $x_0 \in S$. If u is defined by (3.14) and (3.18), then u is continuous at x_0.

Proof: Given $\epsilon > 0$, we wish to produce $\delta > 0$ so that $|u(x) - u(x_0)| < \epsilon$ when $|x - x_0| < \delta$. Let C be the constant in Lemma (3.20), and let $C' = \int_S |K(x, y)| \, d\sigma(y)$ (which is finite since K is a kernel of order $n - 2$: cf. the proof of (3.10).) Choose $\eta > 0$ so that $|\phi(y)| < \epsilon/3(C + C')$ when $y \in B_\eta = \{z \in S : |z - x_0| < \eta\}$. Then

$$|u(x) - u(x_0)| \leq \int_{B_\eta} (|\partial_{\nu_y} N(x, y)| + |\partial_{\nu_y} N(x_0, y)|) |\phi(y)| \, d\sigma(y)$$

$$+ \int_{S \setminus B_\eta} |\partial_{\nu_y} N(x, y) - \partial_{\nu_y} N(x_0, y)||\phi(y)| \, d\sigma(y).$$

(Of course $\partial_{\nu_y} N = K$ when both of its arguments are on S.) The first integral on the right is less than $2\epsilon/3$. Moreover, if $|x - x_0| < \tfrac{1}{2}\eta$ the integrand of the third term is bounded and continuous on $S \setminus B_\eta$ and tends uniformly to zero as $x \to x_0$. We can therefore choose $\delta < \tfrac{1}{2}\eta$ small enough so that the third term is less than $\epsilon/3$ when $|x - x_0| < \delta$. and we are done. (For future reference we note that δ depends only on η and the uniform norm of ϕ.) ∎

We are now ready for the main theorem of this section. If u is defined by (3.14), we define the functions u_t on S for small $t \neq 0$ by

$$u_t(x) = u(x + t\nu(x)).$$

Thus u_t is, in effect, the restriction of u to a surface parallel to S at a distance t from S.

(3.22) Theorem.
Suppose $\phi \in C(S)$ and u is defined by (3.14). The restriction of u to Ω has a continuous extension to $\overline{\Omega}$, and the restriction of u to Ω' has a continuous extension to $\overline{\Omega'}$. More precisely, the functions u_t converge uniformly on S to continuous limits u_- and u_+ as t approaches zero from below and above, respectively. u_- and u_+ are given by

$$u_-(x) = \tfrac{1}{2}\phi(x) + \int_S K(x,y)\phi(y)\,d\sigma(y),$$

$$u_+(x) = -\tfrac{1}{2}\phi(x) + \int_S K(x,y)\phi(y)\,d\sigma(y),$$

that is,

$$u_- = \tfrac{1}{2}\phi + T_K\phi, \qquad u_+ = -\tfrac{1}{2}\phi + T_K\phi.$$

Proof: If $x \in S$ and $t < 0$ is sufficiently small, then $x + t\nu(x) \in \Omega$, so by Proposition (3.19),

$$u_t(x) = \phi(x)\int_S \partial_{\nu_y} N(x + t\nu(x), y)\,d\sigma(y)$$

$$+ \int_S \partial_{\nu_y} N(x + t\nu(x), y)[\phi(y) - \phi(x)]\,d\sigma(y)$$

$$= \phi(x) + \int_S \partial_{\nu_y} N(x + t\nu(x), y)[\phi(y) - \phi(x)]\,d\sigma(y).$$

By Lemma (3.21), the second integral is continuous in t as $t \to 0$, so by Proposition (3.19) again,

$$\lim_{t \to 0} u_t(x) = \phi(x) + \int_S K(x,y)\phi(y)\,d\sigma(y) - \phi(x)\int_S K(x,y)\,d\sigma(y)$$

$$= \tfrac{1}{2}\phi(x) + \int_S K(x,y)\phi(y)\,d\sigma(y).$$

If $t > 0$, the argument is the same except that

$$\phi(x)\int_S \partial_{\nu_y} N(x + t\nu(x), y)\,d\sigma(y) = 0.$$

The uniformity of the convergence follows from the proof of Lemma (3.21): we have only to observe that since S is compact, for any $\epsilon > 0$ we can choose $\eta > 0$ so that $|\phi(x) - \phi(y)| < \epsilon/3(C + C')$ for all $x, y \in S$ such that $|x - y| < \eta$. ∎

(3.23) Corollary.
$\phi = u_- - u_+$.

Theorem (3.22) may be interpreted as follows. If for small $t \neq 0$ we define the function K_t on $S \times S$ by

$$K_t(x, y) = \partial_{\nu_y} N(x + t\nu(x), y),$$

then for each $x \in S$,

$$\lim_{t<0,\ t\to 0} K_t(x, \cdot) = \tfrac{1}{2}\delta_x + K(x, \cdot), \qquad \lim_{t>0,\ t\to 0} K_t(x, \cdot) = -\tfrac{1}{2}\delta_x + K(x, \cdot),$$

where δ_x is the point mass at x and both sides are interpreted as distributions (or measures) on S.

D. Single Layer Potentials

We now consider the single layer potential

$$(3.24) \qquad\qquad u(x) = \int_S N(x, y)\phi(y)\, d\sigma(y)$$

with moment ϕ, where $\phi \in C(S)$. As with the double layer potentials, it is clear that u is harmonic on $\mathbb{R}^n \setminus S$ and that $|u(x)| = O(|x|^{2-n})$ as $x \to \infty$ when $n > 2$, so that u is harmonic at infinity for $n > 2$. (If $n = 2$, in general we only have $|u(x)| = O(\log|x|)$ as $x \to \infty$; we shall say more about this later.) Moreover, the restriction of N to $S \times S$ is clearly a continuous kernel of order $n - 2$, so u is also well defined on S.

(3.25) Proposition.
If $\phi \in C(S)$ and u is defined by (3.24), then u is continuous on $\mathbb{R}^n$.

Proof: We need only show continuity on S, and the proof is very similar to that of Lemma (3.21). Given $x_0 \in S$ and $\delta > 0$, let us set

$B_\delta = \{y \in S : |x_0 - y| < \delta\}$. Then

$$|u(x) - u(x_0)| \leq \int_{B_\delta} |N(x, y)\phi(y)|\, d\sigma(y) + \int_{B_\delta} |N(x_0, y)\phi(y)|\, d\sigma(y)$$

$$+ \int_{S \setminus B_\delta} |N(x, y) - N(x_0, y)||\phi(y)|\, d\sigma(y).$$

Since ϕ is bounded and $N(x, y) = O(|x - y|^{2-n})$ (or $O(\log |x - y|^{-1})$) if $n = 2$), an integration in polar coordinates shows that the first two terms on the right are $O(\delta)$ (or $O(\delta \log \delta^{-1})$ if $n = 2$). Given $\epsilon > 0$, then, we can make these terms each less than $\epsilon/3$ by choosing δ small enough. If we now require that $|x - x_0| < \frac{1}{2}\delta$, the integrand in the third term is bounded on $S \setminus B_\delta$ and tends uniformly to zero as $x \to x_0$, so by choosing $|x - x_0|$ small enough we can make the third term less than $\epsilon/3$ also.   ∎

Now let us consider the normal derivative of u. Let V be the tubular neighborhood of S given by Proposition (0.2). Recall that we have defined the normal derivative ∂_ν on V by formula (0.3); thus for $x \in V \setminus S$ we have

$$(3.26) \qquad \partial_\nu u(x) = \int_S \partial_{\nu_x} N(x, y)\phi(y)\, d\sigma(y).$$

This looks just like a double layer potential except that ∂_ν is applied to N with respect to x instead of y. In fact, since $N(x, y) = N(y, x)$, $\partial_{\nu_x} N(x, y)$ is just $\partial_{\nu_y} N$ evaluated at (y, x). In particular, if we set

$$K^*(x, y) = K(y, x),$$

the right hand side of (3.26) makes sense for $x \in S$ if we interpret it as

$$(3.27) \qquad \int_S K^*(x, y)\phi(y)\, d\sigma(y) = T_{K^*}\phi(x).$$

Since K is a continuous kernel of order $n - 2$, so is K^*; thus (3.27) defines a continuous function on S by Proposition (3.12). Moreover, since K is real-valued, it is easily checked that T_{K^*} is the adjoint of T_K as an operator on $L^2(S)$; hence (3.27) is just $T_K^*\phi(x)$.

As might be expected, there is a jump discontinuity between the quantities defined by (3.26) on $V \setminus S$ and by (3.27) on S. Indeed, we have the following theorem.

(3.28) Theorem.
Suppose $\phi \in C(S)$ and u is defined on $\mathbb{R}^n$ by (3.24). Then the restriction of u to $\overline{\Omega}$ (resp. $\overline{\Omega'}$) is in $C_\nu(\Omega)$ (resp. $C_\nu(\Omega')$), and for $x \in S$ we have

$$\partial_{\nu_-} u(x) = -\tfrac{1}{2}\phi(x) + \int_S K(y, x)\phi(y)\, d\sigma(y),$$

$$\partial_{\nu_+} u(x) = \tfrac{1}{2}\phi(x) + \int_S K(y, x)\phi(y)\, d\sigma(y),$$

that is,

$$\partial_{\nu_-} u = -\tfrac{1}{2}\phi + T_K^*\phi, \qquad \partial_{\nu_+} u = \tfrac{1}{2}\phi + T_K^*\phi.$$

Proof: We already know that u is everywhere continuous. Consider the double layer potential

$$v(x) = \int_S \partial_{\nu_y} N(x, y)\phi(y)\, d\sigma(y)$$

on $\mathbb{R}^n \setminus S$, and define the function f on the tubular neighborhood V of S by

$$f(x) = \begin{cases} v(x) + \partial_\nu u(x) & (x \in V \setminus S), \\ T_K\phi(x) + T_K^*\phi(x) & (x \in S). \end{cases}$$

We claim that f is continuous on V. The restrictions of f to $V \setminus S$ and S are continuous, so it suffices to show that if $x_0 \in S$ and $x = x_0 + t\nu(x_0)$ then $f(x) - f(x_0) \to 0$ as $t \to 0$, the convergence being uniform in x_0. But

$$f(x) - f(x_0)$$
$$= \int_S \left[\partial_{\nu_x} N(x, y) + \partial_{\nu_y} N(x, y) - \partial_{\nu_x} N(x_0, y) - \partial_{\nu_y} N(x_0, y)\right] \phi(y)\, d\sigma(y).$$

We proceed as in the proof of Lemma (3.21): write this expression as an integral over $B_\delta = \{y \in S : |x_0 - y| < \delta\}$ plus an integral over $S \setminus B_\delta$. As before, the integral over $S \setminus B_\delta$ tends uniformly to zero as $x \to x_0$. On the other hand, the integral over B_δ is bounded by

$$\|\phi\|_\infty \int_{B_\delta} |\partial_{\nu_x} N(x, y) + \partial_{\nu_y} N(x, y)|\, d\sigma(y)$$

plus the same expression with x replaced by x_0. Thus it suffices to show that for all x on the normal line through x_0,

$$\int_{B_\delta} |\partial_{\nu_x} N(x, y) + \partial_{\nu_y} N(x, y)|\, d\sigma(y)$$

can be made arbitrarily small by taking δ sufficiently small, independent of x and x_0. Now

$$\partial_{\nu_x} N(x, y) = \frac{(x - y) \cdot \nu(x_0)}{\omega_n |x - y|^n}, \qquad \partial_{\nu_y} N(x, y) = \frac{-(x - y) \cdot \nu(y)}{\omega_n |x - y|^n},$$

so

$$\partial_{\nu_x} N(x, y) + \partial_{\nu_y} N(x, y) = \frac{(x - y) \cdot [\nu(x_0) - \nu(y)]}{\omega_n |x - y|^n}.$$

But $|\nu(x_0) - \nu(y)| = O(|x_0 - y|)$ since ν is C^1, and $|x - y| \geq C|x_0 - y|$ since x is on the normal through x_0. Hence

$$|\partial_{\nu_x} N(x, y) + \partial_{\nu_y} N(x, y)| \leq C' |x_0 - y|^{2-n},$$

and the integral is dominated by

$$\int_0^\delta r^{2-n} r^{n-2} \, dr = \delta.$$

Thus $f = v + \partial_\nu u$ extends continuously across S. Therefore, by Theorem (3.22), for all $x \in S$ we have

$$T_K \phi(x) + T_K^* \phi(x) = v_-(x) + \partial_{\nu_-} u(x) = \tfrac{1}{2}\phi(x) + T_K \phi(x) + \partial_{\nu_-} u(x),$$

so that

$$\partial_{\nu_-} u(x) = -\tfrac{1}{2}\phi(x) + T_K^* \phi(x);$$

and also

$$T_K \phi(x) + T_K^* \phi(x) = v_+(x) + \partial_{\nu_+} u(x) = -\tfrac{1}{2}\phi(x) + T_K \phi(x) + \partial_{\nu_+} u(x),$$

so that

$$\partial_{\nu_+} u(x) = \tfrac{1}{2}\phi(x) + T_K^* \phi(x).$$

The convergence of $\partial_\nu u(x + t\nu(x))$ to $\partial_{\nu_\pm} u(x)$ is uniform in x since the same is true of v and $v + \partial_\nu u$. ∎

(3.29) Corollary.
$\phi = \partial_{\nu_+} u - \partial_{\nu_-} u.$

We conclude the discussion of single layer potentials with three lemmas that will be needed in the next section.

(3.30) Lemma.

If $\phi \in C(S)$ and $\frac{1}{2}\phi + T_K^* \phi = f$, then $\int_S \phi = \int_S f$.

Proof: By Proposition (3.19) and Fubini's theorem,

$$\int_S f(x)\,d\sigma(x) = \frac{1}{2}\int_S \phi(x)\,d\sigma(x) + \int_S\int_S K(y,x)\phi(y)\,d\sigma(y)\,d\sigma(x)$$

$$= \frac{1}{2}\int_S \phi(x)\,d\sigma(x) + \frac{1}{2}\int_S \phi(y)\,d\sigma(y) = \int_S \phi(x)\,d\sigma(x). \quad\blacksquare$$

(3.31) Lemma.

Suppose $n = 2$. If $\phi \in C(S)$, the single layer potential u with moment ϕ is harmonic at infinity if and only if $\int_S \phi = 0$, in which case u vanishes at infinity.

Proof: We have

$$u(x) = \frac{1}{2\pi}\int_S (\log|x - y| - \log|x|)\phi(y)\,d\sigma(y) + \frac{\log|x|}{2\pi}\int_S \phi(y)\,d\sigma(y).$$

Since $\log|x-y| - \log|x| \to 0$ uniformly for $y \in S$ as $x \to \infty$, the first integral vanishes as $x \to \infty$. In view of Proposition (2.74), the result follows. $\quad\blacksquare$

(3.32) Lemma.

Suppose $n = 2$. If $\phi \in C(S)$, $\int_S \phi = 0$, and the single layer potential u with moment ϕ is constant on $\overline{\Omega}$, then $\phi = 0$ (and hence $u = 0$).

Proof: By Lemma (3.31), u is harmonic at infinity, so if $u = c$ on $\overline{\Omega}$, u solves the exterior Dirichlet problem with $f = c$. But the solution to this problem (unique, by Proposition (3.2)) is $u = c$. Thus $u = c$ everywhere, so $\phi = 0$ by Corollary (3.29) (or because $0 = \Delta u = \phi\,d\sigma$). $\quad\blacksquare$

EXERCISES

1. Let $\Omega = B_R(0)$, $S = S_R(0)$. Show that the single layer potential $u(x)$ with moment $\phi \equiv 1$ is given by

$$\frac{[\max(|x|, R)]^{2-n}}{2 - n} \quad (n > 2), \qquad \log[\max(|x|, R)] \quad (n = 2).$$

That is, the potential generated by a uniform distribution of charge on a spherical shell is constant inside the shell, and outside the shell it behaves as if the charge were all concentrated at the origin. (Hint: Show that u is radial and that the asserted formula for u is asymptotically correct as $x \to \infty$, and apply Corollary (2.3).)

2. Use the result of Exercise 1 to show that if $f \in L^1(\mathbb{R}^n)$ is radial and $u = f * N$ is the associated potential, then

$$u(x) = N(x) \int_{|y|<|x|} f(y) \, dy + \int_{|y|>|x|} f(y) \, dy.$$

E. Solution of the Problems

We can now apply the Fredholm theory to solve the Dirichlet and Neumann problems. For $f \in C(S)$, consider the integral equations

(3.33)
$$\frac{1}{2}\phi + T_K\phi = f, \qquad -\frac{1}{2}\phi + T_K\phi = f,$$
$$\frac{1}{2}\phi + T_K^*\phi = f, \qquad -\frac{1}{2}\phi + T_K^*\phi = f,$$

where K is defined by (3.16). By Proposition (3.13) (with K and f replaced by $\pm 2K$ and $\pm 2f$), the solutions ϕ will be continuous on S, if they exist. Therefore, by Theorems (3.22) and (3.28), the double layer potential u with moment ϕ will solve the interior (resp. exterior) Dirichlet problem if ϕ satisfies $\frac{1}{2}\phi + T_K\phi = f$ (resp. $-\frac{1}{2}\phi + T_K\phi = f$), and the single layer potential u with moment ϕ will solve the interior (resp. exterior) Neumann problem if ϕ satisfes $-\frac{1}{2}\phi + T_K^*\phi = f$ (resp. $\frac{1}{2}\phi + T_K^*\phi = f$). There is one exception: if $n = 2$, in general the single layer potential will grow like $\log|x|$ at infinity and so will not be harmonic at infinity. However, for $n = 2$ there is an extra necessary condition for the solution of the exterior Neumann problem given by Proposition (3.6), and this condition is equivalent to harmonicity at infinity by Lemmas (3.30) and (3.31). Thus if the integral equations are solvable and the necessary conditions are satisfied, the boundary value problems are solvable. (We shall soon see that the Dirichlet problems are solvable in any case.)

We proceed to study the solvability of the equations (3.33). By Corollary (0.42), this means identifying the eigenspaces

$$\mathcal{V}_+ = \left\{\phi : T_K\phi = \tfrac{1}{2}\phi\right\}, \qquad \mathcal{V}_- = \left\{\phi : T_K\phi = -\tfrac{1}{2}\phi\right\},$$
$$\mathcal{W}_+ = \left\{\phi : T_K^*\phi = \tfrac{1}{2}\phi\right\}, \qquad \mathcal{W}_- = \left\{\phi : T_K^*\phi = -\tfrac{1}{2}\phi\right\}.$$

Here ϕ is allowed to range over either $L^2(S)$ or $C(S)$; the result is the same, by Proposition (3.13).

Recall that Ω has components $\Omega_1, \ldots, \Omega_m$, and that Ω' has components $\Omega'_0, \ldots, \Omega'_{m'}$, where Ω'_0 is unbounded. We define functions $\alpha_1, \ldots, \alpha_m$ and $\alpha'_1, \ldots, \alpha'_{m'}$ on S by

$$\alpha_j(x) = \begin{cases} 1 & \text{if } x \in \partial\Omega_j, \\ 0 & \text{otherwise,} \end{cases}$$

$$\alpha'_j(x) = \begin{cases} 1 & \text{if } x \in \partial\Omega'_j, \\ 0 & \text{otherwise.} \end{cases}$$

(3.34) Proposition.
$\alpha_j \in \mathcal{V}_+$ for $j = 1, \ldots, m$, and $\alpha'_j \in \mathcal{V}_-$ for $j = 1, \ldots, m'$.

Proof: Since

$$T_K \alpha_j = \int_{\partial\Omega_j} K(x, y)\, d\sigma(y), \qquad T_K \alpha'_j = \int_{\partial\Omega'_j} K(x, y)\, d\sigma(y),$$

the result follows by applying Proposition (3.19) with Ω replaced by Ω_j or Ω'_j. The sign is reversed for α'_j because the normal ν points into Ω'_j. ∎

Clearly $\alpha_1, \ldots, \alpha_m$ are linearly independent, so $\dim \mathcal{V}_+ \geq m$, and by Theorem (0.38b), $\dim \mathcal{W}_+ = \dim \mathcal{V}_+$. On the other hand, suppose $\beta \in \mathcal{W}_+$, and let w be the single layer potential with moment β. Since $\partial_{\nu_-} w = 0$ by Theorem (3.28), w is constant on each Ω_j by Proposition (3.3), so we can define a linear map from $\mathcal{W}_+$ to $\mathbb{C}^m$ by

(3.35) $$\beta \rightarrow (w|\Omega_1, \ldots, w|\Omega_m).$$

If $n > 2$, this map is injective. Indeed, if $w|\Omega = 0$, then w solves the exterior Dirichlet problem with $f = 0$, so $w|\Omega' = 0$ by Proposition (3.2). Hence $w = 0$ everywhere, so $\beta = 0$ by Corollary (3.29). It follows that $\dim \mathcal{W}_+ \leq m$, so $\dim \mathcal{W}_+ = m$ and the map (3.35) is an isomorphism.

If $n = 2$, the preceding argument breaks down because w need not be harmonic at infinity. (A counterexample to the injectivity of (3.35) is provided by Exercise 1 of §3D, with $R = 1$. In that situation, $\mathcal{V}_+ = \mathcal{W}_+$ because $K(x, y) = K(y, x)$ by explicit calculation; the function $\phi = 1$ on S belongs to $\mathcal{V}_+ = \mathcal{W}_+$; and the associated potential u vanishes on Ω.) However, if we set

$$\mathcal{W}_+^0 = \left\{ \beta \in \mathcal{W}_+ : \int_S \beta = 0 \right\},$$

in view of Lemma (3.31) the preceding argument shows that the restriction of the map (3.35) to $\mathcal{W}_+^0$ is injective, and by Lemma (3.32) its range does not contain the vector $(1, 1, \ldots, 1)$. Hence $\dim \mathcal{W}_+^0 \leq m-1$. But $\dim \mathcal{W}_+ \geq \dim \mathcal{W}_+ - 1$ since $\mathcal{W}_+^0$ is defined by the vanishing of one linear functional, so again it follows that $\dim \mathcal{W}_+ = m$. In short, we have proved the following result.

(3.36) Proposition.
The spaces $\mathcal{V}_+$ and $\mathcal{W}_+$ have dimension m. Moreover:
  a. *If $n > 2$, for each $(a_1, \ldots, a_m) \in \mathbb{C}^m$ there is a unique $\beta \in \mathcal{W}_+$ such that the single layer potential w with moment β satisfies $w|\Omega_j = a_j$ for $j = 1, \ldots, m$.*
  b. *If $n = 2$, there is an $(m-1)$-dimensional subspace X of $\mathbb{C}^m$ such that:*
    i. *$\mathbb{C}^m = X \oplus \mathbb{C}(1, 1, \ldots, 1)$;*
    ii. *for each $(a_1, \ldots, a_m) \in X$ there is a unique $\beta \in \mathcal{W}_+^0$ such that the single layer potential w with moment β satisfies $w|\Omega_j = a_j$ for $j = 1, \ldots, m$.*

A similar argument applies to $\mathcal{V}_-$ and $\mathcal{W}_-$. Indeed, we have $\dim \mathcal{V}_- = \dim \mathcal{W}_- \geq m'$ by Theorem (0.38b) and Proposition (3.34). If $\beta \in \mathcal{W}_-$ and w is the associated single layer potential, then w is constant on each Ω_j', and $w = 0$ on Ω_0' since w vanishes at infinity. (This is true even when $n = 2$. Indeed, every $\beta \in \mathcal{W}_-$ satisfies $\int_S \beta = 0$ by Lemma (3.30), so its single layer potential vanishes at infinity by Lemma (3.31).) If $w = 0$ on Ω' then $w = 0$ on Ω by the uniqueness for the interior Dirichlet problem, so $\beta = 0$ by Corollary (3.29). Thus the map

$$\beta \to (w|\Omega_1, \ldots, w|\Omega_{m'})$$

from $\mathcal{W}_-$ to $\mathbb{C}^{m'}$ is injective and hence an isomorphism. Therefore:

(3.37) Proposition.
The spaces $\mathcal{V}_-$ and $\mathcal{W}_-$ have dimension m'. For each $(a_1, \ldots, a_{m'}) \in \mathbb{C}^{m'}$ there is a unique $\beta \in \mathcal{W}_-$ such that the single layer potential w with moment β satisfies $w|\Omega_j' = a_j$ for $j = 1, \ldots, m'$ and $w|\Omega_0' = 0$.

We need one more technical result.

(3.38) Proposition.
$L^2(S) = \mathcal{V}_+^\perp \oplus \mathcal{W}_+ = \mathcal{V}_-^\perp \oplus \mathcal{W}_-$. (The direct sums are not necessarily orthogonal.)

Proof: By Proposition (3.35), $\mathcal{V}_+^\perp$ is a closed subspace of codimension m and $\dim \mathcal{W}_+ = m$, so for the first equality it is enough to show that $\mathcal{V}_+^\perp \cap \mathcal{W}_+ = \{0\}$. Suppose $\phi \in \mathcal{V}_+^\perp \cap \mathcal{W}_+$; then $T_K^* \phi = \frac{1}{2}\phi$, and by Corollary (0.42), $\phi = -\frac{1}{2}\psi + T_K^*\psi$ for some $\psi \in L^2(S)$. By Proposition (3.13), ϕ and ψ are continuous. Let u and v be the single layer potentials with moments ϕ and ψ. Then by Theorem (3.28),

$$\partial_{\nu_-} u = 0, \qquad \partial_{\nu_-} v = \phi = \tfrac{1}{2}\phi + T_K^*\phi = \partial_{\nu_+} u.$$

Multiplying the first equation by v and the second by u, subtracting, and integrating over S, we obtain

$$\int_S (u\partial_{\nu_-} v - v\partial_{\nu_-} u) = \int_S u\partial_{\nu_+} u.$$

By Green's identities, the left hand side equals

$$\int_\Omega (u\Delta v - v\Delta u) = 0,$$

while the right hand side equals

$$-\int_{\Omega'} (u\Delta u + |\nabla u|^2) = -\int_{\Omega'} |\nabla u|^2.$$

(The application of Green's identity on Ω' needs some justification, which we shall give below.) Therefore u is locally constant on Ω', so $\phi = \partial_{\nu_+} u = 0$.

The proof that $L^2(S) = \mathcal{V}_-^\perp \oplus \mathcal{W}_-$ is much the same: again it suffices to show that if $\phi \in \mathcal{V}_-^\perp \cap \mathcal{W}_-$ then $\phi = 0$. But for such a ϕ we have $T_K^* \phi = -\frac{1}{2}\phi$ and $\phi = \frac{1}{2}\psi + T_K^*\psi$ for some ψ, so if we let u and v be the single layer potentials with moments ϕ and ψ, it follows that $\partial_{\nu_+} u = 0$ and $\partial_{\nu_+} v = \phi = \partial_{\nu_-} u$. Hence

$$\int_S (u\partial_{\nu_+} v - v\partial_{\nu_+} u) = \int_S u\partial_{\nu_-} u.$$

By Green's idenities,

$$0 = -\int_{\Omega'} (u\Delta v - v\Delta u) = \int_\Omega |\nabla u|^2,$$

so u is locally constant on Ω and thus $\phi = \partial_{\nu_-} u = 0$.

To justify these uses of Green's identities on the unbounded region Ω' we replace Ω' by $\Omega' \cap B_r(0)$ and let $r \to \infty$ as in the proof of Proposition (3.4). To make this work it is enough to know that u is harmonic at infinity, so that u and its radial derivative satisfy the estimates of Propositions (2.74) and (2.75). Harmonicity at infinity is automatic when $n > 2$, and for $n = 2$ it is equivalent to the condition $\int_S \phi = 0$ by Lemma (3.31). The latter condition is valid when $\phi \in \mathcal{V}_+^\perp$ by Proposition (3.34) since $\int_S \phi = \sum_1^m \langle \phi \,|\, \alpha_j \rangle$, and it is valid when $\phi \in \mathcal{W}_-$ by Lemma (3.30).   ∎

(3.39) Corollary.
$L^2(S) = \text{Range}(-\frac{1}{2}I + T_K) \oplus \mathcal{V}_+ = \text{Range}(\frac{1}{2}I + T_K) \oplus \mathcal{V}_-.$

Proof: Since $\text{Range}(-\frac{1}{2}I + T_K) = \mathcal{W}_+^\perp$ and $\text{Range}(\frac{1}{2}I + T_K) = \mathcal{W}_-^\perp$ by Corollary (0.42), as in the proof of Proposition (3.38) it suffices to show that $\mathcal{W}_+^\perp \cap \mathcal{V}_+ = \mathcal{W}_-^\perp \cap \mathcal{V}_- = \{0\}$. Suppose $\phi \in \mathcal{W}_+^\perp \cap \mathcal{V}_+$. By Proposition (3.38) we can write $\phi = \phi_1 + \phi_2$ where $\phi_1 \in \mathcal{W}_+$ and $\phi_2 \in \mathcal{V}_+^\perp$. But $\langle \phi \,|\, \phi_1 \rangle = 0$ since $\phi \in \mathcal{W}_+^\perp$ and $\langle \phi \,|\, \phi_2 \rangle = 0$ since $\phi \in \mathcal{V}_+$; hence $\langle \phi \,|\, \phi \rangle = 0$, so $\phi = 0$. Likewise $\mathcal{W}_-^\perp \cap \mathcal{V}_- = \{0\}$. ∎

Finally, we come to the theorem for which this whole chapter has been preparing.

(3.40) Theorem.
With the notation and terminology of §3A, we have:
a. *The interior Dirichlet problem has a unique solution for every $f \in C(S)$.*
b. *The exterior Dirichlet problem has a unique solution for every $f \in C(S)$.*
c. *The interior Neumann problem for $f \in C(S)$ has a solution if and only if $\int_{\partial\Omega_j} f = 0$ for $j = 1, \ldots, m$. The solution is unique modulo functions which are constant on each Ω_j.*
d. *The exterior Neumann problem for $f \in C(S)$ has a solution if and only if $\int_{\partial\Omega_j'} f = 0$ for $j = 1, \ldots, m'$ and also for $j = 0$ in case $n = 2$. The solution is unique modulo functions which are constant on $\Omega_1', \ldots, \Omega_{m'}'$ and also on Ω_0' in case $n = 2$.*

Proof: We have already proved the uniqueness and the necessity of the conditions on f in Propositions (3.1–6), so all that remains is existence.

For (c) we simply observe that $\int_{\partial\Omega_j} f = \langle f \,|\, \alpha_j \rangle$, so these integrals vanish if and only if $f \in \mathcal{V}_+^\perp$. By Corollary (0.42), this is necessary and sufficient to solve the integral equation $-\frac{1}{2}\phi + T_K^*\phi = f$. If ϕ is a solution, then ϕ is continuous by Proposition (3.13), so by Theorem (3.28) the single layer potential with moment ϕ solves the interior Neumann problem.

Similarly, for (d) we have $\int_{\partial\Omega_j'} f = \langle f \,|\, \alpha_j' \rangle$ for $j = 1, \ldots, m'$, so these integrals vanish if and only if $f \in \mathcal{V}_-^\perp$, in which case we can solve the equation $\frac{1}{2}\phi + T_K^*\phi = f$ and then solve the Neumann problem with the single layer potential with moment ϕ. In case $n = 2$, by Lemmas (3.30) and (3.31) this potential is harmonic at infinity if and only if $\int_{\partial\Omega_0'} f = 0$, since $\int_{\partial\Omega_j'} f$ is already assumed to vanish for $j \geq 1$.

Now consider (a). By Corollary (3.39) and Propositions (3.34) and (3.37) we can write

$$f = \tfrac{1}{2}\phi + T_K\phi + \sum_1^{m'} a_j\alpha'_j \qquad (a_j \in \mathbb{C}),$$

where ϕ is continuous since $f - \sum a_j\alpha'_j$ is (Proposition (3.13)). By Theorem (3.22), the double layer potential v with moment ϕ solves the interior Dirichlet problem with f replaced by $\tfrac{1}{2}\phi + T_K\phi$. Moroever, by Proposition (3.37) there exists $\beta \in \mathcal{W}_-$ such that the single layer potential w with moment β satisfies $w|\Omega'_j = a_j$ for $j \geq 1$ and $w|\Omega'_0 = 0$. But then $w|S = \sum_1^{m'} a_j\alpha'_j$ since w is continuous on S (Proposition (3.25)), so the solution of the original Dirichlet problem is $u = v + w$.

When $n > 2$, the proof of (b) is exactly the same, with the roles of Ω and Ω' interchanged and Proposition (3.36) replacing Proposition (3.37). For the case $n = 2$, the argument needs to be modified as follows. As above, we can write

$$f = -\tfrac{1}{2}\phi + T_K\phi + \sum_1^{m} a_j\alpha_j \qquad (\phi \in C(S),\ a_j \in \mathbb{C}).$$

The double layer potential v with moment ϕ solves the exterior Dirichlet problem with f replaced by $-\tfrac{1}{2}\phi + T_K\phi$. Moreover, since $\sum \alpha_j = 1$ on S, with notation as in Proposition (3.36) we can write

$$\sum_1^{m} a_j\alpha_j = \sum_1^{m} b_j\alpha_j + c \qquad ((b_1, \ldots, b_m) \in X,\ c \in \mathbb{C}),$$

and there exists $\beta \in \mathcal{W}^0_+$ such that the single layer potential with moment β satisfies $w|\Omega_j = b_j$. w is also harmonic at infinity by Lemma (3.31), so w solves the exterior Dirichlet problem with f replaced by $\sum b_j\alpha_j$. Finally, the constant function c solves the exterior Dirichlet problem with f replaced by c, so the solution to the original Dirichlet problem is $v + w + c$. $\blacksquare$

F. Further Remarks

The classic treatise on potential theory, which has retained its value after more than a half-century in print, is Kellogg [31]. The reader may consult this work for more information on the behavior of single and double layer potentials, as well as various other aspects of the subject not discussed here. Our treatment also owes much to the lucid exposition in Mikhlin [36].

The results of this chapter extend, with no essential change, to the somewhat more general case where S is assumed only to be of class $C^{1+\alpha}$ for some $\alpha > 0$; see Mikhlin [36]. Some of the arguments involving tubular neighborhoods need some technical elaboration, but the main difference is that K is a kernel of order $n-1-\alpha$ rather than of order $n-2$. (The reason is that the conclusion of Lemma (3.15) becomes $|(x-y) \cdot \nu(y)| \leq c|x-y|^{1+\alpha}$.) In the limiting case $\alpha = 0$, where S is assumed to be only C^1 or Lipschitz, the kernel K is of a sufficiently singular nature that the theory of §2B no longer applies. Indeed, only recently has the method of layer potentials been extended to these cases — by Fabes, Jodeit, and Rivière [13] when S is C^1, and by Verchota [54] when S is Lipschitz. These authors obtain sharp theorems for the Dirichlet and Neumann problems with boundary data in L^p; their work relies on some deep results of Calderón, Coifman, Macintosh, and Meyer on singular integrals. See also Jerison and Kenig [28].

There are other methods for solving the Dirichlet problem with continuous boundary data that yield results under minimal regularity hypotheses on S, of which the most popular is Perron's method of subharmonic functions. An exposition of this theory can be found in John [30], Treves [52], and Jerison and Kenig [28]; see also Kellogg [31] for some related material.

The techniques of integral equations can also be applied to solve other boundary value problems for the Laplacian. (Here and in what follows we shall assume that S is at least of class C^2.) For example, if $b \in C^\alpha(S)$ and $f \in C^\alpha(S)$ for some $\alpha > 0$, consider the problem

$$\Delta u = 0 \text{ on } \Omega, \qquad \partial_{\nu_-} u + bu = f \text{ on } S,$$

which arises in the theory of stationary heat flow. Provided $b > 0$ on S, a unique solution to this problem exists in the form of a single layer potential with moment ϕ, where ϕ satisfies

$$\tfrac{1}{2}\phi + T_{K'}\phi = f, \qquad K'(x,y) = K(x,y) + b(x)N(x,y).$$

Even if b is not positive, this equation can be solved provided f satisfies a finite number of compatibility conditions. See Kellogg [31, §XI.12].

More generally, one can consider the **oblique derivative problem**

$$\Delta u = 0 \text{ on } \Omega, \qquad \partial_\mu u + bu = f \text{ on } S.$$

Here $\partial_\mu u = \nabla u \cdot \mu$ where μ is a smooth vector field on a neighborhood of S with $\mu \cdot \nu > 0$ on S, and $b, f \in C^\alpha(S)$ for some $\alpha > 0$. Again, one attempts to find u in the form of a single layer potential, but the kernel $\partial_{\mu_x} N(x, y)$ which arises is no longer of order $n - 2$. Rather, it satisfies only $|\partial_{\mu_x} N(x, y)| \le C|x - y|^{1-n}$, so it is not integrable as a function of either x or y on S. However, it has certain cancellation properties which guarantee that the integral

$$\int_S \partial_{\mu_x} N(x, y)\phi(y) \, d\sigma(y)$$

is well defined for $\phi \in C^\alpha(S)$ $(\alpha > 0)$ if it is interpreted in a suitable principal-value sense. (The idea is much the same as in Theorem (2.29).) The "singular integral operator" defined in this way is not compact. Nonetheless, one can show that there is another singular integral operator which inverts it modulo a compact operator, so the problem is again reduced to Fredholm theory.

These results were obtained by G. Giraud in a long series of papers in the 1930's, in which he dealt with the Dirichlet, Neumann, and oblique derivative problems for general second-order elliptic operators with variable coefficients. For a description of Giraud's work and references to the literature, see Miranda [37]. (This remarkable book contains an enormous amount of information about elliptic equations, often with sketchy or nonexistent proofs, and it concludes with a seventy-page bibliography.)

More recently, the singular integral operators used by Giraud were studied systematically by Calderón and Zygmund and then incorporated into the theory of pseudodifferential operators. The theory of layer potentials has found a new incarnation in a powerful method, due to Calderón and others, for reducing boundary value problems for quite general differential operators on smoothly bounded domains to the study of pseudodifferential equations on the boundary. (The constructions in this chapter and in the work of Giraud are special cases of this method, although they also yield detailed information that is not available in in the general setting.) See Hörmander [27, vol. III] and Taylor [48].

One may expect that if we impose additional smoothness conditions on the boundary data, the solution will have corresponding smoothness properties near the boundary. This is indeed the case, and it can be proved by the techniques of this chapter. In Chapter 7 we shall obtain some results along these lines, by different methods, for the Dirichlet, Neumann, and oblique derivative problems for a general class of second-order elliptic operators.

Chapter 4
THE HEAT OPERATOR

We now turn our attention to the **heat operator**

$$\partial_t - \Delta = \partial_t - \sum_1^n \partial_j^2$$

on $\mathbb{R}^n \times \mathbb{R}$ with coordinates (x, t). It has the following physical interpretation: the temperature $u(x, t)$ at position x and time t in a homogeneous isotropic medium with unit coefficient of thermal diffusivity satisfies the **heat equation** $\partial_t u - \Delta u = 0$. (See Folland [17, Appx. 1] for a brief derivation.) The same equation also governs other diffusion processes, such as the mixing of two fluids by Brownian motion.

A few caveats about the use of the heat equation in physics: First, the heat equation says nothing about the microscopic physical processes that actually produce heat flow. It describes a limiting situation in which the size of atoms can be considered as infinitesimal, or — statistically speaking — in which it is legitimate to pass to the limit in the central limit theorem. It does not recognize the existence of absolute zero, since if u is a solution then so is $u + c$ for any constant c. And, of course, it takes no account of convection effects in fluids. Nonetheless, the heat equation is very useful in many physical situations, and it is of great mathematical importance.

The heat operator is the prototype of the class of **parabolic** operators. These are the operators of the form $\partial_t + \sum_{|\alpha| \leq 2m} a_\alpha(x, t) \partial_x^\alpha$ where the sum satisfies the strong ellipticity condition discussed in §7A, namely, $(-1)^m \operatorname{Re} \sum_{|\alpha|=2m} a_\alpha(x, t) \xi^\alpha \geq c|\xi|^{2m}$ for some $c > 0$. For information about general parabolic operators, we refer the reader to Friedman [20], [21], Treves [52], and Protter and Weinberger [40].

A. The Gaussian Kernel

We begin by considering the initial value problem

(4.1) $\partial_t u - \Delta u = 0$ on $\mathbb{R}^n \times (0, \infty)$, $u(x, 0) = f(x)$.

Physically, this is a reasonable problem: given the temperature at time $t = 0$, find the temperature at subsequent times. It is also reasonable mathematically, since the heat equation is first order in t. (The Cauchy problem for the hyperplane $t = 0$ is certainly overdetermined, since this hyperplane is everywhere characteristic.)

We can quickly obtain a solution by taking the Fourier transform of (4.1) with respect to the x variables. Indeed, assuming for the moment that f is in the Schwartz class $\mathcal{S}$, and denoting the Fourier transform of $u(x, t)$ with respect to x by $\hat{u}(\xi, t)$, we have

$$\partial_t \hat{u}(\xi, t) + 4\pi^2 |\xi|^2 \hat{u}(\xi, t) = 0, \qquad \hat{u}(\xi, 0) = \hat{f}(\xi).$$

This is an initial value problem for a simple ordinary differential equation in t, and the solution is

$$\hat{u}(\xi, t) = \hat{f}(\xi) e^{-4\pi^2 |\xi|^2 t} \qquad (t > 0).$$

Thus $u(x, t) = f * K_t(x)$, where $\hat{K}_t(\xi) = e^{-4\pi^2 |\xi|^2 t}$. By Theorem (0.25), this means that

(4.2) $K_t(x) \equiv K(x, t) = (4\pi t)^{-n/2} e^{-|x|^2/4t} \qquad (t > 0)$.

The function K defined on $\mathbb{R}^n \times (0, \infty)$ by (4.2) is called the **Gaussian kernel** (or **Gauss-Weierstrass kernel** or **heat kernel**). We note that

$$K_t(x) = t^{-n/2} K_1(t^{-1/2} x), \qquad \int K_t(x) \, dx = \hat{K}_t(0) = 1.$$

Thus by Theorem (0.13), $\{K_t\}_{t>0}$ is an approximation to the identity. (Make the substitution $\epsilon = t^{1/2}$ to obtain the usual formulation.) We therefore have:

(4.3) Theorem.
Suppose $f \in L^p(\mathbb{R}^n)$, $1 \le p \le \infty$. Then $u(x, t) = f * K_t(x)$ satisfies $\partial_t u - \Delta u = 0$ on $\mathbb{R}^n \times (0, \infty)$. If f is bounded and continuous, then u is continuous on $\mathbb{R}^n \times [0, \infty)$ and $u(x, 0) = f(x)$. If $f \in L^p$ where $p < \infty$, then $u(\cdot, t)$ converges to f in the L^p norm as $t \to 0$.

Actually, since $K_t(x)$ decays very rapidly as $x \to \infty$, $f * K_t$ makes sense for $t < T$ provided only that $|f(x)| \leq Ce^{|x|^2/4T}$. Under this condition, one easily verifies that $f * K_t$ satisfies the heat equation (by differentiating under the integral) and that it approaches f uniformly on bounded sets as $t \to 0$ provided f is continuous.

Moreover, since all derivatives of $K(x,t)$ decrease rapidly as $x \to \infty$, we can differentiate under the integral as often as we please and conclude that u is C^∞. Thus the heat equation takes arbitrary initial data and immediately smooths them out. In particular, we cannot expect to obtain a solution for $t < 0$ unless the initial data are already very smooth: for the heat equation, time is irreversible. (This is related to the second law of thermodynamics.)

How about uniqueness for (4.1)? As for the Dirichlet problem for Δ in a half-space (§2G), we must impose some conditions at infinity, but here the conditions are much weaker.

(4.4) Theorem.
Suppose u is continuous on $\mathbb{R}^n \times [0, \infty)$ and C^2 on $\mathbb{R}^n \times (0, \infty)$, and that u satisfies $\partial_t u - \Delta u = 0$ $(t > 0)$ and $u(x, 0) = 0$. If for every $\epsilon > 0$ there exists $C > 0$ such that

$$|u(x,t)| \leq Ce^{\epsilon|x|^2}, \qquad |\nabla_x u(x,t)| \leq Ce^{\epsilon|x|^2},$$

then u is identically zero.

Proof: We first note that if f and g are C^2 functions on a domain in $\mathbb{R}^n \times \mathbb{R}$,

$$g(\partial_t f - \Delta f) + f(\partial_t g + \Delta g) = \sum_1^n \partial_j(f\partial_j g - g\partial_j f) + \partial_t(fg) = \nabla_{x,t} \cdot F,$$

where

$$F = (f\partial_1 g - g\partial_1 f, \ldots, f\partial_n g - g\partial_n f, \ fg).$$

Given $x_0 \in \mathbb{R}^n$ and $t_0 > 0$, let us take $f(x,t) = u(x,t)$ and $g(x,t) = K(x - x_0, t_0 - t)$. Then $\partial_t f - \Delta f = 0$ for $t > 0$ and $\partial_t g + \Delta g = 0$ for $t < t_0$. We apply the divergence theorem on the region

$$\Omega = \{(x,t) : |x| < r, \ a < t < b\} \qquad (0 < a < b < t_0),$$

obtaining (by virtue of (0.1))

$$0 = \int_{\partial\Omega} F \cdot \nu$$

$$= \int_{|x|\leq r} u(x,b)K(x-x_0, t_0-b)\, dx - \int_{|x|\leq r} u(x,a)K(x-x_0, t_0-a)\, dx$$

$$+ \sum_{1}^{n} \int_{a}^{b} \int_{|x|=r} [u(x,t)\partial_j K(x,t) - K(x,t)\partial_j u(x,t)]\frac{x_j}{r}\, d\sigma(x)\, dt.$$

Now let $r \to \infty$. The last sum vanishes by our assumptions on u, and since K is even in x we obtain

$$0 = [u(\cdot,b) * K_{t_0-b}](x_0) - [u(\cdot,a) * K_{t_0-a}](x_0).$$

As $b \to t_0$, the first term on the right approaches $u(x_0,t_0)$, and as $a \to 0$, the second term approaches $u(\cdot,0) * K_{t_0}(x_0) = 0$. Hence $u(x_0,t_0) = 0$ for all $x_0 \in \mathbb{R}^n$ and $t_0 > 0$.   ∎

That uniqueness does not hold for the initial value problem (4.1) without some conditions at infinity can be seen as follows. For simplicity we take $n = 1$. If $g(t)$ is any C^∞ function on $\mathbb{R}$, then formally (disregarding questions of convergence) the series

$$u(x,t) = \sum_{0}^{\infty} \frac{g^{(k)}(t)x^{2k}}{(2k)!}$$

satisfies the heat equation. To produce a nonzero solution of (4.1) with $f = 0$, then, it suffices to produce a nonzero g satisfying $g^{(k)}(0) = 0$ for all k such that this series converges nicely on $\mathbb{R}^n \times [0,\infty)$. The function $g(t) = \exp(-t^{-2})$, for example, will do the job; see John [30] for more details.

There is another theorem due to Widder, however, to the effect that if $f \geq 0$, then $u(x,t) = f * K_t(x)$ is the only nonnegative solution of (4.1); see John [30]. This gives a satisfying uniqueness theorem for physical problems since temperatures on the Kelvin scale are always nonnegative.

The Gaussian kernel can also be used to solve the inhomogeneous heat equation $\partial_t u - \Delta u = f$. Let us extend K to $\mathbb{R}^n \times \mathbb{R}$ by setting

$$(4.5) \qquad K(x,t) = \begin{cases} (4\pi t)^{-n/2}e^{-|x|^2/4t} & (t > 0), \\ 0 & (t \leq 0). \end{cases}$$

We note that K is locally integrable on $\mathbb{R}^n \times \mathbb{R}$ — in fact, integrable on any region whose projection on the t-axis is bounded above, since $\int |K(x,t)|\, dx = 1$ for $t > 0$.

(4.6) Theorem.
The kernel K defined by (4.5) is a fundamental solution for the heat operator.

Proof: Given $\epsilon > 0$, set $K_\epsilon(x, t) = K(x, t)$ if $t > \epsilon$ and $K_\epsilon(x, t) = 0$ otherwise. Clearly $K_\epsilon \to K$ in the topology of distributions as $\epsilon \to 0$ (by the dominated convergence theorem), so we must show that $\partial_t K_\epsilon - \Delta K_\epsilon \to \delta$, that is, for any $\phi \in C_c^\infty$,

$$\langle K_\epsilon, -\partial_t \phi - \Delta \phi \rangle \to \phi(0, 0) \text{ as } \epsilon \to 0.$$

An elementary integration by parts shows that

$$\langle K_\epsilon, -\partial_t \phi - \Delta \phi \rangle = \int_\epsilon^\infty \int_{\mathbb{R}^n} K(x, t)[(-\partial_t - \Delta)\phi(x, t)] \, dx \, dt$$

$$= \int_\epsilon^\infty \int_{\mathbb{R}^n} [(\partial_t - \Delta)K(x, t)]\phi(x, t) \, dx \, dt + \int_{\mathbb{R}^n} K(x, \epsilon)\phi(x, \epsilon) \, dx$$

$$= 0 + \int_{\mathbb{R}^n} K(-x, \epsilon)\phi(x, \epsilon) \, dx,$$

since K is even in x. But this is just the convolution

$$[K_\epsilon * \phi(\cdot, \epsilon)](0) = [K_\epsilon * \phi(\cdot, 0)](0) + [K_\epsilon * [\phi(\cdot, \epsilon) - \phi(\cdot, 0)]](0).$$

The first term on the right tends to $\phi(0, 0)$ as $\epsilon \to 0$ by Theorem (4.3), and the second is bounded by

$$\sup_x |\phi(x, \epsilon) - \phi(x, 0)| \, \|K_\epsilon\|_1 = \sup_x |\phi(x, \epsilon) - \phi(x, 0)|,$$

which tends to zero as $\epsilon \to 0$.   ∎

(4.7) Corollary.
The heat operator is hypoelliptic.

Proof: K is C^∞ except at $(0, 0)$, since $K(x, t)$ vanishes to infinite order as t decreases to zero when $x \neq 0$. Hence Theorem (1.58) applies.   ∎

Remark: The analytic version of this theorem is false, and the fundamental solution K, which satisfies $\partial_t K - \Delta K = 0$ away from the origin but is not analytic on the hyperplane $t = 0$, is a counterexample.

We can now solve the inhomogeneous heat equation $\partial_t u - \Delta u = f$ for any compactly supported distribution f by taking $u = f * K$ (convolution on $\mathbb{R}^n \times \mathbb{R}$). The hypothesis of compact support can be relaxed; here is a representative result.

(4.8) Theorem.
If $f \in L^1(\mathbb{R}^n \times \mathbb{R})$, then $u = f * K$ is well defined almost everywhere and is a distribution solution of $\partial_t u - \Delta u = f$.

Proof: We have

$$u(x,t) = \int_{-\infty}^{t} \int_{\mathbb{R}^n} f(y,s)K(x-y, t-s)\, dy\, ds = \int_{-\infty}^{t} f(\cdot, s) * K_{t-s}(x)\, ds,$$

the convolution here being on $\mathbb{R}^n$. Since $f(\cdot, s) \in L^1(\mathbb{R}^n)$ for almost every s and $\|K_{t-s}\|_1 = 1$, by Young's inequality we have

$$\sup_t \int_{-\infty}^{t} \int_{\mathbb{R}^n} |f(\cdot, s) * K_{t-s}(x)|\, dx\, ds \le \int_{-\infty}^{\infty} \int_{\mathbb{R}^n} |f(y,s)|\, dy\, ds = \|f\|_1.$$

It follows that the integral defining $u(x,t)$ converges absolutely for almost every (x,t) and that u is locally integrable. The proof that $\partial_t u - \Delta u = f$ now proceeds just like the proof of Theorem (2.21); details are left to the reader.   ∎

Remark: An argument similar to the proof of Theorem (2.28) shows that if $f \in C^{k+\alpha}$ for some $k \ge 0$ and $0 < \alpha < 1$, then u is $C^{k+2+\alpha}$ in x and $C^{k+1+\alpha}$ in t and is a classical solution of $\partial_t u - \Delta u = f$.

One of the most amusing applications of the Gaussian kernel is Weierstrass's original proof of his celebrated approximation theorem. To obtain the full strength of this theorem, we shall assume the following technical result: If S is a compact set in $\mathbb{R}^n$ and f is a continuous function on S, then f can be extended to be continuous with compact support on $\mathbb{R}^n$. In general, this follows from the Tietze extension theorem, but if S is reasonably nice one can usually construct a solution explicitly. (For example, if S is a C^2 hypersurface, extend f to the tubular neighborhood V of S given by (0.2) by setting $f(x + t\nu(x)) = \phi(t)f(x)$ where $\phi \in C(\mathbb{R})$, $\phi(0) = 1$, and $\phi(t) = 0$ for $|t| > \frac{1}{2}\epsilon$, and then set $f = 0$ outside V. A similar construction works if S is the closure of a domain with C^2 boundary.)

(4.9) The Weierstrass Approximation Theorem.
If S is a compact subset of $\mathbb{R}^n$, the restrictions of polynomials to S are dense in $C(S)$ in the uniform norm.

Proof: Given $f \in C(S)$, extend f to be continuous with compact support on $\mathbb{R}^n$. Given $\epsilon > 0$, by Theorem (4.3) we can find $t > 0$ so that

$$\sup_{x \in \mathbb{R}^n} |f * K_t(x) - f(x)| < \tfrac{1}{2}\epsilon.$$

But

$$f * K_t(x) = (4\pi t)^{-n/2} \int_{\text{supp } f} f(y) e^{-|x-y|^2/4t}\, dy.$$

Since the power series for e^z converges uniformly on compact sets, we can replace $e^{-|x-y|^2/4t}$ by a suitable partial sum with an error less than $(4\pi t)^{n/2}\epsilon/2\|f\|_1$ for $x \in S$ and $y \in \text{supp } f$. Thus,

$$\sup_{x \in S} |f * K_t(x) - g(x)| < \tfrac{1}{2}\epsilon,$$

where

$$g(x) = (4\pi t)^{-n/2} \int_{\text{supp } f} f(y) \sum_0^K \frac{(-1)^k}{k!} \left[\frac{\sum_1^n (x_j - y_j)^2}{4t} \right]^k dy.$$

But g is clearly a polynomial of degree $2K$. ∎

EXERCISES

1. Complete the proof of Theorem (4.8).

2. Suppose $u_1(y,t), \ldots, u_n(y,t)$ are solutions of the one-dimensional heat equation $\partial_t u = \partial_y^2 u$. Show that $v(x,t) = \prod_1^n u_j(x_j, t)$ satisfies the n-dimensional heat equation. (What is special about the heat equation that makes this work? E.g., there is no analogous result for solutions of Laplace's equation.)

3. A formal use of the Fourier transform on $\mathbb{R}^n \times \mathbb{R}$ suggests that one should obtain a fundamental solution for $\partial_t - \Delta$ by taking the inverse Fourier transform of $G(\xi, \tau) = (2\pi i\tau + 4\pi^2 |\xi|^2)^{-1}$. Prove that this works, as follows.
   a. Show that G is locally integrable on $\mathbb{R}^n \times \mathbb{R}$.
   b. Show that $G = \widehat{K}$ where K is defined by (4.5). (Hint: Take the Fourier transform of K first in x, then in t. Why is this legitimate?)

4. Solve the one-dimensional heat equation $\partial_t u = \partial_x^2 u$ on the region $x > 0$, $t > 0$ subject to the initial condition $u(x, 0) = f(x)$ $(x > 0)$ and either the boundary condition (a) $u(0, t) = 0$ or (b) $\partial_x u(0, t) = 0$. (Hint: consider the odd or even extension of f to $\mathbb{R}$.)

B. Functions of the Laplacian

The Gaussian kernel has many other interesting applications in analysis and probability. We shall limit ourselves to one of them, namely, the computation of convolution kernels for functions of the Laplacian.

We begin by observing that $(-\Delta f)\widehat{\ }(\xi) = 4\pi^2 |\xi|^2 \hat{f}(\xi)$ for $f \in S(\mathbb{R}^n)$. It follows that if P is a polynomial in one variable,

$$[P(-\Delta)f]\widehat{\ }(\xi) = P(4\pi^2 |\xi|^2)\hat{f}(\xi),$$

and this suggests the following general construction of functions of $-\Delta$. Suppose ψ is a function on $(0, \infty)$ such that $\xi \to \psi(4\pi^2 |\xi|^2)$ is a tempered distribution on $\mathbb{R}^n$. Then we can define an operator $\psi(-\Delta) : S \to S'$ by

$$[\psi(-\Delta)f]\widehat{\ }(\xi) = \psi(4\pi^2 |\xi|^2)\hat{f}(\xi), \quad \text{i.e.,} \quad \psi(-\Delta)f = [\psi(4\pi^2 |\cdot|^2)\hat{f}]^\vee.$$

$\psi(-\Delta)$ can also be expressed as a convolution operator: $\psi(-\Delta)f = f * \kappa_\psi$ where κ_ψ is the inverse Fourier transform of the tempered distribution $\xi \to \psi(4\pi^2 |\xi|^2)$. For example, if $\psi(s) = e^{-ts}$ with $t > 0$, then κ_ψ is just the Gaussian kernel K_t by Theorem (0.25); thus

$$f * K_t = e^{t\Delta}f,$$

which is just what one would expect by formally solving the heat equation $\partial_t u = \Delta u$ with initial data $u(\cdot, 0) = f$ as an ordinary differential equation in t. (This is essentially what we did in deriving Theorem (4.3).)

With this information in hand, we have a method for computing the convolution kernel κ_ψ whenever ψ can be expressed nicely in terms of exponential functions. More specifically, suppose

$$(4.10) \qquad \psi(s) = \int_0^\infty \phi(\tau)e^{-s\omega(\tau)}\,d\tau$$

for some functions ϕ and ω on $(0, \infty)$ with $\omega > 0$. Then, formally,

$$\psi(-\Delta) = \int_0^\infty \phi(\tau)e^{\omega(\tau)\Delta}\,d\tau,$$

so the kernel k_ψ should be given by

$$k_\psi(x) = \int_0^\infty \phi(\tau) K(x, \omega(\tau)) \, d\tau.$$

So far this is just a heuristic procedure which needs some justification. The interested reader may wish to formulate a theorem along these lines that encompasses a general class of ϕ's and ω's, but here we shall content ourselves with working out some important specific examples.

Our first example is the negative fractional powers of $-\Delta$, $\psi(s) = s^{-\beta}$ where $0 < \operatorname{Re}\beta < \frac{1}{2}n$. (The restriction $\operatorname{Re}\beta < \frac{1}{2}n$ is necessary to make $\psi(4\pi^2|\xi|^2)$ integrable near the origin.) We have

$$s^{-\beta} = \frac{1}{\Gamma(\beta)} \int_0^\infty \tau^{\beta-1} e^{-s\tau} \, d\tau,$$

so the convolution kernel of $(-\Delta)^{-\beta}$ should be

$$\frac{1}{\Gamma(\beta)} \int_0^\infty \tau^{\beta-1} K(x, \tau) \, d\tau.$$

We can compute this integral by making the substitution $\sigma = |x|^2/4\tau$:

$$\frac{1}{\Gamma(\beta)} \int_0^\infty \tau^{\beta-1} K(x, \tau) \, d\tau = \frac{1}{\Gamma(\beta)(4\pi)^{n/2}} \int_0^\infty \tau^{\beta-1-(n/2)} e^{-|x|^2/4\tau} \, d\tau$$

$$= \frac{1}{\Gamma(\beta) 4^\beta \pi^{n/2} |x|^{n-2\beta}} \int_0^\infty \sigma^{(n/2)-\beta-1} e^{-\sigma} \, d\sigma$$

$$= \frac{\Gamma(\frac{1}{2}n - \beta)}{\Gamma(\beta) 4^\beta \pi^{n/2} |x|^{n-2\beta}}.$$

It is not hard to see that this function is indeed the inverse Fourier transform of $(4\pi^2|\xi|^2)^{-\beta}$; see Exercise 1. (It is not quite trivial either, since neither of these functions is globally integrable, so one must interpret the Fourier transform in the sense of distributions.)

It is customary to set $\beta = \frac{1}{2}\alpha$ and to define

$$(4.11) \qquad R_\alpha(x) = \frac{\Gamma(\frac{1}{2}(n - \alpha))}{\Gamma(\frac{1}{2}\alpha) 2^\alpha \pi^{n/2} |x|^{n-\alpha}} \qquad (0 < \operatorname{Re}\alpha < n),$$

so that

$$(-\Delta)^{-\alpha/2} f = f * R_\alpha.$$

R_α is called the **Riesz potential** of order α. In particular, when $\alpha = 2$ (and $n > 2$) we see that

$$R_2(x) = \frac{\Gamma(\frac{1}{2}n - 1)}{4\pi^{n/2}|x|^{n-2}} = \frac{\Gamma(\frac{1}{2}n)}{(n-2)2\pi^{n/2}|x|^{n-2}} = \frac{1}{(n-2)\omega_n|x|^{n-2}},$$

which is the fundamental solution for $-\Delta$ given by (2.18).

We have thus rederived the fundamental solution for Δ by Fourier analysis, at least when $n > 2$. When $n = 2$ this procedure breaks down: as $\alpha \to 2$, the coefficient $\Gamma(\frac{1}{2}(2 - \alpha))$ in $R_\alpha(x)$ becomes infinite, and the function $(4\pi^2|\xi|^2)^{-\alpha/2}$ ceases to be integrable at the origin. However, one can proceed as follows. When $0 < \alpha < 2$, the Riesz potential R_α satisfies $(-\Delta)^{\alpha/2}R_\alpha = \delta$. But so does $R_\alpha - c_\alpha$ for any constant c_α, and we can choose c_α so as to obtain a finite limit as $\alpha \to 2$. Namely, let

$$(4.12) \qquad Q_\alpha(x) = R_\alpha(x) - \frac{\Gamma(1 - \frac{1}{2}\alpha)}{\Gamma(\frac{1}{2}\alpha)2^\alpha\pi} = \frac{\Gamma(1 - \frac{1}{2}\alpha)}{\Gamma(\frac{1}{2}\alpha)2^\alpha\pi}(|x|^{\alpha-2} - 1).$$

Since $\Gamma(1 - \frac{1}{2}\alpha) = 2\Gamma(2 - \frac{1}{2}\alpha)/(2 - \alpha)$, an application of l'Hôpital's rule shows that

$$Q_2(x) = \lim_{\alpha \to 2} Q_\alpha(x) = -\frac{1}{2\pi}\log|x|,$$

which is the fundamental solution for $-\Delta$. Roughly speaking, Q_2 is obtained from R_2 by subtracting off an infinite constant. Correspondingly, its Fourier transform is obtained from $(4\pi^2|\xi|^2)^{-1}$ by subtracting off an infinite multiple of δ; see Exercise 2.

Our next example is a modification of the preceding one. It is sometimes convenient to consider powers of $I - \Delta$ rather than $-\Delta$ so as to avoid confronting the singularity of $|\xi|^{-\alpha}$ at $\xi = 0$. Accordingly, we observe that

$$(1 + s)^{-\beta} = \frac{1}{\Gamma(\beta)}\int_0^\infty \tau^{\beta-1}e^{-(1+s)\tau}\,d\tau \qquad (\text{Re}\,\beta > 0),$$

so that the convolution kernel of $(I - \Delta)^{-\beta}$ should be

$$\frac{1}{\Gamma(\beta)}\int_0^\infty \tau^{\beta-1}e^{-\tau}K(x, \tau)\,d\tau.$$

Again we set $\beta = \frac{1}{2}\alpha$ and define

$$(4.13) \qquad B_\alpha(x) = \frac{1}{\Gamma(\frac{1}{2}\alpha)(4\pi)^{n/2}}\int_0^\infty \tau^{(\alpha-n)/2-1}e^{-\tau-|x|^2/4\tau}\,d\tau.$$

for $\operatorname{Re}\alpha > 0$. (The integral converges for all $x \neq 0$.) B_α is called the **Bessel potential** of order α. (The name comes from the fact that B_α is a Bessel function; in fact, $B_\alpha(x)$ is a constant multiple of $|x|^{(\alpha-n)/2}K_{(n-\alpha)/2}(|x|)$, where K_μ is the modified Bessel function of the second kind of order μ.) B_α is in $L^1(\mathbb{R}^n)$, for

$$\|B_\alpha\|_1 \leq \frac{1}{|\Gamma(\frac{1}{2}\alpha)|} \int_0^\infty \int_{\mathbb{R}^n} \tau^{\operatorname{Re}(\alpha/2)-1} e^{-\tau} K(x,\tau)\,dx\,d\tau$$

$$= \frac{1}{|\Gamma(\frac{1}{2}\alpha)|} \int_0^\infty \tau^{\operatorname{Re}(\alpha/2)-1} e^{-\tau}\,d\tau = \frac{\Gamma(\frac{1}{2}\operatorname{Re}\alpha)}{|\Gamma(\frac{1}{2}\alpha)|}.$$

Because of this, it is easy to use Fubini's theorem to show that the Fourier transform of B_α is indeed $(1 + 4\pi^2|\xi|^2)^{-\alpha/2}$ (Exercise 3), so

$$(I - \Delta)^{-\alpha/2}f = f * B_\alpha \qquad (\operatorname{Re}\alpha > 0).$$

We shall meet this operator again in Chapter 6.

For our final example, we rederive the solution of the Dirichlet problem for the Laplacian in a half-space (see §2G) by Fourier analysis. We recall that the problem is to solve

$$\partial_t^2 u + \Delta_x u = 0 \text{ for } t > 0, \qquad u(x,0) = f(x).$$

As in §4A, we apply the Fourier transform in x to obtain an ordinary differential equation in t:

$$\partial_t^2 \widehat{u} - 4\pi^2|\xi|^2\widehat{u} = 0 \text{ for } t > 0, \qquad \widehat{u}(\xi,0) = \widehat{f}(\xi).$$

The general solution of this differential equation is a linear combination of $e^{2\pi|\xi|t}$ and $e^{-2\pi|\xi|t}$. We reject $e^{2\pi|\xi|t}$ since it is not tempered for $t > 0$, and the initial condition then yields

$$\widehat{u}(\xi,t) = e^{-2\pi|\xi|t}\widehat{f}(\xi), \quad \text{or} \quad u(x,t) = e^{-t\sqrt{-\Delta}}f(x).$$

We therefore have $u(x,t) = f * P_t(x)$ where P_t is the inverse Fourier transform of $e^{-2\pi|\xi|t}$. The magic formula that enables us to compute this is as follows.

(4.14) Lemma.
If $\beta \geq 0$,

$$e^{-\beta} = \int_0^\infty \frac{e^{-\tau}}{\sqrt{\pi\tau}} e^{-\beta^2/4\tau}\,d\tau.$$

Proof: A standard application of the residue theorem yields

$$e^{-\beta} = \frac{1}{\pi} \int_{-\infty}^{\infty} \frac{e^{i\beta s}}{1+s^2}\, ds,$$

and obviously

$$\frac{1}{1+s^2} = \int_0^{\infty} e^{-(1+s^2)\tau}\, d\tau.$$

Hence, by Fubini's theorem and Theorem (0.25),

$$e^{-\beta} = \frac{1}{\pi} \int_0^{\infty} \int_{-\infty}^{\infty} e^{i\beta s} e^{-\tau s^2} e^{-\tau}\, ds\, d\tau$$

$$= 2 \int_0^{\infty} \int_{-\infty}^{\infty} e^{2\pi i \beta \sigma} e^{-4\tau\pi^2\sigma^2} e^{-\tau}\, d\sigma\, d\tau$$

$$= \int_0^{\infty} \frac{1}{\sqrt{\pi\tau}} e^{-\beta^2/4\tau} e^{-\tau}\, d\tau. \qquad\blacksquare$$

Now if we set $\beta = t\sqrt{s}$ in Lemma (4.14), we obtain a formula of the form (4.10),

$$e^{-t\sqrt{s}} = \int_0^{\infty} \frac{e^{-\tau}}{\sqrt{\pi\tau}} e^{-t^2 s/4\tau}\, d\tau,$$

so the inverse Fourier transform P_t of $e^{-2\pi|\xi|t}$ should be given by

$$P_t(x) = \int_0^{\infty} \frac{e^{-\tau}}{\sqrt{\pi\tau}} K(x, t^2/4\tau)\, d\tau.$$

To verify this and evaluate the integral, simply take $\beta = 2\pi|\xi|t$ in Lemma (4.14) and use Fubini's theorem and Theorem (0.25):

$$P_t(x) = \int_0^{\infty} \int_{\mathbb{R}^n} e^{2\pi i x \cdot \xi} \frac{e^{-\tau}}{\sqrt{\pi\tau}} e^{-\pi^2|\xi|^2 t^2/\tau}\, d\xi\, d\tau$$

$$= \int_0^{\infty} \frac{e^{-\tau}}{\sqrt{\pi\tau}} \left[\frac{\tau}{\pi t^2}\right]^{n/2} e^{-|x|^2\tau/t^2}\, d\tau$$

$$= \frac{1}{\pi^{(n+1)/2} t^n} \int_0^{\infty} \tau^{(n-1)/2} e^{-\tau(1+|x|^2/t^2)}\, d\tau$$

$$= \frac{1}{\pi^{(n+1)/2} t^n (1+|x|^2/t^2)^{(n+1)/2}} \int_0^{\infty} \sigma^{(n-1)/2} e^{-\sigma}\, d\sigma$$

$$= \frac{\Gamma((n+1)/2)}{\pi^{(n+1)/2}} \frac{t}{(t^2+|x|^2)^{(n+1)/2}}.$$

Since $\Gamma((n+1)/2)/\pi^{(n+1)/2} = 2/\omega_{n+1}$, we have recovered the formula (2.43) for the Poisson kernel, and we have proved:

(4.15) Theorem.
If P_t is the Poisson kernel defined by (2.43), then $\widehat{P}_t(\xi) = e^{-2\pi t|\xi|}$.

This result gives an easy new proof of the semigroup property $P_{t+s} = P_t * P_s$, for it is obvious that $\widehat{P}_{t+s} = \widehat{P}_t \widehat{P}_s$. It also makes clear how the Dirichlet and Neumann data of the harmonic function $u(x,t) = f * P_t(x)$ determine each other. Indeed, we have $u(\cdot, 0) = f$, and

$$[-\partial_t u]\widehat{}(\xi, 0) = 2\pi|\xi|e^{-2\pi|\xi|t}\widehat{f}(\xi)\big|_{t=0} = 2\pi|\xi|\widehat{f}(\xi).$$

That is, if $g = -\partial_t u(\cdot, 0)$ then $g = (-\Delta_x)^{1/2}f$; in other words, $f = (-\Delta_x)^{-1/2}g = g * R_1$ where R_1 is the Riesz potential given by (4.11).

EXERCISES

1. By Theorems (0.25) and (0.26), for any $\phi \in S(\mathbb{R}^n)$ and $\tau > 0$ we have

$$\int e^{-\pi\tau|x|^2}\widehat{\phi}(x)\,dx = \tau^{-n/2}\int e^{-\pi|\xi|^2/\tau}\phi(\xi)\,d\xi.$$

For $0 < \operatorname{Re}\alpha < n$, multiply both sides by $\tau^{(n-\alpha)/2-1}\,d\tau$ and integrate from 0 to ∞ to obtain

$$\frac{\Gamma((n-\alpha)/2)}{\pi^{(n-\alpha)/2}}\int |x|^{\alpha-n}\widehat{\phi}(x)\,dx = \frac{\Gamma(\alpha/2)}{\pi^{\alpha/2}}\int |\xi|^{-\alpha}\phi(\xi)\,d\xi.$$

Conclude that if R_α is given by (4.11), then $\widehat{R}_\alpha(\xi) = (2\pi|\xi|)^{-\alpha}$, the Fourier transform being interpreted in the distibutional sense.

2. From Exercise 1 and the final paragraph of §0E, if Q_α $(0 < \alpha < 2)$ is given on $\mathbb{R}^2$ by (4.12) we have

$$\widehat{Q}_\alpha(\xi) = (2\pi|\xi|)^{-\alpha} - \frac{\Gamma(1-\frac{1}{2}\alpha)}{\Gamma(\frac{1}{2}\alpha)2^\alpha\pi}\delta(\xi).$$

Show that

$$\frac{\Gamma(1-\frac{1}{2}\alpha)}{\Gamma(\frac{1}{2}\alpha)2^\alpha\pi} - \int_{|\xi|<1}\frac{d\xi}{(2\pi|\xi|)^\alpha}$$

approaches a finite limit c as α approaches 2 from below, and conclude that the Fourier transform of $Q_2(x) = (-2\pi)^{-1}\log|x|$ is the distribution given by

$$\langle\widehat{Q}_2, \phi\rangle = \int_{|\xi|<1}\frac{\phi(\xi) - \phi(0) - \xi\cdot\nabla\phi(0)}{(2\pi|\xi|)^2}\,d\xi + \int_{|\xi|>1}\frac{\phi(\xi)}{(2\pi|\xi|)^2}\,d\xi - c\phi(0).$$

(The first integral is absolutely convergent because the numerator is $O(|\xi|^2)$ by Taylor's theorem. The term $\xi \cdot \nabla\phi(0)/(2\pi|\xi|)^2$ is really a phantom, since it is an odd function and so its integral over the unit ball vanishes formally. Thus, intuitively we have $\widehat{Q}_2(\xi) = (2\pi|\xi|)^{-2} - C\delta(\xi)$ where C is the infinite constant $\int_{|\xi|<1}(2\pi|\xi|)^{-2}\,d\xi + c$.)

3. Show that if $\operatorname{Re}\alpha > 0$ and B_α is given by (4.13), then $\widehat{B}_\alpha(\xi) = (1 + 4\pi^2|\xi|^2)^{-\alpha/2}$.

C. The Heat Equation in Bounded Domains

If one wishes to study heat flow in a bounded region of space $\Omega \subset \mathbb{R}^n$ over a time interval $0 \leq t \leq T \leq \infty$, it is appropriate not only to specify the initial temperature $u(x, 0)$ $(x \in \Omega)$ but also to prescribe a boundary condition on $\partial\Omega \times [0, T]$. For example:

i. $u = g$ (the temperature on the boundary is specified);

ii. $\partial_\nu u = 0$ (the boundary is insulated; there is no heat flow into or out of Ω);

iii. $\partial_\nu u + c(u - u_0) = 0$ (Newton's law of cooling: outside Ω the temperature is maintained at u_0, and the rate of heat flow across the boundary is proportional to $u - u_0$).

The first basic result concerning such problems is the following **maximum principle**:

(4.16) Theorem.
Let Ω be a bounded domain in $\mathbb{R}^n$ and $0 < T < \infty$. Suppose u is a real-valued continuous function on $\overline{\Omega} \times [0, T]$ that satisfies $\partial_t u - \Delta u = 0$ on $\Omega \times (0, T)$ (and hence is C^∞ there). Then u assumes its maximum either on $\Omega \times \{0\}$ or on $\partial\Omega \times [0, T]$.

Proof: Given $\epsilon > 0$, set $v(x, t) = u(x, t) + \epsilon|x|^2$. Then $\partial_t v - \Delta v = -2n\epsilon < 0$. Suppose $0 < T' < T$. If the maximum value of v on $\overline{\Omega} \times [0, T']$ occurs at an interior point, the first derivatives of v vanish there and the pure second derivatives $\partial_j^2 v$ are nonpositive. In particular, $\partial_t v = 0$ and $\Delta v \leq 0$, which contradicts $\partial_t v - \Delta v < 0$. Likewise, if the maximum occurs on $\Omega \times \{T'\}$, the t-derivative must be nonnegative and the pure second x-derivatives must be nonpositive, so $\partial_t v \geq 0$ and $\Delta v \leq 0$, which again contradicts $\partial_t v - \Delta v < 0$. Therefore,

$$\max_{\overline{\Omega} \times [0,T']} u \leq \max_{\overline{\Omega} \times [0,T']} v = \max_{(\Omega \times \{0\}) \cup (\partial\Omega \times [0,T'])} v$$

$$\leq \max_{(\Omega \times \{0\}) \cup (\partial\Omega \times [0,T'])} u + \epsilon \max_{\overline{\Omega}} |x|^2.$$

Letting $\epsilon \to 0$ and $T' \to T$, we obtain the desired result. ∎

Replacing u by $-u$, we see that the minimum is also achieved on either $\Omega \times \{0\}$ or $\partial\Omega \times [0,T]$. Therefore:

(4.17) Corollary.
There is at most one continuous function u on $\overline{\Omega} \times [0,T]$ which agrees with a given continuous function on $\Omega \times \{0\}$ and on $\partial\Omega \times [0,T]$ and satisfies $\partial_t u - \Delta u = 0$ on $\Omega \times (0,T)$.

(In particular, note that the "Dirichlet problem" is overdetermined: the boundary values on $\Omega \times \{T\}$ cannot be specified in advance.)

Let us look more closely at the problem obtained by holding the boundary at a constant temperature, which we may assume to be zero by adjusting our temperature scale:

$$(4.18) \quad \begin{aligned} &\partial_t u - \Delta u = 0 \text{ on } \Omega \times (0,\infty), \\ &u(x,0) = f(x) \quad (x \in \Omega), \qquad u(x,t) = 0 \quad (x \in \partial\Omega, \ t \in (0,\infty)). \end{aligned}$$

This problem can be solved by the method of separation of variables. That is, we begin by looking for solutions of the differential equation of the form $u(x,t) = F(x)G(t)$. For such a function we have $\partial_t u - \Delta u = FG' - (\Delta F)G$, so the heat equation implies that $G'/G = (\Delta F)/F$. But G'/G is a function of t alone, and $(\Delta F)/F$ is a function of x alone, so both these quantities must be some constant λ. Thus $G' = \lambda G$, so $G(t) = Ce^{\lambda t}$, and $\Delta F = \lambda F$.

Suppose we can find an orthonormal basis $\{F_j\}$ for $L^2(\Omega)$ such that $\Delta F_j = \lambda_j F_j$ and $F_j(x) = 0$ for $x \in \partial\Omega$. We can then solve (4.18) by expanding f in terms of the F_j's, say $f = \sum a_j F_j$, and taking

$$(4.19) \qquad u(x,t) = \sum a_j F_j(x) e^{\lambda_j t}.$$

Provided that the λ_j's are nonpositive, this series will converge in $L^2(\Omega)$ for each $t \geq 0$ and will satisfy $u(x,t) = 0$ for $x \in \partial\Omega$ as best it can. (For example, if $f \in H_1^0(\Omega)$ as defined in §2F, then u will vanish on $\partial\Omega \times (0,\infty)$ in the sense of Corollary (2.40).) Also, one easily verifies that u is a distribution solution of the heat equation (Exercise 1), hence a classical solution by Corollary (4.7).

The problem (4.18) is therefore reduced to finding an eigenbasis for the Dirichlet problem for the Laplacian. We have done this in §3H when Ω is a ball. We will show in §7E that such an eigenbasis exists for a general bounded domain Ω with smooth boundary and that the eigenvalues are all negative.

Similar considerations apply to the insulated-boundary problem

$$\partial_t u - \Delta u = 0 \text{ on } \Omega \times (0, \infty),$$
$$u(x, 0) = f(x) \quad (x \in \Omega), \qquad \partial_\nu u(x, t) = 0 \quad (x \in \partial\Omega, \ t \in (0, \infty)).$$

This problem boils down to finding an orthonormal basis $\{H_j\}$ for $L^2(\Omega)$ such that $\Delta H_j = \mu_j H_j$ ($\mu_j \leq 0$) and $\partial_\nu H_j = 0$ on $\partial\Omega$. Again, we have seen how to do this when Ω is a ball, and we will prove the existence of such an eigenbasis for a general Ω in §7E.

Finally, we solve the initial value problem for the heat equation on the circle (physically: an insulated loop made of homogeneous material of negligible diameter). We think of the circle as the set of complex numbers of modulus one and identify it with the unit interval by the correspondence $x \leftrightarrow e^{2\pi i x}$. We then wish to solve the problem

$$(4.20) \qquad \begin{aligned} &\partial_t u - \partial_x^2 u = 0 \text{ for } t > 0, \\ &u(x, 0) = f(x), \qquad u(x + 1, t) = u(x, t), \end{aligned}$$

where $f(x + 1) = f(x)$. The natural tool here is Fourier series, i.e., expansions with respect to the orthonormal basis $\{e^{2\pi i k x}\}_{-\infty}^{\infty}$ for $L^2(0, 1)$. (One would be led to this in any event by separation of variables.) Indeed, if we look for a solution of (4.20) in the form $u(x, t) = \sum_{-\infty}^{\infty} C_k(t) e^{2\pi i k x}$, the heat equation implies that $C_k'(t) = -4\pi^2 k^2 C_k(t)$, and the initial condition means that $C_k(0)$ must be the kth Fourier coefficient $\widehat{f}(k)$ of f. Therefore $C_k(t) = \widehat{f}(k) e^{-4\pi^2 k^2 t}$, and

$$\begin{aligned} u(x, t) &= \sum_{-\infty}^{\infty} \widehat{f}(k) e^{2\pi i k x - 4\pi^2 k^2 t} = \sum_{-\infty}^{\infty} \int_0^1 f(y) e^{2\pi i k(x-y) - 4\pi^2 k^2 t} \, dy \\ &= [f * \vartheta(\cdot, t)](x), \end{aligned}$$

where

$$\vartheta(x, t) = \sum_{-\infty}^{\infty} e^{2\pi i k x - 4\pi^2 k^2 t}$$

and $*$ means convolution on the circle group $\mathbb{R}/\mathbb{Z}$. (The sums are rapidly convergent for $t > 0$ because of the factors $e^{-4\pi^2 k^2 t}$, so the formal manipulations are justified. In particular, ϑ and u are C^∞ functions for $t > 0$ and satisfy the heat equation.)

The function ϑ therefore plays the same role for the heat equation on the circle as the Gaussian kernel K does on $\mathbb{R}^n$. Like K, ϑ has a significance which reaches far beyond the study of heat flow: it is essentially one of the Jacobi theta functions, which have deep connections with elliptic functions and number theory. (More precisely, in the notation of the Bateman Manuscript Project [4], $\vartheta(x,t) = \theta_3(x|4\pi it)$.)

EXERCISES

1. Suppose $\{F_j\}$ is an orthonormal basis for $L^2(\Omega)$, $\lambda_j \leq 0$, and $\sum |a_j|^2 < \infty$.

    a. Show that the series (4.19) converges in the topology of distributions on $\Omega \times (0, \infty)$.

    b. Show that if $\Delta F_j = \lambda_j F_j$, the series (4.19) satisfies the heat equation on $\Omega \times (0, \infty)$ in the sense of distributions. (This is almost immediate from part (a).)

2. Suppose $\{F_j\}$ is an orthonormal basis for $L^2(\Omega)$ such that $\Delta F_j = \lambda_j F_j$ ($\lambda_j \leq 0$) and $F_j = 0$ on $\partial\Omega$. Solve the inhomogeneous heat equation

    $$\partial_t u - \Delta u = g(x,t) \text{ on } \Omega \times (0, \infty),$$
    $$u(x,0) = 0 \quad (x \in \Omega), \qquad u(x,t) = 0 \quad (x \in \partial\Omega, \ t \in (0, \infty))$$

    in terms of the basis $\{F_j\}$. (The solution will look like (4.19), but with $e^{\lambda_j t}$ replaced by the solution of an inhomogeneous ordinary differential equation.)

3. Suppose Ω is a bounded domain with C^1 boundary, and suppose u is a C^1 function on $\overline{\Omega} \times (0, T)$ that satisfies $\partial_t u - \Delta u = 0$ on $\Omega \times (0, T)$ and either of the boundary conditions $u = 0$ or $\partial_\nu u = 0$ on $\partial\Omega \times (0, T)$. Show that $\int_\Omega |u(x,t)|^2 \, dx$ is a decreasing function of t. (Hint: Observe that $\overline{u}(\partial_t u - \Delta u) = \frac{1}{2}\partial_t(|u|^2) - \nabla \cdot (\overline{u}\nabla u) + |\nabla u|^2$ and integrate over Ω.)

Chapter 5
THE WAVE OPERATOR

The last of the three great second-order operators is the **wave operator** or **d'Alembertian**

$$\partial_t^2 - \Delta = \partial_t^2 - \sum_1^n \partial_j^2$$

on $\mathbb{R}^n \times \mathbb{R}$. As the name suggests, the **wave equation**

$$\partial_t^2 u - \Delta u = 0$$

is satisfied by waves with unit speed of propagation in homogeneous isotropic media. (Actually, in most cases such as water waves or vibrating strings or membranes, the wave equation gives only an approximation to the correct physics that is valid for vibrations of small amplitude. However, it is an easy consequence of Maxwell's equations that the wave equation is satisfied exactly by the components of the classical electromagnetic field in a vacuum.) The characteristic variety of the wave operator,

$$\Sigma = \left\{ (\xi, \tau) \neq (0,0) \in \mathbb{R}^n \times \mathbb{R} : \tau^2 = |\xi|^2 \right\},$$

plays an important role in the theory. It is called the **light cone**, and the two nappes $\{(\xi, \tau) \in \Sigma : \tau > 0\}$ and $\{(\xi, \tau) \in \Sigma : \tau < 0\}$ are called the **forward** and **backward** light cones.

The wave operator is the prototype of the class **hyperbolic** operators. (There are several related definitions of hyperbolicity for general differential operators, of which perhaps the most widely useful is the following. An operator $L = \sum_{|\alpha|+j \leq k} a_{\alpha j}(x, t) \partial_x^\alpha \partial_t^j$ on $\mathbb{R}^n \times \mathbb{R}$ is called **strictly hyperbolic** if, for every $(x, t) \in \mathbb{R}^n \times \mathbb{R}$ and every nonzero $\xi \in \mathbb{R}^n$, the polynomial $P(\tau) = \sum_{|\alpha|+j=k} a_\alpha(x, t) \xi^\alpha \tau^j$ has k distinct real roots.) There is an extensive literature on hyperbolic equations, of which we mention only a few basic references: John [30] for a brief introduction; Courant and Hilbert [10], Hörmander [26], [27, vol. II], John [29], and Bers, John, and Schechter [7] for accounts of various aspects of the classical theory; and Gårding [22], Treves [53, vol.II], Hörmander [27, vols. III and IV], and Taylor [48] for more recent developments using the machinery of pseudodifferential operators and Fourier integral operators.

A. The Cauchy Problem

The basic boundary value problem for the wave equation is the Cauchy problem. We know from our analysis in §1C that the initial hypersurface S should be non-characteristic in order to produce reasonable results. However, for $n \geq 2$ this condition is not sufficient. Indeed, recall the Hadamard example

$$u(x_1, x_2) = e^{-\sqrt{k}}(\sin kx_1)(\sinh kx_2)$$

which shows that the Cauchy problem for Δ in $\mathbb{R}^2$ on the line $x_2 = 0$ is badly behaved. (See (1.46).) If we think of u as a function on $\mathbb{R}^n \times \mathbb{R}$ ($n \geq 2$) that is independent of $x_3, \ldots, x_n$ and t, then u satisfies $\partial_t^2 u - \Delta u = 0$ and the initial conditions

$$u(x_1, 0, x_3, \ldots, x_n, t) = 0, \qquad \partial_2 u(x_1, 0, x_3, \ldots, x_n, t) = ke^{-\sqrt{k}} \sin kx_1$$

on the non-characteristic hyperplane $x_2 = 0$. As $k \to \infty$, the Cauchy data tend uniformly to zero along with all their derivatives, but u blows up for $x_2 \neq 0$.

We can generalize this example. Let S be the hyperplane $\nu' \cdot x + \nu_0 t = 0$ through the origin with normal vector

$$\nu = (\nu', \nu_0) = (\nu_1, \ldots, \nu_n, \nu_0).$$

Suppose there is a vector $(\mu_1, \ldots, \mu_n, \mu_0) = (\mu', \mu_0)$ such that

(5.1) $$f(x, t) = \exp[i(\mu' \cdot x + \mu_0 t) + \nu' \cdot x + \nu_0 t]$$

satisfies $\partial_t^2 f - \Delta f = 0$. Since $\partial_t^2 - \Delta$ is real, the imaginary part

$$\sin(\mu' \cdot x + \mu_0 t) \exp(\nu' \cdot x + \nu_0 t)$$

also satisfies the wave equation. Moreover, since the wave operator is invariant under the transformation $(x, t) \to (-x, -t)$, the even part

$$\sin(\mu' \cdot x + \mu_0 t) \sinh(\nu' \cdot x + \nu_0 t)$$

still satisfies the wave equation. Finally, for any $k > 0$,

$$u(x, t) = e^{-\sqrt{k}} \sin k(\mu' \cdot x + \mu_0 t) \sinh k(\nu' \cdot x + \nu_0 t)$$

will satisfy the wave equation with Cauchy data

$$u(x, t) = 0, \qquad \partial_\nu u(x, t) = k e^{-\sqrt{k}} \sin k(\mu' \cdot x + \mu_0 t)$$

on S. As $k \to \infty$ we obtain the same pathology as before. So when can this happen?

(5.2) Proposition.
There exists $(\mu', \mu_0) \neq (0, 0) \in \mathbb{R}^n \times \mathbb{R}$ such that (5.1) satisfies the wave equation if and only if
i. $\nu_0 = \pm\nu_1$ (i.e., the line $\nu_0 t + \nu_1 x = 0$ is characteristic), if $n = 1$;
ii. $|\nu_0| \leq |\nu'|$, if $n \geq 2$.

Proof: (5.1) satisfies the wave equation precisely when

$$(\nu_0 + i\mu_0)^2 - \sum_1^n (\nu_j + i\mu_j)^2 = 0.$$

If $n = 1$, this just means that $\nu_0 + i\mu_0 = \pm(\nu_1 + i\mu_1)$, i.e., that $\nu_0 = \pm\nu_1$ and $\mu_0 = \pm\mu_1$. If $n \geq 2$, we take real and imaginary parts:

$$\nu_0^2 - \mu_0^2 - |\nu'|^2 + |\mu'|^2 = 0, \qquad \mu_0\nu_0 - \mu' \cdot \nu' = 0.$$

In case $\nu_0 = 0$ (so in particular $|\nu_0| \leq |\nu'|$), we can choose $\mu_0 = 0$ and μ' any n-vector of length $|\nu'|$ perpendicular to ν'. If $\nu_0 \neq 0$, we have $\mu_0 = \nu_0^{-1}\mu' \cdot \nu'$, so

$$\nu_0^2 - |\nu'|^2 = \mu_0^2 - |\mu'|^2 = \nu_0^{-2}(\mu' \cdot \nu')^2 - |\mu'|^2$$
$$\leq \nu_0^{-2}|\mu'|^2|\nu'|^2 - |\mu'|^2 = \nu_0^{-2}|\mu'|^2(|\nu'|^2 - \nu_0^2).$$

This forces $\nu_0^2 - |\nu'|^2 \leq 0$, i.e., $|\nu_0| \leq |\nu'|$. On the other hand, if this condition is satisfied, we can take $\mu_0 = 1$ and μ' any vector of length $(1 + |\nu'|^2 - \nu_0^2)^{1/2}$ making an angle with ν' equal to

$$\arccos \frac{\nu_0}{|\nu'|(1 + |\nu'|^2 - \nu_0^2)^{1/2}}.$$

(This works since the argument of arccos lies in $[-1, 1]$.)   ∎

This leads to the following definition. The hypersurface S in $\mathbb{R}^n \times \mathbb{R}$ is called **space-like** if its normal vector $\nu = (\nu', \nu_0)$ satisfies $|\nu_0| > |\nu'|$ at every point of S, that is, if ν lies inside the light cone. The preceding argument suggests that for $n \geq 2$, the Cauchy problem on S will be well behaved if and only if S is space-like, and this turns out to be so. We shall restrict our attention to the most important special case, where S is the hyperplane $t = 0$, and make some remarks about more general S at the end of the section.

We remark to begin with that the wave operator (unlike the heat operator) is invariant under time reversal $(x, t) \to (x, -t)$. It therefore suffices to consider solutions in the half-space $t > 0$, as similar results may be obtained for $t < 0$ be replacing t by $-t$. Our first result is a uniqueness theorem.

(5.3) Theorem.
Suppose $u(x, t)$ is C^2 in the strip $0 \leq t \leq T$ and that $\partial_t^2 - \Delta u = 0$. Suppose also that $u = \partial_t u = 0$ on the ball

$$B = \big\{(x, 0) : |x - x_0| \leq t_0\big\}$$

in the hyperplane $t = 0$, where $x_0 \in \mathbb{R}^n$ and $0 < t_0 \leq T$. Then u vanishes in the region

$$\Omega = \big\{(x, t) : 0 \leq t \leq t_0 \text{ and } |x - x_0| \leq t_0 - t\big\}.$$

(Note that Ω is the [truncated] cone with base B and vertex (x_0, t_0), or in other words, the region in the strip $0 \leq t \leq t_0$ that is inside the backward light cone with vertex at (x_0, t_0). See Figure 5.1.)

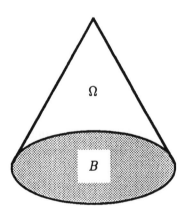

Figure 5.1. The regions in Theorem (5.3).

Proof: By considering real and imaginary parts, we may assume that u is real. For $0 \leq t \leq t_0$, let

$$B_t = \{x : |x - x_0| \leq t_0 - t\}.$$

We consider the integral

$$E(t) = \tfrac{1}{2} \int_{B_t} |\nabla_{x,t} u(x,t)|^2 \, dx = \tfrac{1}{2} \int_{B_t} \left[\left(\frac{\partial u}{\partial t} \right)^2 + \sum \left(\frac{\partial u}{\partial x_j} \right)^2 \right] \, dx,$$

which represents the energy of the wave in the region B_t at time t. The rate of change of $E(t)$ is

$$\frac{dE}{dt} = \int_{B_t} \left[\frac{\partial u}{\partial t} \frac{\partial^2 u}{\partial t^2} + \sum \frac{\partial u}{\partial x_j} \frac{\partial^2 u}{\partial x_j \partial t} \right] dx - \tfrac{1}{2} \int_{\partial B_t} |\nabla_{x,t} u|^2 \, d\sigma.$$

(The second term comes from the change in the region B_t; see Exercise 1.) Now we observe that

$$\frac{\partial}{\partial x_j} \left[\frac{\partial u}{\partial x_j} \frac{\partial u}{\partial t} \right] = \frac{\partial u}{\partial x_j} \frac{\partial^2 u}{\partial x_j \partial t} + \frac{\partial^2 u}{\partial x_j^2} \frac{\partial u}{\partial t},$$

so by the divergence theorem,

$$\int_{B_t} \left[\frac{\partial u}{\partial t} \frac{\partial^2 u}{\partial t^2} + \sum \frac{\partial u}{\partial x_j} \frac{\partial^2 u}{\partial x_j \partial t} \right] dx$$

$$= \int_{B_t} \frac{\partial u}{\partial t} \left[\frac{\partial^2 u}{\partial t^2} - \sum \frac{\partial^2 u}{\partial x_j^2} \right] dx + \int_{\partial B_t} \sum \frac{\partial u}{\partial t} \frac{\partial u}{\partial x_j} \nu_j \, d\sigma,$$

where ν is the normal to B_t in $\mathbb{R}^n$. The first integrand on the right vanishes, and for the second we have the estimate

$$\sum \frac{\partial u}{\partial t} \frac{\partial u}{\partial x_j} \nu_j \leq \left| \frac{\partial u}{\partial t} \right| \left(\sum \left| \frac{\partial u}{\partial x_j} \right|^2 \right)^{1/2}$$

$$\leq \frac{1}{2} \left(\left| \frac{\partial u}{\partial t} \right|^2 + \sum \left| \frac{\partial u}{\partial x_j} \right|^2 \right) = \tfrac{1}{2} |\nabla_{x,t} u|^2,$$

by the Schwarz inequality and the fact that $2ab \leq a^2 + b^2$. Therefore,

$$\frac{dE}{dt} = \int_{\partial B_t} \left(\sum \frac{\partial u}{\partial t} \frac{\partial u}{\partial x_j} \nu_j - \tfrac{1}{2} |\nabla_{x,t} u|^2 \right) d\sigma \leq 0.$$

But clearly $E \geq 0$, and $E(0) = 0$ because the Cauchy data vanish. Hence $E(t) = 0$, so $\nabla_{x,t} u = 0$ on $\Omega = \{(x,t) : x \in B_t\}$. But $u(x,0) = 0$, so $u = 0$ on Ω.   ∎

This is a very strong result. It shows that the value of a solution u of the wave equation at a point (x_0, t_0) depends only on the Cauchy data of u on the ball $\{x : |x - x_0| \leq t_0\}$ cut out of the initial hyperplane by the backward light cone with vertex at (x_0, t_0). (This expresses the fact that waves propagate with unit speed.) Conversely, the Cauchy data on a region R in the initial hyperplane influence only those points inside the forward light cones issuing from points of R.

Similar results hold when the hyperplane $t = 0$ is replaced by a space-like hypersurface $S = \{(x, t) : t = \phi(x)\}$. (The condition that S is space-like means precisely that $|\nabla \phi| < 1$.) Indeed, the change of variable $t' = t - \phi(x)$ transforms the Cauchy problem for the wave equation on S to the Cauchy problem for another differential equation $Lu = 0$ on the hyperplane $t' = 0$. The fact that S is space-like guarantees that L is still strictly hyperbolic, and one can apply the theory of general hyperbolic operators; cf. the references in the introduction to this chapter. See also Exercise 3 for the case where S is a hyperplane.

EXERCISES

1. Show that if f is a continuous function on $\mathbb{R}^n$ and $x \in \mathbb{R}^n$,

$$\frac{d}{dr} \int_{B_r(x)} f(y)\, dy = \int_{S_r(x)} f(y)\, d\sigma(y).$$

2. Suppose u is a C^2 solution of the wave equation in $\mathbb{R}^n \times \mathbb{R}$. Show that if $u(\cdot, t_0)$ has compact support in $\mathbb{R}^n$ for some t_0 then $u(\cdot, t)$ has compact support in $\mathbb{R}^n$ for all t, and adapt the proof of Theorem (5.3) to show that the energy integral $E = \int_{\mathbb{R}^n} |\nabla_{x,t} u|^2\, dx$ is independent of t.

3. Suppose $\nu = (\nu', \nu_0)$ is a unit vector in $\mathbb{R}^n \times \mathbb{R}$ with $|\nu'| < |\nu_0|$, so that the hyperplane $S = \{(x, t) : \nu' \cdot x + \nu_0 t = 0\}$ is space-like.
   a. Show that there is a linear transformation of $\mathbb{R}^n \times \mathbb{R}$ that maps S onto the hyperplane $t = 0$ and has the form $T = T_2 T_1$ with

   $T_1(x, t) = (Rx, t)$, where R is a rotation of $\mathbb{R}^n$;

   $T_2(x, t) = (x_1', x_2, \ldots, x_n, t')$, where

   $$x_1' = x_1 \cosh \theta + t \sinh \theta, \quad t' = x_1 \sinh \theta + t \cosh \theta \ (\theta \in \mathbb{R}).$$

   b. Use Theorem (2.1) and Exercise 2 of §2A to show that if T is as in part (a) then $(\partial_t^2 - \Delta)(u \circ T) = [(\partial_t^2 - \Delta)u] \circ T$.
   c. Conclude that the Cauchy problem for the hyperplane S can be reduced to the Cauchy problem for the hyperplane $t = 0$ by composition with the transformation T.

B. Solution of the Cauchy Problem

In this section we shall construct the solution of the Cauchy problem

$$
(5.4) \qquad
\begin{aligned}
\partial_t^2 u - \Delta u &= 0, \\
u(x,0) &= f(x), \qquad \partial_t u(x,0) = g(x).
\end{aligned}
$$

We start with the one-dimensional case, which is very simple. First, we observe that if ϕ is an arbitrary locally integrable function of one real variable, the functions $u_\pm(x,t) = \phi(x \pm t)$ satisfy the wave equation, for $\partial_t^2 u_\pm = \partial_x^2 u_\pm = \phi''(x \pm t)$. (More precisely, if ϕ is C^2 then $u_\pm$ is a classical solution of the wave equation, while if ϕ is merely locally integrable, u is a distribution solution.) Conversely, it is easy to see — at least on the formal level — that any solution of the one-dimensional wave equation is of the form $\phi(x+t) + \psi(x-t)$ where ϕ and ψ are functions of one variable. Indeed, if we make the change of variables $\xi = x + t$, $\eta = x - t$, the chain rule gives $\partial_x = \partial_\xi + \partial_\eta$ and $\partial_t = \partial_\xi - \partial_\eta$, so that $\partial_t^2 - \partial_x^2 = -4\partial_\xi \partial_\eta$, and the wave equation becomes $\partial_\xi \partial_\eta u = 0$. To solve this, we integrate in η, obtaining $\partial_\xi u = \Phi(\xi)$, where Φ is an arbitrary function, and then integrate in ξ, obtaining $u = \phi(\xi) + \psi(\eta)$ where $\phi' = \Phi$ and ψ is again arbitrary.

With this in mind, it is easy to solve the Cauchy problem (5.4) when $n = 1$. We look for a solution of the form $u(x,t) = \phi(x+t) + \psi(x-t)$. Setting $t = 0$, we must have $\phi + \psi = f$ and $\phi' - \psi' = g$. Hence $\phi' = \frac{1}{2}(f' + g)$ and $\psi' = \frac{1}{2}(f' - g)$, so

$$
\phi(x) = \tfrac{1}{2}f(x) + \tfrac{1}{2}\int_0^x g(s)\,ds + C_1, \qquad
\psi(x) = \tfrac{1}{2}f(x) - \tfrac{1}{2}\int_0^x g(s)\,ds + C_2,
$$

where $C_1 + C_2 = 0$ since $\phi + \psi = f$. Therefore,

$$
(5.5) \qquad u(x,t) = \tfrac{1}{2}[f(x+t) + f(x-t)] + \tfrac{1}{2}\int_{x-t}^{x+t} g(s)\,ds.
$$

It is a simple exercise to verify that this formula really works. To be precise, if f is C^2 and g is C^1 on $\mathbb{R}^n$, then u is C^2 on $\mathbb{R}^n \times \mathbb{R}$ and satisfies (5.4) in the classical sense; if f and g are merely locally integrable, u satisfies the wave equation in the sense of distributions and the initial conditions pointwise. In fact, one can take f and g to be arbitrary distributions on $\mathbb{R}$; $\int_{x-t}^{x+t} g(s)\,ds$ is then to be interpreted as the distribution $g * \chi_t$ where χ is the characteristic function of $[-t,t]$. We leave the details to the reader (Exercise 1) and summarize our results briefly:

(5.6) Theorem.
If $n = 1$, the solution of the Cauchy problem (5.4) is given by (5.5).

The situation in space dimensions $n > 1$ is a good deal more subtle. We shall construct the solution when n is odd by using a clever device to reduce the problem to the one-dimensional case, and then obtain the even-dimensional solutions by modifying the odd-dimensional ones. As in the one-dimensional case, we shall first proceed on the classical level, assuming that all functions in question have lots of derivatives, and then observe that the results also work in the setting of distributions.

If ϕ is a continuous function on $\mathbb{R}^n$, $x \in \mathbb{R}^n$, and $r > 0$, we define the **spherical mean** $M_\phi(x, r)$ to be the average value of ϕ on $S_r(x)$:

$$M_\phi(x, r) = \frac{1}{r^{n-1}\omega_n} \int_{|z-x|=r} \phi(z) \, d\sigma(z).$$

The substitution $z = x + ry$ turns this onto

$$(5.7) \qquad M_\phi(x, r) = \frac{1}{\omega_n} \int_{|y|=1} \phi(x + ry) \, d\sigma(y),$$

which makes sense for all $r \in \mathbb{R}$. Accordingly, we regard M_ϕ as the function on $\mathbb{R}^n \times \mathbb{R}$ defined by (5.7). As such, it is even in r, as one sees by making the substitution $y \to -y$, and it is C^k in both x and r if ϕ is C^k, as one sees by differentiating under the integral. Moreover, $M_\phi(\cdot, 0) = \phi$.

M_ϕ satisfies an interesting differential equation which may be derived as follows. If T is any rotation of $\mathbb{R}^n$, by Theorem (2.1) we have

$$\Delta_x[\phi(x + Ty)] = [\Delta\phi](x + Ty) = \Delta_y[\phi(x + Ty)],$$

where Δ_x and Δ_y denote the Laplacian acting in the variables x and y. We now average these quantities over all rotations T. Since the average of $\phi(x + Ty)$ over all rotations is $M_\phi(x, |y|)$, we obtain

$$\Delta_x M_\phi(x, |y|) = \Delta_y M_\phi(x, |y|).$$

Therefore, by Proposition (2.2),

$$(5.8) \qquad \Delta_x M_\phi(x, r) = \left[\partial_r^2 + \frac{n-1}{r}\partial_r\right] M_\phi(x, r).$$

To make this argument complete we should explain carefully what it means to average over all rotations; instead, we shall give an alternative derivation that finesses this problem.

(5.9) Proposition.
If ϕ is a C^2 function on $\mathbb{R}^n$, then M_ϕ satisfies (5.8) on $\mathbb{R}^n \times \mathbb{R}$.

Proof: It suffices to consider $r > 0$, since both sides of (5.8) are even functions of r. First, by (0.1) and the divergence theorem,

$$\partial_r M_\phi(x, r) = \frac{1}{\omega_n} \int_{|y|=1} \sum y_k \partial_{y_k} \phi(x + ry)\, d\sigma(y)$$

$$= \frac{1}{\omega_n} \int_{|y| \le 1} r\Delta\phi(x + ry)\, dy$$

$$= \frac{1}{r^{n-1}\omega_n} \int_{|z| \le r} \Delta\phi(x + z)\, dz.$$

Multiplying by r^{n-1} and expressing the integral in polar coordinates, we obtain

$$r^{n-1}\partial_r M_\phi(x, r) = \frac{1}{\omega_n} \int_0^r \int_{|y|=1} \Delta\phi(x + \rho y)\rho^{n-1}\, d\sigma(y)\, d\rho,$$

so

$$\partial_r[r^{n-1}\partial_r M_\phi(x, r)] = \frac{1}{\omega_n} \int_{|y|=1} \Delta\phi(x + ry)r^{n-1}\, d\sigma(y) = r^{n-1}\Delta_x M_\phi(x, r).$$

The desired result follows by working out the derivative on the left and dividing by r^{n-1}.   ∎

(5.10) Corollary.
Suppose $u(x, t)$ is a C^2 function on $\mathbb{R}^n \times \mathbb{R}$, and let $M_u(x, r, t)$ denote the spherical mean of the function $x \to u(x, t)$. Then u satisfies the wave equation if and only if M_u satisfies

(5.11) $$\left[\partial_r^2 + \frac{n-1}{r}\partial_r\right] M_u(x, r, t) = \partial_t^2 M_u(x, r, t)$$

for each $x \in \mathbb{R}^n$.

Proof: We have merely to apply Proposition (5.9) after observing that $\Delta_x M_u = M_{\Delta u}$ and $\partial_t^2 M_u = M_{\partial_t^2 u}$.   ∎

When n is odd, the differential equation (5.11) can be reduced to the one-dimensional wave equation by means of the following identity.

(5.12) Lemma.
If $k \geq 1$ and $\phi \in C^{k+1}(\mathbb{R})$, then

$$(5.13) \qquad \partial_r^2(r^{-1}\partial_r)^{k-1}[r^{2k-1}\phi(r)] = (r^{-1}\partial_r)^k[r^{2k}\phi'(r)].$$

Proof: It is easy to verify directly that (5.13) holds when $\phi(r) = r^m$ and hence when ϕ is any polynomial, and that both sides of (5.13) vanish at r_0 when ϕ and its derivatives of order $\leq k+1$ vanish at r_0. But by Taylor's theorem, for any r_0 we can write $\phi = P + R$ where P is a polynomial and R vanishes to order $k + 1$ at r_0, so (5.13) follows in general. ∎

The right side of (5.13) equals

$$(r^{-1}\partial_r)^{k-1}[r^{2k-1}\phi''(r) + 2kr^{2k-2}\phi'(r)],$$

so if we define the differential operator T_k by

$$T_k\phi(r) = (r^{-1}\partial_r)^{k-1}[r^{2k-1}\phi(r)],$$

(5.13) says that

$$\partial_r^2 T_k\phi = T_k[(\partial_r^2 + 2kr^{-1}\partial_r)\phi].$$

Thus, if $n = 2k + 1$, T_k intertwines the operator occurring on the left of (5.11) with ∂_r^2, so applying T_k to both sides of (5.11) converts (5.11) into the one-dimensional wave equation.

For future reference we make one more observation about the operator T_k. In $T_k\phi$ there are $2k - 1$ powers of r in the numerator and $k - 1$ in the denominator, and $k - 1$ derivatives. Expand $T_k\phi$ by the product rule: if j derivatives hit ϕ, then $k - 1 - j$ derivatives must hit the powers of r, leaving a factor of r to the power $(2k - 1) - (k - 1) - (k - 1 - j) = j + 1$. Therefore,

$$(5.14) \qquad T_k\phi(r) = \sum_0^{k-1} c_j r^{j+1}\phi^{(j)}(r), \text{ where}$$

$$c_0 r = (r^{-1}\partial_r)^{k-1}r^{2k-1} = 1 \cdot 3 \cdots (2k - 1)r.$$

At last we are in a position to solve the Cauchy problem (5.4) when the space dimension n is odd and > 1. We shall start by assuming that a solution exists and deriving a formula for it, and then we shall show that the formula always gives a solution. Suppose, then, that u satisfies (5.4), and suppose for the moment that u, f, and g are smooth — at least of class $C^{(n+3)/2}$. By Corollary (5.10), the spherical mean M_u satisfies the differential equation (5.11) with initial conditions

$$M_u(x, r, 0) = M_f(x, r), \qquad \partial_t M_u(x, r, 0) = M_g(x, r).$$

Thus, if we set

$$\tilde{u} = TM_u, \quad \tilde{f} = TM_f, \quad \tilde{g} = TM_g,$$

where

$$T(\cdot) = T_{(n-1)/2}(\cdot) = (r^{-1}\partial_r)^{(n-3)/2}[r^{n-2}(\cdot)],$$

the remarks following Lemma (5.12) show that

$$\partial_r^2 \tilde{u}(x, r, t) = \partial_t^2 \tilde{u}(x, r, t),$$
$$\tilde{u}(x, r, 0) = \tilde{f}(x, r), \qquad \partial_t \tilde{u}(x, r, 0) = \tilde{g}(x, r).$$

The solution to this problem is given by (5.5),

$$\tilde{u}(x, r, t) = \tfrac{1}{2}[\tilde{f}(x, r + t) + \tilde{f}(x, r - t)] + \tfrac{1}{2} \int_{r-t}^{r+t} \tilde{g}(x, s) \, ds,$$

so it remains to recover u from $\tilde{u}$. In principle, M_u is obtained from $\tilde{u}$ by undoing the operator T, i.e., by successive integrations, and then u is obtained from M_u by setting $r = 0$. However, one can take a shortcut by using (5.14) with $\phi(r) = M_u(x, r, t)$ and $k = \tfrac{1}{2}(n - 1)$:

$$u(x, t) = M_u(x, 0, t) = \lim_{r \to 0} \frac{\tilde{u}(x, r, t)}{c_0 r},$$

where $c_0 = 1 \cdot 3 \cdots (n - 2)$. Moreover, since M_f and M_g are even functions of r, $\tilde{f}$ and $\tilde{g}$ are odd, and $\partial_r \tilde{f}$ is even; hence, by l'Hôpital's rule,

$$
\begin{aligned}
u(x, t) &= \lim_{r \to 0} \frac{1}{2c_0 r} \left[\tilde{f}(x, r + t) + \tilde{f}(x, r - t) + \int_{r-t}^{r+t} \tilde{g}(x, s) \, ds \right] \\
&= \frac{1}{2c_0} \left[(\partial_r \tilde{f})(x, r)\big|_{r=t} + (\partial_r \tilde{f})(x, r)\big|_{r=-t} + \tilde{g}(x, t) - \tilde{g}(x, -t) \right] \\
&= \frac{1}{c_0} \left[(\partial_r \tilde{f})(x, r)\big|_{r=t} + \tilde{g}(x, t) \right].
\end{aligned}
$$

If we unravel the definitions of $\tilde{f}$, $\tilde{g}$, and c_0, we obtain the promised formula for u in terms of f and g, which we state explicitly in the following theorem.

(5.15) Theorem.

Suppose n is odd and $n \geq 3$. If $f \in C^{(n+3)/2}(\mathbb{R}^n)$ and $g \in C^{(n+1)/2}(\mathbb{R}^n)$, the function

(5.16)

$$u(x,t) = \frac{1}{1 \cdot 3 \cdots (n-2)\omega_n} \left[\partial_t (t^{-1}\partial_t)^{(n-3)/2} \left(t^{n-2} \int_{|y|=1} f(x+ty)\, d\sigma(y) \right) \right.$$

$$\left. + (t^{-1}\partial_t)^{(n-3)/2} \left(t^{n-2} \int_{|y|=1} g(x+ty)\, d\sigma(y) \right) \right]$$

solves the Cauchy problem (5.4).

Proof: Up to a constant factor, the second term on the right of (5.16) is

$$v(x,t) = (t^{-1}\partial_t)^{(n-3)/2}[t^{n-2} M_g(x,t)],$$

and by Corollary (5.10) and Lemma (5.12) we have

$$\begin{aligned}
\Delta_x v(x,t) &= (t^{-1}\partial_t)^{(n-3)/2}[t^{n-2}\Delta_x M_g(x,t)]\\
&= (t^{-1}\partial_t)^{(n-3)/2}[t^{n-2}\partial_t^2 M_g(x,t) + (n-1)t^{n-3}\partial_t M_g(x,t)]\\
&= (t^{-1}\partial_t)^{(n-1)/2}[t^{n-1}\partial_t M_g(x,t)]\\
&= \partial_t^2 (t^{-1}\partial_t)^{(n-3)/2}[t^{n-2} M_g(x,t)]\\
&= \partial_t^2 v(x,t),
\end{aligned}$$

so this term satifies the wave equation. Likewise, the function

$$w(x,t) = (t^{-1}\partial_t)^{(n-3)/2}[t^{n-2} M_f(x,t)]$$

satisfies the wave equation, and hence so does $\partial_t w(x,t)$, which is the first term on the right of (5.16). As for the initial conditions, by (5.14) we have

$$u(x,t) = \partial_t \left[t M_f(x,t) + \frac{c_1}{c_0} t^2 \partial_t M_f(x,t) + O(t^3) \right] + t M_g(x,t) + O(t^2)$$

$$= M_f(x,t) + \frac{c_0 + 2c_1}{c_0} t \partial_t M_f(x,t) + t M_g(x,t) + O(t^2).$$

Hence

$$u(x,0) = M_f(x,0) = f(x),$$

and

$$\partial_t u(x,0) = \frac{2(c_0 + c_1)}{c_0} \partial_t M_f(x,0) + M_g(x,0) = g(x),$$

because $M_f(x,t)$ is even in t and so its derivative vanishes at $t = 0$. The proof is complete. ∎

The solution of the Cauchy problem (5.4) for even n is readily derived from the solution for odd n by the "method of descent." This is just the trivial observation that if u is a solution of the wave equation in $\mathbb{R}^{n+1} \times \mathbb{R}$ that does not depend on x_{n+1}, then u satisfies the wave equation in $\mathbb{R}^n \times \mathbb{R}$. Thus to solve (5.4) in $\mathbb{R}^n \times \mathbb{R}$ with n even, we think of f and g as functions on $\mathbb{R}^{n+1}$ which are independent of x_{n+1}, write down the solution (5.16) of the Cauchy problem, and check that it does not depend on x_{n+1}. The result is as follows.

(5.17) Theorem.
Suppose n is even. If $f \in C^{(n+4)/2}(\mathbb{R}^n)$ and $g \in C^{(n+2)/2}(\mathbb{R}^n)$, the function
(5.18)

$$u(x,t) = \frac{2}{1 \cdot 3 \cdots (n-1)\omega_{n+1}} \left[\partial_t (t^{-1}\partial_t)^{(n-2)/2} \left(t^{n-1} \int_{|y| \leq 1} \frac{f(x+ty)}{\sqrt{1-|y|^2}} \, dy \right) \right.$$
$$\left. + (t^{-1}\partial_t)^{(n-2)/2} \left(t^{n-1} \int_{|y| \leq 1} \frac{g(x+ty)}{\sqrt{1-|y|^2}} \, dy \right) \right]$$

solves the Cauchy problem (5.4).

Proof: Denote a point in $\mathbb{R}^{n+1}$ by (x, x_{n+1}) where $x = (x_1, \ldots, x_n)$. If we think of f and g as functions on $\mathbb{R}^{n+1}$ that depend only on x, the solution to (5.4) in $\mathbb{R}^{n+1} \times \mathbb{R}$ is given by (5.16) with n replaced by $n+1$, which involves the integral

$$\int_{|y|^2 + y_{n+1}^2 = 1} f(x+ty) \, d\sigma(y, y_{n+1})$$

and a similar integral with f replaced by g. We parametrize the upper and lower hemispheres of the sphere $|y|^2 + y_{n+1}^2 = 1$ by $y_{n+1} = \pm\phi(y)$ where $\phi(y) = \sqrt{1 - |y|^2}$, $|y| < 1$. By a standard formula of advanced calculus, the element of surface area is given by

$$d\sigma(y, y_{n+1}) = \sqrt{1 + |\nabla\phi|^2} \, dy = \frac{dy}{\sqrt{1-|y|^2}},$$

so we obtain (5.18). (The factor of 2 is there because we integrate over both the upper and lower hemispheres.)   ∎

The formulas (5.5), (5.16), and (5.18) give the solution to the Cauchy problem (5.4) in arbitrary dimensions. We note that these formulas agree with the uniqueness theorem (5.3): for $t_0 > 0$, the value of u at (x_0, t_0) depends only on the values of f and g on the ball $|x - x_0| \leq t_0$ cut out of the initial hyperplane by the backward light cone with vertex at (x_0, t_0). Actually, when n is odd we can say more. If $n = 1$, $u(x_0, t_0)$ depends on the values of g on $|x - x_0| \leq t_0$, but on the values of f only at $x_0 = \pm t_0$. If n is odd and $n \geq 3$, $u(x_0, t_0)$ depends only on the values of f and g and their first few derivatives on the *sphere* $|x - x_0| = t_0$ — or, so to speak, on the values of f and g on an infinitesimal neighborhood of the sphere $|x - x_0| = t_0$. This fact is known as the **Huygens principle**, and we shall refer to the fact that it holds only in odd dimensions as the **Huygens phenomenon**.

Physically, the Huygens phenomenon can be understood as follows. Suppose you are in a dark room at position x_0, and someone at the origin sets off a flash bulb at time $t = 0$. At time $t_0 = |x_0|$ you see a flash of light, then darkness again. In an even-dimensional world you would see a burst of light at time t_0, but instead of disappearing promptly it would gradually fade away. This effect can be observed in two-dimensional water waves by dropping a pebble into a pond and watching the ripples.

We should say something about the role of the differentiability hypotheses in Theorems (5.15) and (5.17). The conditions on f and g in these theorems are designed to ensure that $u \in C^2$, so that u is a classical solution of the wave equation. Additional smoothness of f and g clearly implies additional smoothness of u: to be precise, if $f \in C^{k+(n-1)/2}$ and $g \in C^{k+(n-3)/2}$ (n odd) or $f \in C^{k+(n/2)}$ and $g \in C^{k-1+(n/2)}$ (n even) then $u \in C^k$. On the other hand, if we assume weaker conditions on f and g and interpret the formulas (5.16) and (5.18) properly, we obtain distribution solutions of (5.4).

In fact, for odd $n \geq 3$ let us observe that

$$\frac{1}{\omega_n} \int_{|y|=1} f(x + ty) \, d\sigma(y) = \frac{1}{\omega_n} \int_{|y|=1} f(x - ty) \, d\sigma(y) = f * \Sigma_t(x),$$

where Σ_t is the distribution

(5.19)
$$\langle \Sigma_t, \psi \rangle = \frac{1}{\omega_n} \int_{|y|=1} \psi(ty) \, d\sigma(y).$$

(That is, Σ_t is surface measure on the sphere $S_{|t|}(0)$, normalized so that the measure of $S_{|t|}(0)$ is 1 and considered as a distribution on $\mathbb{R}^n$.) The map $t \to \Sigma_t$ is a C^∞ (in fact, analytic) function of $t \in \mathbb{R}$ with values in the space of compactly supported distributions on $\mathbb{R}^n$, and hence so is

(5.20)
$$\Phi_t = \frac{1}{1 \cdot 3 \cdots (n-2)} (t^{-1} \partial_t)^{(n-3)/2} [t^{n-2} \Sigma_t].$$

(Φ_t is smooth even at $t = 0$, by (5.14).) (5.16) then says that

$$(5.21) \qquad u(\cdot, t) = f * \partial_t \Phi_t + g * \Phi_t.$$

When rewritten this way, (5.16) makes sense when f and g are arbitrary distibutions on $\mathbb{R}^n$ and defines $u(\cdot, t)$ as a C^∞ function of t with values in the space of distributions on $\mathbb{R}^n$. (In particular, u may be regarded as a distribution on $\mathbb{R}^n \times \mathbb{R}$.) A standard approximation argument (consider f and g as limits of smooth functions) shows that u satisfies (5.4) in the appropriate sense. Similarly, when n is even, (5.21) holds with

$$(5.22) \qquad \begin{aligned} \Phi_t &= \frac{1}{1 \cdot 3 \cdots (n-1)} (t^{-1} \partial_t)^{(n-2)/2} [t^{n-1} \Upsilon_t], \\ \langle \Upsilon_t, \psi \rangle &= \frac{1}{\omega_{n+1}} \int_{|y|=1} \frac{\psi(ty)}{\sqrt{1 - |y|^2}} \, dy. \end{aligned}$$

In fact, the solution for $n = 1$ also has the form (5.21) where

$$(5.23) \qquad \langle \Phi_t, \psi \rangle = \tfrac{1}{2} \int_{-t}^{t} \psi(s) \, ds.$$

Indeed, $\partial_t \int_{-t}^{t} \psi(s) \, ds = \psi(t) + \psi(-t)$, so $\partial_t \Phi_t(s) = \frac{1}{2}[\delta(s-t) + \delta(s+t)]$ when Φ_t is given by (5.23).

The loss of continuous differentiability in passing from the Cauchy data to the solution in dimensions $n \geq 2$ is unavoidable. Intuitively, it happens because "weak" singularites in the initial data at different points will propagate along light rays and collide at later times, possibly creating "stronger" singularities. The remarkable thing, however, is that this can happen only to a limited extent. That is, although for any $\epsilon > 0$ there can be a loss of roughly $\frac{1}{2}n$ continuous derivatives in passing from $u(x, 0)$ to $u(x, \epsilon)$, once these derivatives are lost there is no further loss in passing from $u(x, \epsilon)$ to $u(x, t)$ for $t > \epsilon$! Moreover, if one considers "L^2 derivatives" instead of "continuous derivatives," there is no loss at all. We shall say more about this in §5D.

Incidentally, one consequence of all this, which the reader may have noticed already, is that the wave operator is not hypoelliptic: Solutions of the homogeneous equation $\partial_t^2 u - \Delta u = 0$ can be arbitrarily rough.

EXERCISES

1. Show that if f and g are locally integrable functions on $\mathbb{R}$, the function $u(x, t)$ defined by (5.5) is a distribution solution of the one-dimensional

wave equation. In what sense do we have $\lim_{t\to 0} u(\cdot, t) = f$ and $\lim_{t\to 0} \partial_t u(\cdot, t) = g$? More generally, if f and g are distributions on $\mathbb{R}$, show how to interpret (5.5) as a smooth $\mathcal{D}'(\mathbb{R})$-valued function of t that satisfies (5.4) in an appropriate sense.

2. The differential equation (5.11), i.e., $[\partial_r^2 + (n-1)r^{-1}\partial_r]w = \partial_t^2 w$, is the n-dimensional wave equation for radial functions (Proposition (2.2)). As explained in the text, when n is odd this equation can be reduced to the one-dimensional wave equation. In this problem we consider the case $n = 3$.

  a. Show that the general radial solution of the 3-dimensional wave equation is

$$u(x, t) = r^{-1}[\phi(r + t) + \psi(r - t)] \qquad (r = |x|),$$

  where ϕ and ψ are functions on $\mathbb{R}$.

  b. Solve the Cauchy problem with radial initial data $(n = 3)$,

$$\partial_t^2 u - \Delta u = 0, \qquad u(x, 0) = f(|x|), \quad \partial_t u(x, 0) = g(|x|),$$

  in a form similar to (5.5). (Hint: Extend f and g to be even functions on $\mathbb{R}$.)

  c. Let u, f, g be as in part (b). Show that $u(0, t) = f(t) + tf'(t) + tg(t)$, so that u is generally no better than $C^{(k)}$ if $f \in C^{(k+1)}$ and $g \in C^{(k)}$.

C. The Inhomogeneous Equation

We now consider the Cauchy problem for the inhomogeneous wave equation:

(5.24)
$$\partial_t u - \Delta u = w(x, t),$$
$$u(x, 0) = f(x), \qquad \partial_t u(x, 0) = g(x),$$

which represents waves influenced by a driving force $w(x, t)$. We know how to find a solution u_1 of this problem when w is replaced by 0. If we can also find a solution u_2 when f and g are replaced by 0, then $u = u_1 + u_2$ will be a solution of (5.24). But the latter problem is easily reduced to the former one by a version of the "variation of parameters" method known as **Duhamel's principle:**

(5.25) Theorem.
Suppose $w \in C^{[n/2]+1}(\mathbb{R}^n \times \mathbb{R})$. For each $s \in \mathbb{R}$ let $v(x, t; s)$ be the solution of

$$\partial_t v - \Delta v = 0, \qquad v(x, 0; s) = 0, \qquad \partial_t v(x, 0; s) = w(x, s).$$

Then $u(x, t) = \int_0^t v(x, t - s; s)\, ds$ satisfies (5.24) with $f = g = 0$.

Proof: Clearly $u(x, 0) = 0$. Also,

$$\partial_t u(x, t) = v(x, 0; t) + \int_0^t \partial_t v(x, t - s; s)\, ds = \int_0^t \partial_t v(x, t - s; s)\, ds,$$

so $\partial_t u(x, 0) = 0$. Differentiating once more in t, we see that

$$\partial_t^2 u(x, t) = \partial_t v(x, 0; t) + \int_0^t \partial_t^2 v(x, t - s; s)\, ds$$

$$= w(x, t) + \int_0^t \Delta v(x, t - s; s)\, ds = w(x, t) + \Delta u(x, t),$$

so the proof is complete.   ∎

The problem (5.24) is really of physical interest only for $t > 0$. (If one considers $t < 0$, the Cauchy data f and g are "final conditions" rather than "initial conditions.") If one wants to consider solutions of $\partial_t^2 - \Delta u = w$ for arbitrary times, the following problem may be more natural. Suppose the system is completely at rest in the distant past, and then at some time t_0 the driving force $w(x, t)$ starts to operate. Thus, we assume that $w(x, t) = 0$ for $t \leq t_0$, and we wish to solve $\partial_t^2 u - \Delta u = w$ subject to the condition that $u(x, t) = 0$ for $t \leq t_0$. This problem reduces immediately to Theorem (5.25) if we make the change of variable $t \to t - t_0$, but we can restate the solution in a way that does not mention the starting time t_0:

(5.26) Theorem.
Suppose $w \in C^{[n/2]+1}(\mathbb{R}^n \times \mathbb{R})$, and $w(\cdot, t) = 0$ for $t \ll 0$. For each $s \in \mathbb{R}$ let $v(x, t; s)$ be the solution of

$$\partial_t v - \Delta v = 0, \qquad v(x, 0; s) = 0, \qquad \partial_t v(x, 0; s) = w(x, s).$$

(Thus $v(\cdot, \cdot; s) = 0$ for $s \ll 0$.) Then $u(x, t) = \int_{-\infty}^t v(x, t - s; s)\, ds$ satisfies

$$\partial_t^2 u - \Delta u = w, \qquad u(\cdot, t) = 0 \text{ for } t \ll 0.$$

Proof: Simply repeat the proof of Theorem (5.25).   ∎

We can rewrite the solution u in Theorem (5.26) as follows. By the discussion at the end of §5B, the function v in Theorem (5.26) is given by

$$v(\cdot, t; s) = w(\cdot, s) *_x \Phi_t,$$

where Φ_t is given by (5.19) and (5.20), (5.22), or (5.23) depending on n, and $*_x$ denotes convolution with respect to the space variables. But then

$$u(\cdot, t) = \int_{-\infty}^{t} w(\cdot, s) *_x \Phi_{t-s}\, ds,$$

which is a convolution integral in t if we replace Φ_s by 0 for $s < 0$. In other words, if we define the distibution Φ_+ on $\mathbb{R}^n \times \mathbb{R}$ by

$$\Phi_+(\cdot, t) = \begin{cases} \Phi_t(\cdot) & \text{if } t \geq 0, \\ 0 & \text{if } t < 0, \end{cases}$$

that is,

(5.27) $\langle \Phi_+, \psi \rangle = \int_0^\infty \langle \Phi_t, \psi(\cdot, t) \rangle\, dt \qquad (\phi \in C_c^\infty(\mathbb{R}^n \times \mathbb{R})),$

we have

$$u = w * \Phi_+ \qquad \text{(convolution on } \mathbb{R}^n \times \mathbb{R}).$$

This holds in particular for any $w \in C_c^\infty(\mathbb{R}^n \times \mathbb{R})$, which means that Φ_+ is a fundamental solution for the wave operator. In short, we have:

(5.28) Theorem.
Define the distribution Φ_t on $\mathbb{R}^n$ by (5.19) and (5.20) if n is odd and $n \geq 3$, by (5.22) if n is even, and by (5.23) if $n = 1$, and define the distribution Φ_+ on $\mathbb{R}^n \times \mathbb{R}$ by (5.27). Then Φ_+ is a fundamental solution for the wave operator.

The solution of the inhomogeneous equation in Theorem (5.25) is not given by a convolution integral, since the imposition of the conditions at $t = 0$ breaks the translation invariance in t. We shall investigate Φ_+ further in the next section.

EXERCISES

1. Let u be the solution of $\partial_t^2 u - \Delta u = w$ given by Theorem (5.25). For each (x_0, t_0), determine the region of $\mathbb{R}^n \times \mathbb{R}$ in which the values of w influence the value of u at (x_0, t_0). Do the same for the solution u in Theorem (5.26).

2. Let L be a constant-coefficient differential operator on $\mathbb{R}^n$. Suppose $\{F_t\}_{t>0}$ is a family of distributions depending smoothly on t such that

$$\partial_t^m F_t = L F_t \quad (t > 0), \qquad \lim_{t \to 0} \partial_t^j F_t = \begin{cases} 0 & (j < m - 1), \\ \delta & (j = m - 1), \end{cases}$$

   where δ denotes the point mass at the origin.
   a. Show that if $g \in C_c^\infty(\mathbb{R}^n)$, $u(x, t) = g * F_t(x)$ solves the Cauchy problem

$$\partial_t^m u = Lu \quad (t > 0), \qquad \partial_t^j u(x, 0) = \begin{cases} 0 & (j < m - 1), \\ g(x) & (j = m - 1). \end{cases}$$

   b. Define the distribution F_+ on $\mathbb{R}^n \times \mathbb{R}$ by

$$\langle F_+, \psi \rangle = \int_0^\infty \langle F_t, \psi(\cdot, t) \rangle \, dt.$$

   Show that F_+ is a fundamental solution for $\partial_t^m - L$. (Note that the fundamental solutions of the heat and wave equations given by (4.5) and (5.27) are of this form.)

D. Fourier Analysis of the Wave Operator

We now re-examine the problems solved in the previous two sections by using the Fourier transform. To begin with, we consider the Cauchy problem (5.4) for the homogeneous wave equation. Application of the Fourier transform in the space variables converts this into

$$\partial_t^2 \hat{u} + 4\pi^2 |\xi|^2 \hat{u} = 0, \qquad \hat{u}(\xi, 0) = \hat{f}(\xi), \qquad \partial_t \hat{u}(\xi, t) = \hat{g}(\xi),$$

where $\hat{u}(\cdot, t)$ denotes is the Fourier transform of $u(\cdot, t)$ for each t. The general solution of this ordinary differential equation is a linear combination of $\cos 2\pi|\xi|t$ and $\sin 2\pi|\xi|t$; taking the initial conditions into account, we see that

(5.29) $$\hat{u}(\xi, t) = \hat{f}(\xi) \cos 2\pi|\xi|t + \hat{g}(\xi) \frac{\sin 2\pi|\xi|t}{2\pi|\xi|}.$$

Therefore,

$$u(\cdot, t) = f * \Psi_t + g * \Phi_t,$$

where Ψ and Φ are the inverse Fourier transforms of the functions

$$\widehat{\Phi}_t(\xi) = \frac{\sin 2\pi|\xi|t}{2\pi|\xi|} \qquad \widehat{\Psi}_t(\xi) = \cos 2\pi|\xi|t = \partial_t \widehat{\Phi}_t(\xi).$$

Direct evaluation of these inverse Fourier transforms is not easy, because although $\widehat{\Psi}_t$ and $\widehat{\Phi}_t$ are perfectly nice tempered distributions, they are not L^1 functions. Of course, we already know the answer: if we compare (5.29) with (5.21), we see that Φ_t is the distribution defined by (5.19) and (5.20) when n is odd and ≥ 3, by (5.22) when n is even, and by (5.23) when $n = 1$; and $\Psi_t = \partial_t \Phi_t$. When $n = 1$, the direct calculation of $\widehat{\Phi}_t$ from Φ_t is a triviality:

$$\widehat{\Phi}_t(\xi) = \tfrac{1}{2} \int_{-t}^{t} e^{-2\pi i \xi x} \, dx = \frac{e^{-2\pi i \xi t} - e^{2\pi i \xi t}}{-4\pi i \xi}$$

$$= \frac{\sin 2\pi \xi t}{2\pi \xi} = \frac{\sin 2\pi|\xi|t}{2\pi|\xi|}.$$

The corresponding calculation for $n = 3$, where Φ_t is a scalar multiple of surface measure on $S_{|t|}(0)$, is also quite easy; see Exercise 1. For $n \neq 1, 3$, however, going from Φ_t to $\widehat{\Phi}_t$ is rather arduous.

It is, however, possible to apply the Fourier inversion formula to $\widehat{\Phi}_t$ in a way that readily displays some of the most important qualitative features of Φ_t, although not the exact nature of its singularities on the sphere $S_{|t|}(0)$. Namely, we consider

$$\widehat{\Phi}_t^{\epsilon}(\xi) = e^{-2\pi\epsilon|\xi|} \widehat{\Phi}_t(\xi) = e^{-2\pi\epsilon|\xi|} \frac{\sin 2\pi|\xi|t}{2\pi|\xi|} \qquad (\epsilon > 0).$$

Clearly $\widehat{\Phi}_t^{\epsilon} \to \widehat{\Phi}_t$ uniformly as $\epsilon \to 0$, so that Φ_t will be the limit in the topology of tempered distributions of Φ_t^{ϵ}, the inverse Fourier transform of $\widehat{\Phi}_t^{\epsilon}$. Moreover, $\widehat{\Phi}_t^{\epsilon} \in L^1(\mathbb{R}^n)$, so we can calculate its inverse Fourier transform as an ordinary integral. In fact, we have

$$\widehat{\Phi}_t^{\epsilon}(\xi) = \frac{e^{-2\pi(\epsilon-it)|\xi|} - e^{-2\pi(\epsilon+it)|\xi|}}{4\pi i|\xi|} = \frac{1}{2i} \int_{\epsilon-it}^{\epsilon+it} e^{-2\pi s|\xi|} \, ds,$$

so

$$\Phi_t^{\epsilon}(x) = \frac{1}{2i} \int_{\epsilon-it}^{\epsilon+it} \int_{\mathbb{R}^n} e^{2\pi i \xi \cdot x} e^{-2\pi s|\xi|} \, d\xi \, ds.$$

Now, the inner integral is the inverse Fourier transform of $e^{-2\pi s|\xi|}$, which we calculated in Theorem (4.15):

$$\int_{\mathbb{R}^n} e^{2\pi i \xi \cdot x} e^{-2\pi s|\xi|}\, d\xi = \frac{\Gamma((n+1)/2)}{\pi^{(n+1)/2}} \frac{s}{(s^2 + |x|^2)^{(n+1)/2}}.$$

Actually, Theorem (4.15) pertains to the case $s > 0$, but this formula remains valid for all complex s with $\mathrm{Re}\, s > 0$ provided that $(s^2 + |x|^2)^{1/2}$ is taken to be the square root of $s^2 + |x|^2$ with positive real part. (One can either repeat the proof of Theorem (4.15) with s complex, or argue that both sides of the formula are analytic functions in the half-plane $\mathrm{Re}\, s > 0$ that agree for $s > 0$ and hence everywhere.) Therefore, if $n > 1$,

(5.30)
$$\Phi_t^\epsilon(x) = \frac{\Gamma((n+1)/2)}{2i\pi^{(n+1)/2}} \int_{\epsilon-it}^{\epsilon+it} \frac{s}{(s^2 + |x|^2)^{(n+1)/2}}\, ds$$

$$= \frac{\Gamma((n+1)/2)}{2i(1-n)\pi^{(n+1)/2}} \left[\frac{1}{[(\epsilon+it)^2 + |x|^2]^{(n-1)/2}} - \frac{1}{[(\epsilon-it)^2 + |x|^2]^{(n-1)/2}} \right].$$

(The corresponding calculation for $n = 1$ is left to the reader; see Exercise 2.)

Now we let $\epsilon \to 0$. Clearly $\Phi_t^\epsilon \to 0$ uniformly on compact subsets of $\{x : |x| > |t|\}$, so we recover the fact that $\mathrm{supp}\, \Phi_t \subset \{x : |x| \le |t|\}$. This in turn is equivalent to the facts about the localized dependence of the solution of the Cauchy problem on the initial data that we originally derived from Theorem (5.3). Moreover, if n is odd (and > 1), so that $(n-1)/2$ is an integer, $\Phi_t^\epsilon \to 0$ uniformly on compact subsets of $\{x : |x| < |t|\}$ also. Hence $\mathrm{supp}\, \Phi_t \subset \{x : |x| = |t|\}$, which is a restatement of the Huygens principle. However, for $|x| < |t|$ the quantities $(\epsilon \pm it)^2 + |x|^2$ approach the negative real axis (the branch cut for the square root) from opposite sides, so if n is even the two terms on the right of (5.30) do not cancel out but add up, with the result that Φ_t agrees on the region $\{x : |x| < |t|\}$ with the function

$$\frac{(-1)^{(n/2)-1}\Gamma((n+1)/2)\,\mathrm{sgn}\, t}{(n-1)\pi^{(n+1)/2}} \frac{1}{(t^2 - |x|^2)^{(n-1)/2}}.$$

On the other hand, (5.22) implies that Φ_t agrees on this region with the function

$$\frac{2\,\mathrm{sgn}\, t}{1 \cdot 3 \cdots (n-1)\omega_{n+1}} (t^{-1}\partial_t)^{(n/2)-1}(t^2 - |x|^2)^{-1/2},$$

and it is an elementary exercise to see that these two functions are the same.

One of the main advantages of the Fourier representation (5.29) is that it yields easy answers to questions about L^2 norms. For example, it is obvious from (5.29) and the Plancherel theorem that if $f, g \in L^2(\mathbb{R}^n)$ then $u(\cdot, t) \in L^2(\mathbb{R}^n)$ and $\|u(\cdot, t)\|_2$ is bounded independently of t, so that $u \in L^2(\mathbb{R}^n \times [t_0, t_1])$ whenever $-\infty < t_0 < t_1 < \infty$. (See also Exercise 3.) This result can be refined to take account of smoothness conditions. To do so, we shall have to get ahead of our story a little and introduce an idea that will be explained more fully in Chapter 6.

If k is a positive integer and Ω is an open set in $\mathbb{R}^n$ (or $\mathbb{R}^n \times \mathbb{R}$), we denote by $H_k(\Omega)$ the space of all $f \in L^2(\Omega)$ whose distribution derivatives $\partial^\alpha f$ also belong to $L^2(\Omega)$ for $|\alpha| \le k$. If $\Omega = \mathbb{R}^n$, by the formula $(\partial^\alpha f)\widehat{\ }(\xi) = (2\pi i \xi)^\alpha \widehat{f}(\xi)$ and the Plancherel theorem we see that $f \in H_k(\mathbb{R}^n)$ if and only if $\xi^\alpha \widehat{f} \in L^2$ for $|\alpha| \le k$, or equivalently, $(1 + |\xi|)^k \widehat{f} \in L^2$.

Now, from (5.29) we have

$$(\partial_x^\alpha \partial_t^j u)\widehat{\ }(\xi, t) = (2\pi i \xi)^\alpha (2\pi |\xi|)^j \widehat{f}(\xi) \operatorname{trig} 2\pi |\xi| t$$
$$+ (2\pi i \xi)^\alpha (2\pi |\xi|)^{j-1} \widehat{g}(\xi) \operatorname{trig} 2\pi |\xi| t,$$

where trig denotes one of the functions $\pm \cos$ or $\pm \sin$. From this it is clear that if $f \in H_k(\mathbb{R}^n)$ and $g \in H_{k-1}(\mathbb{R}^n)$ then $\partial_x^\alpha \partial_t^j u(\cdot, t)$ belongs to $L^2(\mathbb{R}^n)$ with L^2 norm bounded independent of t for $|\alpha| + j \le k$. An integration over a finite t-interval then yields the following result, which shows that, unlike the situation with continuous derivatives, solving the wave equation preserves L^2 derivatives.

(5.31) Theorem.
Let u be the solution of the Cauchy problem

$$\partial_t^2 u - \Delta u = 0, \qquad u(x, 0) = f(x), \qquad \partial_t u(x, 0) = g(x).$$

If $f \in H_k(\mathbb{R}^n)$ and $g \in H_{k-1}(\mathbb{R}^n)$ then $u \in H_k(\mathbb{R}^n \times [t_0, t_1])$ whenever $-\infty < t_0 < t_1 < \infty$.

There are also various boundedness theorems for solutions of the wave equation in terms of L^p norms with $p \ne 2$, but these are mostly quite recent and depend on some deep results of Fourier analysis. In particular, Peral [39] has proved an L^p analogue of Theorem (5.31) (but with a small loss of smoothness, depending on p) for $\left|\frac{1}{p} - \frac{1}{2}\right| \le \frac{1}{n-1}$. See also Stein [46, §VIII.5], and the references given there.

We now turn to the question of finding a fundamental solution for the wave equation by Fourier analysis. Here we wish to solve the equation

$$\partial_t^2 u(x, t) - \Delta u(x, t) = \delta(x, t) = \delta(x)\delta(t).$$

Application of the Fourier transform in x turns this into

$$\partial_t^2 \widehat{u}(\xi, t) + 4\pi^2 |\xi|^2 \widehat{u}(\xi, t) = \delta(t).$$

The recipe for solving this ordinary differential equation is as follows. $\widehat{u}$ must be a solution of the homogeneous equation $\partial_t^2 \widehat{u} + 4\pi^2 |\xi|^2 \widehat{u} = 0$ for $t < 0$ and for $t > 0$, so

$$\widehat{u}(\xi, t) = \begin{cases} a(\xi) \cos 2\pi |\xi| t + b(\xi) \sin 2\pi |\xi| t & (t < 0), \\ c(\xi) \cos 2\pi |\xi| t + d(\xi) \sin 2\pi |\xi| t & (t > 0). \end{cases}$$

To obtain the delta function at $t = 0$ on applying ∂_t^2 to $\widehat{u}$, we require that u should be continuous at $t = 0$ but that $\partial_t u$ should have a jump of size 1 at $t = 0$:

$$a(\xi) = c(\xi), \qquad (2\pi |\xi|)[d(\xi) - b(\xi)] = 1.$$

This gives two equations for the four unknowns a, b, c, d; the remaining degrees of freedom can be used to satisfy side conditions. If we think if $u(x, t)$ as the response of a system at rest to a sudden jolt at $x = 0$, $t = 0$, it is reasonable to require $u(x, t) = 0$ for $t < 0$. Hence, we take $a = b = 0$, which yields $c = 0$ and $d = (2\pi |\xi|)^{-1}$. In short,

$$(5.32) \qquad \widehat{u}(\xi, t) = \begin{cases} (2\pi |\xi|)^{-1} \sin 2\pi |\xi| t = \widehat{\Phi}_t(\xi) & (t > 0), \\ 0 & (t < 0), \end{cases}$$

so that u is the fundamental solution Φ_+ given by Theorem (5.28).

It is of interest to compute the full Fourier transform of Φ_+ in both x and t. We expect to get something like $[4\pi^2(|\xi|^2 - \tau^2)]^{-1}$, but the latter function is not locally integrable near points of the light cone and so will need to be interpreted suitably as a temepered distribution. In fact, we can compute the Fourier transform in t of (5.32) by the same device that we used to compute the inverse Fourier transform in x of $\widehat{\Phi}_t$. Namely, consider

$$\widehat{\Phi}_+^\epsilon(\xi, t) = e^{-2\pi\epsilon t} \widehat{\Phi}_+(\xi, t) = \chi_{[0,\infty)}(t) \frac{e^{2\pi i t(|\xi| + i\epsilon)} - e^{2\pi i t(-|\xi| + i\epsilon)}}{4\pi i |\xi|}.$$

The t-Fourier transform of this is

$$\int e^{-2\pi i \tau t} \widehat{\Phi}_+^\epsilon(\xi, t)\, dt = \int_0^\infty \frac{e^{2\pi i t(|\xi| - \tau + i\epsilon)} - e^{2\pi i t(-|\xi| - \tau + i\epsilon)}}{4\pi i |\xi|}\, dt$$

$$= \frac{1}{8\pi^2 |\xi|} \left[\frac{1}{|\xi| - \tau + i\epsilon} + \frac{1}{|\xi| + \tau - i\epsilon} \right]$$

$$= \frac{1}{4\pi^2 [|\xi|^2 - (\tau - i\epsilon)^2]}.$$

Clearly $\widehat{\Phi}_+^\epsilon \to \widehat{\Phi}_+$ uniformly as $\epsilon \to 0$, so the full Fourier transform of Φ_+ is the limit in the topology of tempered distributions of the functions $[(4\pi^2)(|\xi|^2 - (\tau - i\epsilon)^2)]^{-1}$.

The fundamental solution Φ_+ has the property that the solution of $u = w * \Phi_+$ of the inhomogeneous equation $\partial_t^2 u - \Delta u = w$ at time $t = t_0$ depends only on the values of w at times $t \leq t_0$, and it is uniquely determined by this property. It is therefore the natural fundamental solution to use when u is considered as the response of a system to an driving force w. However, there are other fundamental solutions for the wave operator that are of significance in physics. One is the time reversal of Φ_+,

$$\Phi_-(x,t) = \Phi_+(x,-t),$$

whose full Fourier transform is the distribution limit of the functions

$$\mathcal{F}\Phi_-^\epsilon(\xi,\tau) = \frac{1}{4\pi^2[|\xi|^2 - (\tau + i\epsilon)^2]}.$$

(Φ_+ and Φ_- are often called the **retarded** and **advanced Green's functions** in the physics literature.) Another is the **causal Green's function** or **Feynman propagator** Φ_c, the distribution whose full Fourier transform is the distribution limit of the functions

$$\mathcal{F}\Phi_c^\epsilon(\xi,\tau) = \frac{1}{4\pi^2(|\xi|^2 - \tau^2 - i\epsilon)}$$

as $\epsilon \to 0+$. The reader may find an explanation of the role of these Green's functions in quantum field theory in Bogoliubov and Shirkov [8]; here we merely wish to point out that there are a number of natural but quite different fundamental solutions of the wave equation. (In other words, there are a number of natural but quite different distributions that agree with the function $[4\pi^2(|\xi|^2 - \tau^2)]^{-1}$ away from the light cone!)

We refer the reader to Treves [52] for further discussion of the properties of Φ_+.

EXERCISES

1. When $n = 3$, the distribution Φ_t defined by (4.19) and (4.20) is $(4\pi|t|)^{-1}\sigma_{|t|}$, where σ_r is surface measure on $S_r(0)$. Show by a direct calculation that $\widehat{\Phi}_t(\xi) = (2\pi|\xi|)^{-1}\sin 2\pi|\xi|t$. (Hint: Since Φ_t is radial, so is $\widehat{\Phi}_t$; hence it suffices to consider $\xi = (0,0,\rho)$ with $\rho > 0$. Integrate in spherical coordinates.)

2. Perform the calculation (5.30) when $n = 1$ to obtain the formula (5.22) for Φ_t in the one-dimensional case.

3. Use the Fourier representation (5.29) to rederive the result of Exercise 2, §5A. That is, suppose u, f, and g are related by (5.29) and that $f \in H_1(\mathbb{R}^n)$ and $g \in L^2(\mathbb{R}^n)$. Show that for all $t \in \mathbb{R}$,

$$\int |\nabla_{x,t} u(x,t)|^2 \, dx = \int |\nabla_{x,t} u(x,0)|^2 \, dx = \int (|\nabla f(x)|^2 + |g(x)|^2) \, dx.$$

E. The Wave Equation in Bounded Domains

When solving the wave equation in the region $\Omega \times (0, \infty)$, where Ω is a bounded domain in $\mathbb{R}^n$, it is appropriate to specify not only the Cauchy data on $\Omega \times \{0\}$ but also some condition on $\partial\Omega \times (0, \infty)$ to tell the wave what to do when it hits the boundary. The most commonly used conditions are those of Dirichlet and Neumann, i.e., $u = 0$ or $\partial_\nu u = 0$. For either of these the solution is unique, as may be proved by the same method as Theorem (5.3); see Exercise 1.

If the boundary condition on $\partial\Omega \times (0, \infty)$ is independent of t, the method of separation of variables can be used, just as for the heat equation (§4C). That is, we can look for solutions u in the form $u(x,t) = F(x)G(t)$ where f satisfies the boundary condition on $\partial\Omega$. The wave equation $\partial_t^2 u - \Delta u = 0$ is equivalent to the equation $(\Delta F)/F = G''/G$, and since $(\Delta F)/F$ depends only on x and G''/G depends only on t, these quantities must be equal to some constant which we write as $-\lambda^2$ in the hope that it will be real and negative. If we can solve the equation $\Delta F + \lambda^2 F = 0$ on Ω for some $\lambda > 0$ subject to the given boundary condition on $\partial\Omega$, we obtain solutions of the wave equation

$$u(x,t) = F(x)(a \cos \lambda t + b \sin \lambda t)$$

where a and b are constants that can be adjusted to satisfy initial conditions. This represents a **normal mode** of vibration with frequency λ, and with luck we will be able to obtain the general solution as a superposition of normal modes.

In particular, consider the Dirichlet boundary condition $u(x,t) = 0$ for $x \in \partial\Omega$, and assume that $\partial\Omega$ is smooth. We shall see in §7G that $L^2(\Omega)$ admits an orthonormal basis $\{F_j\}$ consisting of eigenfunctions for Δ on Ω with negative eigenvalues $\{-\lambda_j^2\}$ and satisfying $F_j = 0$ on $\partial\Omega$. Hence we can solve the problem

$$\partial_t^2 u - \Delta u = 0 \text{ on } \Omega \times (0, \infty),$$

$$u(x,0) = f(x), \qquad \partial_t u(x,0) = g(x), \qquad u(x,t) = 0 \text{ for } x \in \partial\Omega,$$

by taking

$$u(x,t) = \sum F_j(x)(a_j \cos \lambda_j t + b_j \sin \lambda_j t)$$

where

$$\sum a_j F_j = f, \qquad \sum \lambda_j b_j F_j = g,$$

that is,

$$a_j = \langle f \mid F_j \rangle, \qquad b_j = \lambda_j^{-1} \langle g \mid F_j \rangle.$$

If Ω is a ball, we have already seen how to do this in §2H.

For $n = 1$ and $\Omega = (0, l)$, the problem is

(5.33)
$$\partial_t u - \Delta u = 0 \text{ on } (0,l) \times (0,\infty),$$

$$u(x,0) = f(x), \quad \partial_t u(x,0) = g(x), \quad u(0,t) = u(l,t) = 0.$$

The normalized eigenfunctions are $F_j(x) = \sqrt{2/l} \sin(j\pi x/l)$ and the associated frequencies λ_j are the integer multiples of the fundamental frequency π/l, and we obtain the familiar expansion in "harmonics" of the vibration of a string fixed at both ends:

(5.34)
$$u(x,t) = \sum \left[a_j \cos \frac{j\pi t}{l} + b_j \sin \frac{j\pi t}{l} \right] \sin \frac{j\pi x}{l},$$

$$a_j = \frac{2}{l} \int_0^l f(x) \sin \frac{j\pi x}{l} \, dx, \qquad b_j = \frac{2}{\pi j} \int_0^l g(x) \sin \frac{j\pi x}{l} \, dx.$$

For $n = 2$ and $\Omega = B_R(0)$, corresponding to a vibrating circular membrane fixed at the boundary, the numbers λ_j are zeros of Bessel functions. These numbers are not rational multiples of each other and so do not "harmonize" particularly well. For this reason a violin has better tone quality than a drum.

For more about vibrations and eigenvalue problems, see, e.g., Courant and Hilbert [10, vol. I] or Folland [17].

EXERCISES

1. Suppose Ω is a smoothly bounded domain in $\mathbb{R}^n$, and suppose u is a C^2 function satisfying $\partial_t^2 u - \Delta u = 0$ in $\Omega \times (0,\infty)$ and either $u = 0$ or $\partial_\nu u = 0$ on $\partial\Omega \times (0,\infty)$. Adapt the proof of Theorem (5.3) to show that $E(t) = \int_\Omega |\nabla_{x,t} u|^2 \, dx$ is independent of t. Conclude that if $u(x,0) = \partial_t u(x,0) = 0$ then $u \equiv 0$.

2. Suppose f and g are continuous functions on $[0, l]$ that vanish at both endpoints. Let $\tilde{f}$ and $\tilde{g}$ be the functions on $\mathbb{R}$ obtained by extending f and g to be odd functions on $[-l, l]$ and then extending them to be periodic functions on $\mathbb{R}$ with period $2l$, and let u be the solution of

$$\partial_t^2 u - \Delta u = 0, \qquad u(x, 0) = \tilde{f}(x), \qquad \partial_t u(x, 0) = \tilde{g}(x)$$

on $\mathbb{R} \times \mathbb{R}$ given by (5.5). Show that the restriction of u to $[0, l] \times [0, \infty)$ is the solution of (5.33). (This form of the solution tells you what you see when you look at a vibrating string, but (5.34) tells you what you hear when you listen to it.)

3. Solve the wave equation $\partial_t^2 u - \Delta u = 0$ on $(0, l) \times (0, \infty)$ with initial conditions $u(x, 0) = f(x)$, $\partial_t u(x, 0) = g(x)$, subject to the boundary conditions $u(0, t) = \partial_x u(l, t) = 0$. In particular, show that the frequencies are the odd integer multiples of the fundamental frequency $\pi/2l$. (This problem models vibrations of air in a cylindrical pipe that is open at one end and closed at the other, like a clarinet; u represents the change in air pressure relative to the ambient pressure. On the other hand, (5.33) models vibrations of air in a cylindrical pipe that is open at both ends, like a flute.)

4. Let $\Omega = \{x \in \mathbb{R}^3 : |x| < l\}$. Use separation of variables to solve the wave equation for radial functions in $\Omega \times (0, \infty)$ with Dirichlet boundary conditions. That is, by Proposition (2.2), the problem is to solve

$$\partial_t^2 u(r, t) - \partial_r^2 u(r, t) - 2r^{-1}\partial_r u(r, t) = 0 \text{ for } 0 \le r < l, \ t > 0,$$
$$u(r, 0) = f(r), \quad \partial_t u(r, 0) = g(r), \quad u(l, t) = 0.$$

(Hint: The eigenvalue problem to be solved is $F''(r) + 2r^{-1}F'(r) + \lambda^2 F(r) = 0$, with $F(l) = 0$ and $F(0)$ finite. Reduce this to a more familiar problem by setting $F(r) = r^{-1}G(r)$.) In particular, show that the frequencies are the integer multiples of the fundamental frequency π/l. (The restriction of u to a narrow conical region, say $\{x : \phi(x) < \epsilon\}$ where $\phi(x)$ is the angle from x to the north pole, is a model for vibrations of air in a conical pipe like an oboe, bassoon, or saxophone; cf. Exercise 3.)

F. The Radon Transform

To conclude this chapter we present an elegant method for solving the Cauchy problem (5.4) based on the "Radon transform." We begin by observing that if the Cauchy data depend only on x_1 and not on $x_2, \ldots, x_n$, then the solution u will also be independent of $x_2, \ldots, x_n$, and will be expressed by the one-dimensional formula (5.5):

$$u(x,t) = \tfrac{1}{2}[f(x_1 + t) + f(x_1 - t)] + \tfrac{1}{2} \int_{x_1-t}^{x_1+t} g(s)\,ds.$$

(This is clearly *a* solution; by uniqueness, it is *the* solution.) By performing a rotation in $\mathbb{R}^n$, we see more generally that if f and g are constant on all hyperplanes with a fixed normal vector ω, say $f(x) = F(x \cdot \omega)$ and $g(x) = G(x \cdot \omega)$, then u will have the same property and will be given by

$$u(x,t) = \tfrac{1}{2}[F(x \cdot \omega + t) + F(x \cdot \omega - t)] + \tfrac{1}{2} \int_{x \cdot \omega - t}^{x \cdot \omega + t} G(s)\,ds.$$

In this case u is called a **plane wave** in the direction ω.

The idea is to decompose "arbitrary" functions f and g into sums (or integrals) of functions depending only on $x \cdot \omega$ as ω varies over the unit sphere, and then to express u as the corresponding sum of plane waves. In order for the integrals that arise to be classically convergent, it is necessary to assume that f and g have some degree of smoothness and that they decay reasonably fast at infinity. We shall avoid all technical complications by assuming that f and g are in the Schwartz class $\mathcal{S}$.

Let us denote by S_n the unit sphere in $\mathbb{R}^n$:

$$S_n = \{x \in \mathbb{R}^n : |x| = 1\}.$$

If $f \in \mathcal{S}(\mathbb{R}^n)$, its **Radon transform** Rf is the function on $\mathbb{R} \times S_n$ defined by

$$Rf(s,\omega) = \int_{x \cdot \omega = s} f(x)\,dx,$$

where dx denotes $(n-1)$-dimensional Lebesgue measure on the hyperplane $x \cdot \omega = s$. Clearly $Rf \in C^\infty(\mathbb{R} \times S_n)$ and $Rf(\cdot,\omega) \in \mathcal{S}(\mathbb{R})$ for each $\omega \in S_n$. Moreover, $Rf(s,\omega) = Rf(-s, -\omega)$.

R is closely related to the Fourier transform. In what follows, we shall denote by $\widehat{Rf}$ the Fourier transform of Rf with respect to its first variable.

(5.35) Proposition.
If $f \in \mathcal{S}$, $\rho \in \mathbb{R}$, and $\omega \in S_n$, then $\widehat{Rf}(\rho, \omega) = \hat{f}(\rho\omega)$.

Proof: We have

$$\hat{f}(\rho\omega) = \int_{\mathbb{R}^n} e^{-2\pi i \rho\omega \cdot x} f(x)\,dx = \int_{\mathbb{R}} \int_{x\cdot\omega=s} e^{-2\pi i \rho s} f(x)\,dx\,ds$$

$$= \int_{\mathbb{R}} e^{-2\pi i \rho s} Rf(s,\omega)\,ds = \widehat{Rf}(\rho,\omega). \qquad \blacksquare$$

(5.36) Corollary.
If $f \in \mathcal{S}$,

$$Rf(s,\omega) = \int e^{2\pi i s \rho} \hat{f}(\rho\omega)\,d\rho.$$

Proof: Apply the Fourier inversion theorem to $\widehat{Rf}$. $\qquad \blacksquare$

Proposition (5.35), together with the Fourier inversion theorem and the formula for integration in polar coordinates, yields an inversion formula for the Radon transform. Indeed,

$$f(x) = \int e^{2\pi i x \cdot \xi} \hat{f}(\xi)\,d\xi = \int_{S_n} \int_0^\infty e^{2\pi i \rho x \cdot \omega} \hat{f}(\rho\omega)\rho^{n-1}\,d\rho\,d\sigma(\omega)$$

$$= \int_{S_n} \int_0^\infty e^{2\pi i \rho x \cdot \omega} \widehat{Rf}(\rho,\omega)\rho^{n-1}\,d\rho\,d\sigma(\omega)$$

$$= \int_{S_n} h(x\cdot\omega, \omega)\,d\sigma(\omega),$$

where

$$h(s,\omega) = \int_0^\infty e^{2\pi i \rho s} \widehat{Rf}(\rho,\omega)\rho^{n-1}\,d\rho.$$

But since $h(x\cdot\omega, \omega)$ is integrated over the entire unit sphere, the result is the same if we take only its even part in ω:

(5.37) $$f(x) = \tfrac{1}{2}\int_{S_n} [h(x\cdot\omega, \omega) + h(-x\cdot\omega, -\omega)]\,d\sigma(\omega).$$

Now clearly $\widehat{Rf}(\rho,\omega) = \widehat{Rf}(-\rho,-\omega)$ by Proposition (5.35), so

$$\tfrac{1}{2}[h(s,\omega) + h(-s,-\omega)]$$

$$= \tfrac{1}{2}\int_0^\infty e^{2\pi i \rho s} \widehat{Rf}(\rho,\omega)\rho^{n-1}\,d\rho + \tfrac{1}{2}\int_0^\infty e^{-2\pi i \rho s} \widehat{Rf}(\rho,-\omega)\rho^{n-1}\,d\rho$$

$$= \tfrac{1}{2}\int_0^\infty e^{2\pi i \rho s} \widehat{Rf}(\rho,\omega)\rho^{n-1}\,d\rho + \tfrac{1}{2}\int_{-\infty}^0 e^{2\pi i \rho s} \widehat{Rf}(\rho,\omega)(-\rho)^{n-1}\,d\rho$$

$$= \tfrac{1}{2}\int_{-\infty}^\infty e^{2\pi i \rho s} |\rho|^{n-1} \widehat{Rf}(\rho,\omega)\,d\rho.$$

We call this function the **modified Radon transform** of f and denote it by $\tilde{R}f(s,\omega)$.

We can express $\tilde{R}f$ more directly in terms of Rf. If n is odd, then $|\rho|^{n-1} = \rho^{n-1}$, and we have

$$
\tilde{R}f(s,\omega) = \frac{(-1)^{(n-1)/2}}{2(2\pi)^{n-1}} \int_{-\infty}^{\infty} e^{2\pi i \rho s} (2\pi i \rho)^{n-1} \widehat{Rf}(\rho,\omega)\, d\rho
$$

$$
= \frac{(-1)^{(n-1)/2}}{2(2\pi)^{n-1}} \partial_s^{n-1} Rf(s,\omega).
$$

On the other hand, if n is even,

$$
\tilde{R}f(s,\omega) = \frac{(-1)^{(n-2)/2}}{2(2\pi)^{n-1}} \int_{-\infty}^{\infty} e^{2\pi i \rho s} (-i\,\mathrm{sgn}\,\rho)(2\pi i \rho)^{n-1} \widehat{Rf}(\rho,\omega)\, d\rho
$$

$$
= \frac{(-1)^{(n-2)/2}}{2(2\pi)^{n-1}} H \partial_s^{n-1} Rf(s,\omega),
$$

where H is the **Hilbert transform**, defined for $\phi \in \mathcal{S}(\mathbb{R})$ by

$$
(H\phi)\widehat{\ }(\rho) = (-i\,\mathrm{sgn}\,\rho)\widehat{\phi}(\rho),
$$

or equivalently by

$$
H\phi(s) = \lim_{\epsilon \to 0} \frac{1}{\pi} \int_{|t| > \epsilon} \frac{\phi(s-t)}{t}\, dt.
$$

(For the equivalence of these two formulas, see Exercise 1.) In any event, (5.37) becomes

$$
f(x) = \int_{S_n} \tilde{R}f(x \cdot \omega, \omega)\, d\sigma(\omega),
$$

which is the inversion formula for the Radon transform.

Now we can solve the Cauchy problem (5.4) for $f, g \in \mathcal{S}$. Namely, we write

$$
f(x) = \int_{S_n} \tilde{R}f(x \cdot \omega, \omega)\, d\sigma(\omega), \qquad g(x) = \int_{S_n} \tilde{R}g(x \cdot \omega, \omega)\, d\sigma(\omega),
$$

expressing f and g as superpositions of functions that depend only on $x \cdot \omega$ for various values of ω. Then the solution u is the corresponding superposition of plane waves:

(5.38)
$$
u(x,t) = \frac{1}{2} \int_{S_n} \left[\tilde{R}f(x \cdot \omega + t, \omega) + \tilde{R}f(x \cdot \omega - t, \omega) \right.
$$
$$
\left. + \int_{x \cdot \omega - t}^{x \cdot \omega + t} \tilde{R}g(s,\omega)\, ds \right] d\sigma(\omega).
$$

(See also Exercise 2.)

In this setup, the Huygens phenomenon arises from the difference in the formulas for $\tilde{R}$ when n is even or odd. It is related to the fact that ∂_s is a local operator (that is, $\partial_s \phi(s_0)$ depends only on the values of ϕ near s_0), whereas the Hilbert transform H is not.

For more about the Radon transform and its applications to differential equations, see John [29] and Ludwig [35]. See also Deans [11] for a discussion of some of the many other applications of the Radon transform in astronomy, medicine, and other areas, and Helgason [24] for a development of the Radon transform in a more general geometric setting.

EXERCISES

1. Show that the inverse Fourier transform of the function $h(\rho) = -i \operatorname{sgn} \rho$ on $\mathbb{R}$ is the distribution $h^\vee$ defined by

$$\langle h^\vee, \psi \rangle = \lim_{\epsilon \to 0} \frac{1}{\pi} \int_{|s| > \epsilon} \frac{\psi(s)}{s} \, ds.$$

   Hint: First show that the inverse Fourier transform of $e^{-2\pi\epsilon|\rho|} h(\rho)$ is $s/\pi(s^2 + \epsilon^2)$, then apply Theorem (0.13) with

$$\phi(s) = \frac{s}{\pi(s^2 + 1)} - \frac{\chi_{(-\infty, -1)}(s) + \chi_{(1, \infty)}(s)}{\pi s}.$$

2. Show that $R(\partial_j f)(s, \omega) = \omega_j \partial_s Rf(s, \omega)$, and hence that $R(\Delta f) = \partial_s^2 Rf$. Application of the Radon transform therefore reduces the n-dimensional wave equation to the one-dimensional wave equation; use this fact to derive (5.38).

3. Show that $R(f * g) = Rf *_s Rg$, where $*_s$ denotes the convolution of two functions of (s, ω) with respect to s.

4. Compute Rf where $f(x) = e^{-\pi|x|^2}$.

Chapter 6
THE L² THEORY OF DERIVATIVES

One of the most elegant and useful ways of measuring differentiability properties of functions on $\mathbb{R}^n$ is in terms of L^2 norms. The reason for this is twofold: first, L^2 is a Hilbert space; second, the Fourier transform, which converts differentiation into multiplication by polynomials, is a unitary isomorphism on L^2. The resulting function spaces are known as (L^2) **Sobolev spaces**.

In the first two sections of this chapter we develop the basic properties of Sobolev spaces on $\mathbb{R}^n$. In the next two sections we use them to prove the local regularity theorem for elliptic operators (a result which we shall re-derive, in a refined form and with more sophisticated methods, in Chapter 8) and Hörmander's characterization of constant-coefficient hypoelliptic operators. In the final section we study Sobolev spaces on bounded domains, where the Fourier transform is not directly available but to which the results on $\mathbb{R}^n$ can be applied.

A. Sobolev Spaces on $\mathbf{R}^n$

To begin with, if k is a nonnegative integer, we define the **Sobolev space** $H_k = H_k(\mathbb{R}^n)$ **of order** k to be the set of all $f \in L^2(\mathbb{R}^n)$ whose (distribution) derivatives $\partial^\alpha f$ belong to $L^2(\mathbb{R}^n)$ for $|\alpha| \le k$:

$$H_k = \left\{ f \in L^2 : \partial^\alpha f \in L^2 \text{ for } |\alpha| \le k \right\}.$$

This definition of H_k makes its meaning clear, but there is an equivalent characterization of H_k in terms of the Fourier transform that is easier to work with:

(6.1) Theorem.
$f \in H_k$ if and only if $(1 + |\xi|^2)^{k/2}\widehat{f} \in L^2$, and the norms

$$f \to \left[\sum_{|\alpha| \le k} \|\partial^\alpha f\|_{L^2}^2 \right]^{1/2} \text{ and } f \to \left[\int |\widehat{f}(\xi)|^2 (1 + |\xi|^2)^k \, d\xi \right]^{1/2}$$

are equivalent.

Proof: Since $(\partial^\alpha f)\widehat{}(\xi) = (2\pi i\xi)^\alpha \widehat{f}(\xi)$, the Plancherel theorem implies that

$$\sum_{|\alpha| \le k} \|\partial^\alpha f\|_{L^2}^2 = \sum_{|\alpha| \le k} \int |\widehat{f}(\xi)|^2 |(2\pi\xi)^\alpha|^2 \, d\xi,$$

so the theorem amounts to proving that the quantities $\sum_{|\alpha| \le k} |\xi^\alpha|^2$ and $(1 + |\xi|^2)^k$ are comparable, i.e., that each is bounded by a constant multiple of the other. But clearly $|\xi^\alpha| \le 1$ for $|\xi| \le 1$, and $|\xi^\alpha| \le |\xi|^{|\alpha|} \le |\xi|^k$ for $|\xi| \ge 1$ and $|\alpha| \le k$, so

$$\sum_{|\alpha| \le k} |\xi^\alpha|^2 \le C_1 \max(1, |\xi|^{2k}) \le (1 + |\xi|^2)^k.$$

On the other hand, since $|\xi|^{2k}$ and $\sum_1^n |\xi_j^k|^2$ are both homogeneous of degree k, we have $|\xi|^{2k} \le C_2 \sum_1^n |\xi_j^k|^2$ where C_2 is the reciprocal of the minimum value of $\sum_1^n |\xi_j^k|^2$ on the unit sphere $|\xi| = 1$. It follows that

$$(1 + |\xi|^2)^k \le 2^k \max(1, |\xi|^{2k}) \le 2^k(1 + |\xi|^{2k})$$

$$\le 2^k C_2 \left[1 + \sum_1^n |\xi_j^k|^2 \right] \le 2^k C_2 \sum_{|\alpha| \le k} |\xi^\alpha|^2,$$

which completes the proof.   ∎

This suggests a generalization of H_k in which k is replaced by an arbitrary real number s. Namely, if $u(\xi)$ is a function on $\mathbb{R}^n$ such that $(1 + |\xi|^2)^{s/2}u(\xi) \in L^2$, then $u\phi \in L^1$ for any $\phi \in \mathcal{S}$, so u is a tempered distribution (whose action on $\phi \in \mathcal{S}$ is $\int u\phi$). Since the Fourier transform maps tempered distributions into tempered distributions, we can define the **Sobolev space of order** s:

$$(6.2) \qquad H_s = H_s(\mathbb{R}^n) = \Big\{ f \in \mathcal{S}'(\mathbb{R}^n) : \widehat{f} \text{ is a function and}$$

$$\|f\|_s^2 \equiv \int |\widehat{f}(\xi)|^2 (1 + |\xi|^2)^s \, d\xi < \infty \Big\}.$$

The norm $\| \cdot \|_s$ on H_s thus defined is called the **Sobolev norm of order s**. Theorem (6.1) shows that this definition agrees with the previous one when s is a nonnegative integer. In particular, $H_0 = L^2$, and the norm on L^2 will henceforth be denoted by $\| \cdot \|_0$. (There will be no notational confusion between the Sobolev norms $\| \cdot \|_s$ and the L^p norms, since we shall rarely use the latter for $p \neq 2$, and when we do, we shall henceforth denote them by $\| \cdot \|_{L^p}$.)

The argument that proves Theorem (6.1) also proves the following generalization if it; one has merely to introduce factors of $(1 + |\xi|^2)^s$ in appropriate places:

(6.3) Theorem.
Suppose k is a positive integer, s is a real number, and f is a tempered distribution. Then $f \in H_s$ if and only if $\partial^\alpha f \in H_{s-k}$ for $|\alpha| \leq k$, and the norms

$$\|f\|_s \quad \text{and} \quad \left[\sum_{|\alpha| \leq k} \|\partial^\alpha f\|_{s-k}^2 \right]^{1/2}$$

are equivalent. In particular, ∂^α is a bounded operator from H_s to H_{s-k} for all s when $|\alpha| \leq k$.

H_s is a Hilbert space with inner product

$$\langle f \mid g \rangle_s = \int \widehat{f}(\xi) \overline{\widehat{g}(\xi)} (1 + |\xi|^2)^s \, d\xi,$$

and the Fourier transform is a unitary isomorphism from H_s to $L^2(\mathbb{R}^n, \mu)$ where $d\mu(\xi) = (1 + |\xi|^2)^s \, d\xi$. Standard approximation arguments show that $\mathcal{S}$ (or even C_c^∞) is dense in the latter space, and since the Fourier transform maps $\mathcal{S}$ onto itself, $\mathcal{S}$ is dense in H_s.

If $s \leq t$, it is clear that $H_t \subset H_s$ and that $\| \cdot \|_s \leq \| \cdot \|_t$, so the inclusion map from H_t to H_s is continuous. In particular, $H_s \subset H_0 = L^2$ for $s \geq 0$, so the elements of H_s for $s \geq 0$ are functions. For $s < 0$ they are, in general, more singular objects.

Example 1.
With $n = 1$, let $f(x) = (\pi x)^{-1} \sin 2\pi x$. An elementary calculation shows that f is the inverse Fourier transform of $\chi_{[-1,1]}$, so $f \in H_s$ for all s. It is also easy to verify directly, by induction, that $f^{(k)}$ decays like x^{-1} at as $x \to \infty$ and hence that $f^{(k)} \in L^2$ for all k. (This example shows, however, that elements of Sobolev spaces need not decay very rapidly at infinity.)

Example 2.

Let δ be the point mass at the origin. Since $\widehat{\delta} = 1$ we have

$$\|\delta\|_s^2 = \int (1 + |\xi|^2)^s \, d\xi = \omega_n \int_0^\infty (1 + r^2)^s r^{n-1} \, dr,$$

which is finite if and only if $s < -\frac{1}{2}n$ since the integrand is roughly r^{2s+n-1} for large r. Hence $\delta \in H_s$ if and only if $s < -\frac{1}{2}n$.

Since H_s is a Hilbert space, it is naturally isomorphic to its own dual space — or rather anti-isomorphic, since the identification is conjugate-linear. However, there is an equally natural anti-isomorphism between $(H_s)^*$ and H_{-s}, given not by the inner product $(\cdot \mid \cdot)_s$ but by the ordinary L^2 inner product on the Fourier transform side. Indeed, if $f \in H_s$ and $g \in H_{-s}$, then

$$\widehat{f}\,\overline{\widehat{g}} = [(1 + |\xi|^2)^{s/2}\widehat{f}][(1 + |\xi|^2)^{-s/2}\widehat{g}] \in L^2 \cdot L^2 \subset L^1,$$

so $\langle \widehat{f} \mid \widehat{g} \rangle = \int \widehat{f}\,\overline{\widehat{g}}$ is absolutely convergent, and $f \to \langle \widehat{f} \mid \widehat{g} \rangle$ defines a bounded linear functional on H_s. Indeed,

$$|\langle \widehat{f} \mid \widehat{g} \rangle| \le \left[\int |\widehat{f}(\xi)|^2 (1 + |\xi|^2)^s \, d\xi \right]^{1/2} \left[\int |\widehat{g}(\xi)|^2 (1 + |\xi|^2)^{-s} \, d\xi \right]^{1/2}$$
$$= \|f\|_s \|g\|_{-s},$$

with equality if $\widehat{f}(\xi) = (1 + |\xi|^2)^{-s}\widehat{g}(\xi)$, and the self-duality of L^2 implies that every bounded linear functional on H_s is of this form. In short, the map taking $g \in H_{-s}$ to the functional $f \to \langle \widehat{f} \mid \widehat{g} \rangle$ defines an isometric anti-isomorphism between H_{-s} and $(H_s)^*$. Moreover, if we restrict the latter functional to S, we obtain a tempered distribution which is nothing but $\overline{g}$:

$$\langle \widehat{\phi} \mid \widehat{g} \rangle = \langle \widehat{\phi}, \overline{\widehat{g}} \rangle = \langle \phi, \overline{\widehat{\widehat{g}}} \rangle = \langle \phi, \overline{g} \rangle.$$

Thus, the duality between H_s and H_{-s} is, except for the necessary introduction of a complex conjugate, compatible with the duality between S and S'.

We can express the norm on H_s neatly in terms of the operator

$$\Lambda^s = \left[I - (2\pi)^{-2}\Delta \right]^{s/2} \qquad (s \in \mathbb{R}).$$

Here we are using the functional calculus for the Laplacian discussed in §4B, that is,

$$(6.4) \qquad\qquad (\Lambda^s f)\widehat{\ }(\xi) = (1 + |\xi|^2)^{s/2}\widehat{f}(\xi).$$

Clearly, by the Plancherel theorem,

$$\|f\|_s = \|\Lambda^s f\|_0 = \|\Lambda^s f\|_{L^2}, \qquad \langle f \mid g \rangle_s = \langle \Lambda^s f \mid \Lambda^s g \rangle,$$

and Λ^s is a unitary isomorphism from H_t to H_{t-s} for all $t \in \mathbb{R}$.

Although Sobolev spaces make the manipulation of distribution derivatives very easy, they would be of limited usefulness if we could not relate L^2 derivatives to classical pointwise derivatives. Fortunately, there is a simple and beautiful connection between the two.

(6.5) The Sobolev Lemma.
If $s > k + \frac{1}{2}n$, then $H_s \subset C^k$ and there is a constant $C = C_{s,k}$ such that

(6.6) $$\sup_{|\alpha| \leq k} \sup_{x \in \mathbb{R}^n} |\partial^\alpha f(x)| \leq C\|f\|_s.$$

Proof: By the Fourier inversion theorem, if $(\partial^\alpha f)^\frown \in L^1$ then $\partial^\alpha f$ is continuous and $\sup_x |\partial^\alpha f(x)| \leq \|(\partial^\alpha f)^\frown\|_{L^1}$. Hence, to prove the theorem it is enough to prove that $\|(\partial^\alpha f)^\frown\|_{L^1} \leq \|f\|_s$ when $|\alpha| \leq k$. But $(\partial^\alpha f)^\frown(\xi) = (2\pi i \xi)^\alpha \widehat{f}(\xi)$, and by the Schwarz inequality, for $|\alpha| \leq k$ we have

$$\int |(2\pi i \xi)^\alpha \widehat{f}(\xi)| \, d\xi \leq (2\pi)^k \int (1 + |\xi|^2)^{k/2} |\widehat{f}(\xi)| \, d\xi$$

$$= (2\pi)^k \int (1 + |\xi|^2)^{s/2} |\widehat{f}(\xi)| (1 + |\xi|^2)^{(k-s)/2} \, d\xi$$

$$\leq (2\pi)^k \left[\int (1 + |\xi|^2)^s |\widehat{f}(\xi)|^2 \, d\xi \int (1 + |\xi|^2)^{k-s} \, d\xi \right]^{1/2}.$$

The first integral on the right is $\|f\|_s^2$, so the theorem boils down to the fact that

$$\int (1 + |\xi|^2)^{k-s} \, d\xi = \omega_n \int_0^\infty (1 + r^2)^{k-s} r^{n-1} \, dr < \infty$$

precisely when $s > k + \frac{1}{2}n$. (Cf. Example 2 above.)   ∎

(6.7) Corollary.
If $f \in H_s$ for all $s \in \mathbb{R}$, then $f \in C^\infty$.

(6.8) Corollary.
Every distribution with compact support belongs to H_s for some $s \in \mathbb{R}$.

Proof: If f is a distribution with compact support, it follows from (0.32) and the Sobolev lemma that for some positive integer N,

$$|\langle f, \phi \rangle| \leq C \sum_{|\alpha| \leq N} \sup_x |\partial^\alpha \phi(x)| \leq C'\|\phi\|_s \qquad (\phi \in \mathcal{S}),$$

where $s = N + \frac{1}{2}n + 1$. But this means that f extends to a bounded linear functional on H_s and hence belongs to H_{-s}. ∎

Remark 1: The proof of the Sobolev lemma shows that $H_s \subset BC^k$ for $s > k + \frac{1}{2}n$, where BC^k is the space of C^k functions f such that $\partial^\alpha f$ is bounded for $|\alpha| \leq k$. Conversely, if $H_s \subset BC^k$, the closed graph theorem easily implies that the estimate (6.6) holds. But this means that the functionals $f \to \partial^\alpha f(x)$ ($|\alpha| \leq k$, $x \in \mathbb{R}^n$) are a bounded set of bounded linear functionals on H_s, or in other words, that the distributions $\partial^\alpha \delta(\cdot - x)$ are a bounded subset of H_{-s}. A calculation like the one in Example 1 above shows that this happens precisely when $s > k + \frac{1}{2}n$. (See Exercise 1.)

Remark 2: The Sobolev lemma can be sharpened a bit: if $s = k + \alpha + \frac{1}{2}n$ where $0 < \alpha < 1$, then $H_s \subset C^{k+\alpha}$. See Exercise 2.

A few examples may help to elucidate the meaning of the Sobolev lemma.

Example 3.
Pick $\phi \in C_c^\infty$ with $\phi = 1$ near the origin, and let $f_\lambda(x) = |x|^\lambda \phi(x)$ where λ is positive and not an even integer. Since $\partial^\alpha f_\lambda$ is homogeneous of degree $\lambda - |\alpha|$ near the origin, an integration in polar coordinates shows that $\partial^\alpha f \in L^2$ for $|\alpha| < \lambda + \frac{1}{2}n$, but $\partial^\alpha f$ is continuous only for $|\alpha| < \lambda$. (See Exercise 3.)

Example 4.
Let $f(x) = e^{-2\pi|x|}$. By Theorem (4.15) and the Fourier inversion theorem, $\widehat{f}(\xi)$ is a constant times $(1 + |\xi|^2)^{-(n+1)/2}$. It follows easily that $f \in H_s$ for $s < \frac{1}{2}n + 1$ but $f \notin H_{(n/2)+1}$. The Sobolev lemma says that f should be continuous (which it is), but just barely fails to guarantee that $f \in C^1$ (which it is not).

Example 5.
Suppose $f \in C^k$ has compact support (so $f \in H_k$ also), and let u be the solution of the Cauchy problem $\partial_t^2 u - \Delta u = 0$, $u(x, 0) = f(x)$, $\partial_t u(x, 0) = 0$.

The argument leading to Theorem (5.31) shows that $u(\cdot, t) \in H_k$ for all t, but (5.16) and (5.18) show that $u(\cdot, t)$ may be no better than $C^{k-[n/2]}$, where $[n/2]$ is the greatest integer in $n/2$. This is essentially the maximum discrepancy between L^2 derivatives and continuous derivatives allowed by the Sobolev lemma.

Elements of L^2 are only defined almost everywhere, and in general it does not make sense to evaluate them at a single point. The Sobolev lemma, however, shows that if $s > \frac{1}{2}n$, the evaluation map $f \to f(x)$ ($x \in \mathbb{R}^n$) has a natural meaning for $f \in H_s$. More generally, it turns out that if $k \leq n$ and $s > \frac{1}{2}k$, it makes sense to restrict functions in H_s to submanifolds of codimension k, even though these sets have measure zero. For the case $k = 1$ in particular, this is useful for the study of boundary value problems. We shall prove this for the special case of linear subspaces here; the general case will then follow from some results of the next section.

To be precise, let us regard $\mathbb{R}^n$ as $\mathbb{R}^{n-k} \times \mathbb{R}^k$ with coordinates $y \in \mathbb{R}^{n-k}$, $z \in \mathbb{R}^k$ and dual coordinates η, ζ. We define the restriction map $R : \mathcal{S}(\mathbb{R}^n) \to \mathcal{S}(\mathbb{R}^{n-k})$ by

$$Rf(y) = f(y, 0).$$

(6.9) Theorem.
If $s > \frac{1}{2}k$, the restriction map R extends to a bounded map from $H_s(\mathbb{R}^n)$ to $H_{s-(k/2)}(\mathbb{R}^{n-k})$.

Proof: It suffices to show that $\|Rf\|_{s-(k/2)}$ is dominated by $\|f\|_s$ for $f \in \mathcal{S}$. We have

$$\int e^{2\pi i \eta \cdot y} \widehat{Rf}(\eta) \, d\eta = Rf(y) = f(y, 0) = \iint e^{2\pi i \eta \cdot y} \widehat{f}(\eta, \zeta) \, d\eta \, d\zeta$$

for all y, so $\widehat{Rf}(\eta) = \int \widehat{f}(\eta, \zeta) \, d\zeta$. Therefore,

$$|\widehat{Rf}(\eta)|^2 = \left[\int \widehat{f}(\eta, \zeta)(1 + |\eta|^2 + |\zeta|^2)^{s/2}(1 + |\eta|^2 + |\zeta|^2)^{-s/2} \, d\zeta \right]^2$$

$$\leq \left[\int |\widehat{f}(\eta, \zeta)|^2(1 + |\eta|^2 + |\zeta|^2)^s \, d\zeta \right] \left[\int (1 + |\eta|^2 + |\zeta|^2)^{-s} \, d\zeta \right].$$

Setting $1 + |\eta|^2 = a^2$ and $|\zeta| = r$, the second factor on the right is

$$\omega_k \int_0^\infty (a^2 + r^2)^{-s} r^{k-1} \, dr = a^{k-2s} \omega_k \int_0^\infty (1 + r^2)^{-s} r^{k-1} \, dr.$$

The latter integral is finite when $s > \frac{1}{2}k$, so

$$|\widehat{Rf}(\eta)|^2 \le C_s (1 + |\eta|^2)^{(k/2)-s} \int |\widehat{f}(\eta,\zeta)|^2 (1 + |\eta|^2 + |\zeta|^2)^s \, d\zeta,$$

or

$$|\widehat{Rf}(\eta)|^2 (1 + |\eta|^2)^{s-(k/2)} \le C_s \int |\widehat{f}(\eta,\zeta)|^2 (1 + |\eta|^2 + |\zeta|^2)^s \, d\zeta.$$

Integrating both sides with respect to η, we conclude that $\|Rf\|^2_{s-(k/2)} \le C_s \|f\|^2_s$ as desired.   ∎

Before proceeding further, we present two simple lemmas that will be used a number of times in the sequel.

(6.10) Lemma.
For all $\xi, \eta \in \mathbb{R}^n$ and all $s \in \mathbb{R}$,

$$\left(\frac{1 + |\xi|^2}{1 + |\eta|^2} \right)^s \le 2^{|s|} (1 + |\xi - \eta|^2)^{|s|}.$$

Proof: Since $|\xi| \le |\xi - \eta| + |\eta|$, we have $|\xi|^2 \le 2(|\xi - \eta|^2 + |\eta|^2)$, so

$$1 + |\xi|^2 \le 2(1 + |\xi - \eta|^2)(1 + |\eta|^2).$$

If $s \ge 0$, we have merely to raise both sides to the power s. If $s < 0$, we apply the latter result with ξ and η interchanged and s replaced by $|s| = -s$.   ∎

(6.11) Lemma.
Suppose $r < s < t$. For any $\epsilon > 0$ there exists $C > 0$ such that for all $f \in H_t$,

$$\|f\|^2_s \le \epsilon \|f\|^2_t + C \|f\|^2_r.$$

Proof: Given $\epsilon > 0$, choose $A > 0$ large enough so that $(1 + A^2)^s \le \epsilon(1 + A^2)^t$, and set $C = (1 + A^2)^{s-r}$. Then

$$(1 + |\xi|^2)^s \le \begin{cases} C(1 + |\xi|^2)^r & \text{if } |\xi| \le A \\ \epsilon(1 + |\xi|^2)^t & \text{if } |\xi| \ge A \end{cases} \le \epsilon(1 + |\xi|^2)^t + C(1 + |\xi|^2)^r.$$

It follows immediately that $\|f\|^2_s \le \epsilon \|f\|^2_t + C \|f\|^2_r$ for any $f \in H_t$.   ∎

We now show that multiplication by Schwartz functions preserves the H_s spaces. If s is a positive integer, this is clear from Theorem (6.1) and the product formula for derivatives. For the general case, a different argument is needed.

(6.12) Proposition.
If $\phi \in \mathcal{S}$, the map $f \rightarrow \phi f$ is bounded on H_s for every $s \in \mathbb{R}$.

Proof: The proposition amounts to the assertion that the map $f \rightarrow \Lambda^s \phi \Lambda^{-s} f$ is bounded on $H_0 = L^2$ for every s, where Λ^s is given by (6.4). But since the Fourier transform converts pointwise multiplication into convolution,

$$(\Lambda^s \phi \Lambda^{-s} f)\widehat{\ }(\xi) = \int K(\xi, \eta) \widehat{f}(\eta) \, d\eta$$

where

$$K(\xi, \eta) = (1 + |\xi|^2)^{s/2} \widehat{\phi}(\xi - \eta)(1 + |\eta|^2)^{-s/2}.$$

By Lemma (6.10),

$$|K(\xi, \eta)| \le 2^{|s|/2}(1 + |\xi - \eta|^2)^{|s|/2} |\widehat{\phi}(\xi - \eta)|,$$

so $\int |K(\xi, \eta)| \, d\xi$ and $\int |K(\xi, \eta)| \, d\eta$ are bounded by

$$2^{|s|/2} \int (1 + |\xi|^2)^{|s|/2} |\widehat{\phi}(\xi)| \, d\xi,$$

which is finite since $\widehat{\phi}$ is rapidly decreasing at infinity. The proposition then follows from the generalized Young's inequality (0.10).   ∎

Proposition (6.12) shows that the Sobolev spaces are preserved by multiplication by smooth cutoff functions, so they can be localized. To be precise, if $s \in \mathbb{R}$ and Ω is an open set in $\mathbb{R}^n$, we define the **localized Sobolev space** $H_s^{\text{loc}}(\Omega)$ to be the set of all distributions f on Ω such that for every bounded open Ω_0 with $\overline{\Omega_0} \subset \Omega$, f agrees with an element of H_s on Ω_0. Alternatively, we have the following characterization of $H_s^{\text{loc}}(\Omega)$:

(6.13) Proposition.
$f \in H_s^{\text{loc}}(\Omega)$ if and only if $\phi f \in H_s$ for every $\phi \in C_c^\infty(\Omega)$. Moreover, $H_s \subset H_s^{\text{loc}}(\Omega)$ for every open $\Omega \subset \mathbb{R}^n$.

Proof: Clearly the second statement follows from the first one, in view of Proposition (6.12). If $f \in H_s^{loc}(\Omega)$ and $\phi \in C_c^\infty(\Omega)$, then f agrees with an element of H_s on the support of ϕ, whence $\phi f \in H_s$ by Proposition (6.12). Conversely, suppose $\phi f \in H_s$ for all $\phi \in C_c^\infty(\Omega)$ and Ω_0 is an open set with compact closure in Ω. Choose $\phi \in C_c^\infty(\Omega)$ with $\phi = 1$ on Ω_0 (Theorem (0.17)); then f agrees with $\phi f \in H_s$ on Ω_0. ∎

Roughly speaking, the condition $f \in H_s^{loc}(\Omega)$ means that f has the requisite smoothness for being in H_s on Ω but imposes no global square-integrability conditions on f.

We conclude this section by remarking that there are L^p analogues of the Sobolev spaces for $1 \le p < \infty$. On the simplest level, if k is a positive integer one can consider the space L_k^p of L^p functions f whose distribution derivatives $\partial^\alpha f$ are in L^p for $|\alpha| \le k$; this is a Banach space with norm $\sum_{|\alpha| \le k} \|\partial^\alpha f\|_{L^p}$. If $1 < p < \infty$, it can be shown (although it requires a fair amount of theory to do so) that $f \in L_k^p$ if and only if $\Lambda^k f \in L^p$, where Λ^k is defined by (6.4). One can then define L_s^p for any $s \in \mathbb{R}$ to be the set of tempered distributions f such that $\Lambda^s f \in L^p$, and much of the theory for the L^2 Sobolev spaces can be extended. In particular, the analogue of the Sobolev lemma is that $L_s^p \subset C^k$ if $s > k + (n/p)$; one also has the embedding theorem $L_s^p \subset L_t^q$ when $q > p$ and $p^{-1} - q^{-1} = n^{-1}(s - t)$. See Stein [45], Adams [1], and Nirenberg [38].

EXERCISES

1. Fill in the details of Remark 1 following the Sobolev lemma to show that if $H_s \subset BC^k$ then $s > k + \frac{1}{2}n$.

2. Show that if $s = \frac{1}{2}n + \alpha$ where $0 < \alpha < 1$ then $\|\delta_x - \delta_y\|_{-s} \le C|x - y|^\alpha$ for all $x, y \in \mathbb{R}^n$, where δ_x and δ_y are the point masses at x and y. (Hint: $\widehat{\delta_x}(\xi) = e^{-2\pi i x \cdot \xi}$. Write the integral defining $\|\delta_x - \delta_y\|_{-s}^2$ as the sum of integrals over the regions $|\xi| \le R$ and $|\xi| > R$ where $R = |x - y|^{-1}$, and use the mean value theorem of calculus to estimate $\widehat{\delta_x} - \widehat{\delta_y}$ over the first region.) Conclude that $H_s \subset C^\alpha$, and more generally that if $s = \frac{1}{2}n + k + \alpha$ with $0 < \alpha < 1$ then $H_s \subset C^{k+\alpha}$.

3. In Example 3 following the Sobolev lemma, we implicitly used the fact that the distribution derivatives of $f_\lambda(x) = |x|^\lambda \phi(x)$ coincide with its pointwise derivatives when the latter are integrable functions — namely, when the order of the derivative is less than $\lambda + n$. Prove this. (One way is to approximate f_λ by smooth functions.)

4. Suppose $s > \frac{1}{2}n$. Prove that H_s is an algebra, and more precisely that $\|fg\|_s \leq C_s\|f\|_s\|g\|_s$ for all $f, g \in H_s$. Do this via the following lemmas:

   a. Show that $(1 + |\xi|^2)^{s/2}(fg)\widehat{}(\xi) = \int K(\xi, \eta)u(\xi - \eta)v(\eta)\,d\eta$ where $u, v \in L^2$ and $K(\xi, \eta) = (1 + |\xi|^2)^{s/2}(1 + |\xi - \eta|^2)^{-s/2}(1 + |\eta|^2)^{-s/2}$.

   b. Show that $K(\xi, \eta) \leq C_s(1 + |\eta|^2)^{-s/2}$ if $|\xi - \eta| \geq \frac{1}{2}|\xi|$ and $K(\xi, \eta) \leq C_s(1 + |\xi - \eta|^2)^{-s/2}$ if $|\xi - \eta| \leq \frac{1}{2}|\xi|$.

   c. Show that if $u, v \in L^2$ and $w(\xi) = \int(1 + |\eta|^2)^{-s/2}u(\xi - \eta)v(\eta)\,d\eta$ then $w \in L^2$ and $\|w\|_{L^2} \leq C_s\|u\|_{L^2}\|v\|_{L^2}$.

B. Further Results on Sobolev Spaces

In this section we present a potpourri of theorems about Sobolev spaces. The most important ones are the Rellich compactness theorem, a characterization of H_s in terms of difference quotients, an interpolation theorem, and the local invariance of Sobolev spaces under smooth coordinate changes.

If Ω is an open set in $\mathbb{R}^n$, we define

$$H_s^0(\Omega) = \text{the closure of } C_c^\infty(\Omega) \text{ in } H_s.$$

Thus $H_s^0(\Omega)$ consists of elements of H_s that are supported in $\overline{\Omega}$, although in general not every such element belongs to $H_s^0(\Omega)$. If $\{f_k\}$ is a sequence in $C_c^\infty(\Omega)$ that converges in H_s, it also converges in H_t when $t < s$; hence $H_s^0(\Omega)$ is a subset of $H_t^0(\Omega)$ for $s > t$.

(6.14) Rellich's Theorem.
If Ω is bounded and $s > t$, the inclusion map $H_s^0(\Omega) \to H_t^0(\Omega)$ is compact. In fact, every bounded sequence in $H_s^0(\Omega)$ has a subsequence that converges in H_t for every $t < s$.

   Proof: Suppose $\{f_k\}$ is a sequence in $H_s^0(\Omega)$ with $\|f_k\|_s \leq C < \infty$. Choose $\phi \in C_c^\infty(\mathbb{R}^n)$ with $\phi = 1$ on Ω. Then $f_k = \phi f_k$, so $\widehat{f_k} = \widehat{\phi} * \widehat{f_k}$. Also, since $\widehat{\phi} \in \mathcal{S}$ and $\widehat{f_k}$ is a tempered function, $\widehat{\phi} * \widehat{f_k}$ is a C^∞ function defined pointwise by the usual convolution integral. Thus, by Lemma (6.10) and the Schwarz inequality,

$$(1 + |\xi|^2)^{s/2}|\widehat{f_k}(\xi)|$$
$$\leq 2^{|s|/2}\int|\widehat{\phi}(\xi - \eta)|(1 + |\xi - \eta|^2)^{|s|/2}|\widehat{f_k}(\eta)|(1 + |\eta|^2)^{s/2}\,d\eta$$
$$\leq 2^{|s|/2}\|\phi\|_{|s|}\|f_k\|_s.$$

For the same reason, if $j = 1, \ldots, n$,

$$(1 + |\xi|^2)^{s/2}|\partial_j \widehat{f_k}(\xi)| \le 2^{|s|/2}\|2\pi i x_j \phi\|_{|s|}\|f_k\|_s.$$

Since $\|f_k\|_s \le C$, the functions $\widehat{f_k}$ and their first derivatives are uniformly bounded on compact sets, so by the Arzelà-Ascoli theorem there is a subsequence $\{\widehat{f_{k_j}}\}$ that converges uniformly on compact sets. We claim that $\{f_{k_j}\}$ converges in H_t for $t < s$. Indeed, for any $R > 0$,

$$
\begin{aligned}
\|f_{k_i} - f_{k_j}\|_t^2 &= \int_{|\xi| \le R} (1 + |\xi|^2)^t |\widehat{f_{k_i}} - \widehat{f_{k_j}}|^2(\xi)\, d\xi \\
&\quad + \int_{|\xi| \ge R} (1 + |\xi|^2)^{t-s}(1 + |\xi|^2)^s |\widehat{f_{k_i}} - \widehat{f_{k_j}}|^2(\xi)\, d\xi \\
&\le \left[\sup_{|\xi| \le R} |\widehat{f_{k_i}} - f_{k_j}|^2(\xi)\right] \int_{|\xi| \le R} (1 + |\xi|^2)^t\, d\xi \\
&\quad + (1 + R^2)^{t-s} \int_{|\xi| > R} (1 + |\xi|^2)^s |\widehat{f_{k_i}} - \widehat{f_{k_j}}|^2(\xi)\, d\xi \\
&\le \left[\sup_{|\xi| \le R} |\widehat{f_{k_i}} - f_{k_j}|^2(\xi)\right] \int_{|\xi| \le R} (1 + |\xi|^2)^t\, d\xi \\
&\quad + (1 + R^2)^{t-s}\|f_{k_i} - f_{k_j}\|_s^2.
\end{aligned}
$$

Now, given $\epsilon > 0$, since $t - s < 0$ and $\|f_{k_i} - f_{k_j}\|_s \le 2C$, we can choose R large enough so that the second term in this last expression is less than $\frac{1}{2}\epsilon$ for all i, j. But then for all sufficiently large i and j, the first term will also be less than $\frac{1}{2}\epsilon$. Thus $\{f_{k_j}\}$ is Cauchy in H_t.   ∎

The hypothesis that Ω is bounded can be relaxed when $s \ge 0$ (see Adams [1] and Lair [32]), but some restriction is necessary: see Exercises 1 and 2.

While we are considering the spaces $H_s^0(\Omega)$, here is another useful result.

(6.15) Proposition.
If Ω is bounded and k is a positive integer, the norms $f \to \|f\|_k$ and $f \to \sum_{|\alpha|=k} \|\partial^\alpha f\|_0$ are equivalent on $H_k^0(\Omega)$.

Proof: By Theorem (6.1), it is enough to show that

$$\|\partial^\beta f\|_0 \le C \sum_{|\alpha|=k} \|\partial^\alpha f\|_0 \text{ for } |\beta| < k.$$

Consider first the case $k = 1$. We shall prove a sharper result, namely that if there are constants a and b such that $a < x_n < b$ whenever $x \in \Omega$, then $\|f\|_0 \leq (b-a)\|\partial_n f\|_0$ for all $f \in H_1^0(\Omega)$.

It suffices to assume that $f \in C_c^\infty(\Omega)$. Writing $x = (x', x_n)$ where $x' = (x_1, \ldots, x_{n-1})$, we have $f(x) = \int_a^{x_n} \partial_n f(x', t) \, dt$, so by the Schwarz inequality,

$$|f(x)|^2 \leq \int_a^{x_n} |\partial_n f(x', t)|^2 \, dt \int_a^{x_n} dt \leq (b-a) \int_a^b |\partial_n f(x', t)|^2 \, dt$$

for $x \in \Omega$. Integrating over Ω, we obtain

$$\|f\|_0^2 \leq (b-a) \int_a^b \int_{\mathbb{R}^{n-1}} \int_a^b |\partial_n f(x', t)|^2 \, dt \, dx' \, dx_n$$
$$= (b-a)^2 \|\partial_n f\|_0^2.$$

Now for the case $k > 1$, the above argument shows that $\|\partial^\beta f\|_0 \leq (b-a)\|\partial_n \partial^\beta f\|_0$ for $|\beta| < k$, so the desired result follows by induction. ∎

Our next result is a more precise version of Proposition (6.12) that we shall need later. We recall that if A and B are operators, their commutator $[A, B]$ is defined to be $AB - BA$. In particular, if B is multiplication by a function ϕ, we shall abuse notation slightly and write $[A, B]$ as $[A, \phi]$:

$$[A, \phi]f = A(\phi \cdot f) - \phi \cdot Af.$$

(6.16) Lemma.
If $s \in \mathbb{R}$ and $\sigma > \frac{1}{2}n$ there is a constant $C = C_{s,\sigma}$ such that for all $\phi \in \mathcal{S}$ and $f \in H_{s-1}$,

$$\|[\Lambda^s, \phi]f\|_0 \leq C\|\phi\|_{|s-1|+1+\sigma}\|f\|_{s-1}.$$

Proof: Setting $f = \Lambda^{1-s}g$, what we need to show is that

$$\|[\Lambda^s, \phi]\Lambda^{1-s}g\|_0 \leq C\|\phi\|_{|s-1|+1+\sigma}\|g\|_0$$

for all $g \in H_0 = L^2$. Since the Fourier transform converts multiplication into convolution, we have

$$([\Lambda^s, \phi]\Lambda^{1-s}g)\widehat{\ }(\xi) = \int K(\xi, \eta)\widehat{g}(\eta) \, d\eta,$$

where

$$K(\xi, \eta) = [(1 + |\xi|^2)^{s/2} - (1 + |\eta|^2)^{s/2}]\hat{\phi}(\xi - \eta)|(1 + |\eta|^2)^{(1-s)/2}.$$

We claim that

(6.17)
$$\begin{aligned}&[(1 + |\xi|^2)^{s/2} - (1 + |\eta|^2)^{s/2}]\\ &\le |s|\,|\xi - \eta|[(1 + |\xi|^2)^{(s-1)/2} + (1 + |\eta|^2)^{(s-1)/2}].\end{aligned}$$

Granted this, by Lemma (6.10) we have

$$\begin{aligned}|K(\xi, \eta)| &\le |s|\,|\hat{\phi}(\xi - \eta)|\,|\xi - \eta|\,[(1 + |\xi|^2)^{(s-1)/2}(1 + |\eta|^2)^{(1-s)/2} + 1]\\ &\le |s|2^{|s-1|/2}|\hat{\phi}(\xi - \eta)|\,|\xi - \eta|\,[(1 + |\xi - \eta|^2)^{|s-1|/2} + 1]\\ &\le C_2|\hat{\phi}(\xi - \eta)|(1 + |\xi - \eta|^2)^{(|s-1|+1)/2}.\end{aligned}$$

Thus $\int |K(\xi, \eta)|\,d\xi$ and $\int |K(\xi, \eta)|\,d\eta$ are bounded by

$$C_2 \left[\int |\hat{\phi}(\xi)|^2(1 + |\xi|^2)^{|s-1|+1+\sigma}\,d\xi \right]^{1/2} \left[\int (1 + |\xi|^2)^{-\sigma}\,d\xi \right]^{1/2}$$
$$= C_3 \|\phi\|_{|s-1|+1+\sigma},$$

so the lemma follows from Theorem (0.10).

It remains to prove (6.17). Since

$$\left| \frac{d}{dt}(1 + t^2)^{s/2} \right| = |st|(1 + t^2)^{(s/2)-1} \le |s|(1 + t^2)^{(s-1)/2},$$

by the mean value theorem of calculus we have

$$\begin{aligned}|(1 + a^2)^{s/2} - (1 + b^2)^{s/2}| &\le |a - b| \sup_{a \le t \le b} |s|(1 + t^2)^{(s-1)/2}\\ &\le |s|\,|a - b|\,[(1 + a^2)^{(s-1)/2} + (1 + b^2)^{(s-1)/2}]\end{aligned}$$

for all $a, b \ge 0$. Setting $a = |\xi|$ and $b = |\eta|$ and using the inequality $||\xi| - |\eta|| \le |\xi - \eta|$, we obtain (6.17). ∎

(6.18) Proposition.
If $s \in \mathbb{R}$ and $\sigma > \frac{1}{2}n$ there is a constant $C = C_{s,\sigma}$ such that for all $\phi \in \mathcal{S}$ and $f \in H_s$,

$$\|\phi f\|_s \le \left[\sup_x |\phi(x)| \right] \|f\|_s + C\|\phi\|_{|s-1|+1+\sigma}\|f\|_{s-1}.$$

Proof: Since $\| \cdot \|_0$ is the L^2 norm, by Lemma (6.16) we have

$$\|\phi f\|_s = \|\Lambda^s \phi f\|_0 \leq \|\phi \Lambda^s f\|_0 + \|[\Lambda^s, \phi] f\|_0$$

$$\leq \left[\sup_x |\phi(x)| \right] \|\Lambda^s f\|_0 + C \|\phi\|_{|s-1|+1+\sigma} \|f\|_{s-1}$$

$$\leq \left[\sup_x |\phi(x)| \right] \|f\|_s + C \|\phi\|_{|s-1|+1+\sigma} \|f\|_{s-1} . \qquad \blacksquare$$

In our study of elliptic operators we shall sometimes wish to determine when certain derivatives of a function $f \in H_s$ are also in H_s. For this purpose we shall use Nirenberg's method of difference quotients, which is based on the following simple result.

First, some notation. If ϕ is a function on $\mathbb{R}^n$ and $x \in \mathbb{R}^n$, we define the translate ϕ_x of ϕ by $\phi_x(y) = \phi(y + x)$. If f is a distribution, we then define the translate f_x by the formula $\langle f_x, \phi \rangle = \langle f, \phi_{-x} \rangle$. Next, let $e_1, \ldots, e_n$ be the standard basis for $\mathbb{R}^n$. If f is a distribution, $h \in \mathbb{R} \setminus \{0\}$, and $1 \leq j \leq n$, we define the **difference quotient** $\Delta_h^j f$ by

$$\Delta_h^j f = h^{-1}(f_{he_j} - f).$$

We also introduce the following notation for products of difference quotients. If α is a multi-index and $\mathbf{h} = \{ h_{jk} : 1 \leq j \leq n, \ 1 \leq k \leq \alpha_j \}$ is a set of $|\alpha|$ nonzero real numbers, we define

$$\Delta_{\mathbf{h}}^\alpha = \prod_{j=1}^n \prod_{k=1}^{\alpha_j} \Delta_{h_{jk}}^j .$$

We also define $|\mathbf{h}|$, the "norm" of $\mathbf{h}$, to be

$$|\mathbf{h}| = \sum_{j=1}^n \sum_{k=1}^{\alpha_j} |h_{jk}|.$$

(6.19) Theorem.
Suppose $u \in H_s$ for some $s \in \mathbb{R}$ and α is a multi-index. Then

$$\|\partial^\alpha f\|_s = \limsup_{|\mathbf{h}| \to 0} \|\Delta_{\mathbf{h}}^\alpha f\|_s$$

(whether both sides are finite or infinite). In particular, $\partial_\alpha f \in H_s$ if and only if $\Delta_{\mathbf{h}}^\alpha f$ remains bounded in H_s as $|\mathbf{h}| \to 0$.

Proof: For simplicity we shall present the proof for $|\alpha| = 1$, i.e., $\Delta_{\mathbf{h}}^\alpha = \Delta_h^j$; the argument in general is exactly the same. In the first place,

$$(\Delta_h^j f)\widehat{\ }(\xi) = \frac{e^{2\pi i h \xi_j} - 1}{h} \widehat{f}(\xi) = 2i e^{\pi i h \xi_j} \frac{\sin \pi h \xi_j}{h} \widehat{f}(\xi).$$

Since $|h^{-1} \sin \pi h \xi_j| \leq \pi |\xi_j|$ for all h and ξ_j, clearly

$$\limsup_{h \to 0} \|\Delta_h^j f\|_s^2 \leq \int (1 + |\xi|^2)^s |2\pi i \xi_j \widehat{u}(\xi)|^2 \, d\xi = \|\partial_j u\|_s^2.$$

Moreover, since $h^{-1} \sin \pi h \xi_j \to \pi \xi_j$ as $h \to 0$, it follows from the dominated convergence theorem that equality holds provided $\|\partial_j u\|_s$ is finite. (We can even replace "lim sup" by "lim".) On the other hand, if $\|\partial_j u\|_s = \infty$, given any $N > 0$ we can find $R > 0$ large enough so that

$$\int_{|\xi| \leq R} (1 + |\xi|^2)^s |2\pi i \xi_j \widehat{f}(\xi)|^2 \, d\xi > 2N,$$

which implies that for sufficiently small h,

$$\|\Delta_j^h f\|_s^2 \geq \int_{|\xi| \leq R} (1 + |\xi|^2)^s |2h^{-1} \sin \pi h \xi_j|^2 |\widehat{f}(\xi)|^2 \, d\xi > N.$$

Thus $\|\Delta_h^j f\|_s \to \infty$ as $h \to 0$. ∎

We also need the following fact about difference quotients.

(6.20) Proposition.
If $s \in \mathbb{R}$ and $\phi \in \mathcal{S}$, the operator $[\Delta_{\mathbf{h}}^\alpha, \phi]$ is bounded from H_s to $H_{s-|\alpha|+1}$ with bound independent of $\mathbf{h}$.

Proof: First suppose $|\alpha| = 1$, so $\Delta_{\mathbf{h}}^\alpha = \Delta_h^j$. Clearly

$$[\Delta_h^j, \phi] f = \frac{(\phi f)_{he_j} - \phi f - [\phi f_{he_j} - \phi f]}{h} = (\Delta_h^j \phi) f_{he_j}.$$

Since the translations $f \to f_{he_j}$ are isometries on H_s and $\Delta_h^j \phi$ converges smoothly to $\partial_j \phi$ as $h \to 0$, by Proposition (6.18) we have $\|[\Delta_h^j, \phi] f\|_s \leq C\|f\|_s$ with C independent of h.

If $|\alpha| > 1$, we can commute ϕ through the factors of $\Delta_{\mathbf{h}}^\alpha$ one at a time, yielding $[\Delta_{\mathbf{h}}^\alpha, \phi]$ as a sum of terms of the form $\Delta_{\mathbf{h}'}^{\alpha'} [\Delta_h^j, \phi] \Delta_{\mathbf{h}''}^{\alpha''}$ where $\alpha' + e_j + \alpha'' = \alpha$. But by Theorems (6.3) and (6.19) and the result for $|\alpha| = 1$,

$$\|\Delta_{\mathbf{h}'}^{\alpha'} [\Delta_h^j, \phi] \Delta_{\mathbf{h}''}^{\alpha''} f\|_{s-|\alpha|+1} \leq \|[\Delta_h^j, \phi] \Delta_{\mathbf{h}''}^{\alpha''} f\|_{s-|\alpha|+|\alpha'|+1}$$
$$\leq C\|\Delta_{\mathbf{h}''}^{\alpha''} f\|_{s-|\alpha|+|\alpha'|+1}$$
$$\leq C\|f\|_{s-|\alpha|+|\alpha'|+|\alpha''|+1} = C\|f\|_s. \quad ∎$$

(6.21) Corollary.
If $L \doteq \sum_{|\beta| \leq k} a_\beta \partial^\beta$ is a differential operator of order k with coefficients in $\mathcal{S}$, then for any $s \in \mathbb{R}$ the operator $[\Delta_{\mathfrak{h}}^\alpha, L]$ is bounded from H_s to $H_{s-k-|\alpha|+1}$ with bound independent of $\mathfrak{h}$.

Proof: Simply observe that $\Delta_{\mathfrak{h}}^\alpha$ commutes with ∂^β, so that $[\Delta_{\mathfrak{h}}^\alpha, L] = \sum [\Delta_{\mathfrak{h}}^\alpha, a_\beta] \partial^\beta$, and apply Theorem (6.3) and Proposition (6.20).   ∎

We now present an interpolation theorem for operators on Sobolev spaces. The proof, like the proof of the Riesz-Thorin interpolation theorem for operators on L^p spaces which it closely resembles, is based on the following lemma from complex analysis.

(6.22) The Three Lines Lemma.
Suppose $F(\zeta)$ is a bounded continuous function in the strip $0 \leq \operatorname{Re}\zeta \leq 1$ which is holomorphic for $0 < \operatorname{Re}\zeta < 1$. If $|F(\zeta)| \leq C_0$ for $\operatorname{Re}\zeta = 0$ and $|F(\zeta)| \leq C_1$ for $\operatorname{Re}\zeta = 1$, then $|F(\zeta)| \leq C_0^{1-\sigma} C_1^\sigma$ for $\operatorname{Re}\zeta = \sigma$, $0 < \sigma < 1$.

Proof: For $\epsilon > 0$, the function $G_\epsilon(\zeta) = C_0^{\zeta-1} C_1^{-\zeta} e^{\epsilon(\zeta^2 - \zeta)} F(\zeta)$ satisfies $|G_\epsilon(\zeta)| \leq 1$ for $\operatorname{Re}\zeta = 0$ or $\operatorname{Re}\zeta = 1$, and also $|G_\epsilon(\zeta)| \to 0$ as $|\operatorname{Im}\zeta| \to \infty$ for $0 \leq \operatorname{Re}\zeta \leq 1$. Applying the maximum modulus theorem on the rectangle $0 \leq \operatorname{Re}\zeta \leq 1$, $|\operatorname{Im}\zeta| \leq R$ with R large, we see that $|G_\epsilon(\zeta)| \leq 1$ for $0 \leq \operatorname{Re}\zeta \leq 1$. Letting $\epsilon \to 0$, we conclude that

$$C_0^{\sigma-1} C_1^{-\sigma} |F(\zeta)| = \lim_{\epsilon \to 0} |G_\epsilon(\zeta)| \leq 1 \quad \text{for } \operatorname{Re}\zeta = \sigma,$$

which is the desired result.   ∎

(6.23) Theorem.
Suppose $s_0 < s_1$ and $t_0 < t_1$, and for $\zeta \in \mathbb{C}$ let

$$s(\zeta) = (1-\zeta)s_0 + \zeta s_1, \qquad t(\zeta) = (1-\zeta)t_0 + \zeta t_1.$$

Suppose T is a bounded linear map from H_{s_0} to H_{t_0} whose restriction to H_{s_1} is bounded from H_{s_1} to H_{t_1}. Then the restriction of T to $H_{s(\sigma)}$ is bounded from $H_{s(\sigma)}$ to $H_{t(\sigma)}$ for $0 < \sigma < 1$. More precisely, if $\|Tf\|_{t_0} \leq C_0 \|f\|_{s_0}$ and $\|Tf\|_{t_1} \leq C_1 \|f\|_{s_1}$, then $\|Tf\|_{t(\sigma)} \leq C_0^{1-\sigma} C_1^\sigma \|f\|_{s(\sigma)}$.

Proof: Given $\phi, \psi \in \mathcal{S}$, let

$$F(\zeta) = \langle T\Lambda^{-s(\zeta)}\phi \mid \Lambda^{t(\zeta)}\psi \rangle.$$

Here Λ^z is defined for $z \in \mathbb{C}$ just as in (6.4); for $z = x + iy$ we have $\Lambda^z = \Lambda^x \Lambda^{iy}$ and $\|\Lambda^z \phi\|_s = \|\Lambda^x \phi\|_s = \|\phi\|_{s-x}$ since $|(1 + |\xi|^2)^{iy/2}| \equiv 1$. Also, since $(1 + |\xi|^2)^{z/2}$ is an entire holomorphic function of z, it follows easily that $F(\zeta)$ is an entire holomorphic function of ζ. The hypotheses of the theorem imply that when $\operatorname{Re}\zeta = 0$,

$$|F(\zeta)| \le \|T\Lambda^{-s(\zeta)}\phi\|_{t_0} \|\Lambda^{t(\zeta)}\psi\|_{-t_0} \le C_0 \|\Lambda^{-s(\zeta)}\phi\|_{s_0} \|\Lambda^{t(\zeta)}\psi\|_{-t_0}$$
$$= C_0 \|\Lambda^{-s_0}\phi\|_{s_0} \|\Lambda^{t_0}\psi\|_{-t_0} = C_0 \|\phi\|_0 \|\psi\|_0,$$

and likewise for $\operatorname{Re}\zeta = 1$,

$$|F(\zeta)| \le \|T\Lambda^{-s(\zeta)}\phi\|_{t_1} \|\Lambda^{t(\zeta)}\psi\|_{-t_1} \le C_1 \|\phi\|_0 \|\psi\|_0.$$

Moreover, for $\operatorname{Re}\zeta = \sigma$, $0 \le \sigma \le 1$,

$$|F(\zeta)| \le \|T\Lambda^{-s(\zeta)}\phi\|_{t_0} \|\Lambda^{t(\zeta)}\psi\|_{-t_0} \le C_0 \|\Lambda^{-s(\zeta)}\phi\|_{s_0} \|\Lambda^{t(\zeta)}\psi\|_{-t_0}$$
$$= C_0 \|\phi\|_{\sigma(s_0 - s_1)} \|\psi\|_{\sigma(t_1 - t_0)} \le C_0 \|\phi\|_0 \|\psi\|_{t_1 - t_0}.$$

Thus, by the three lines lemma,

$$|F(\sigma)| \le C_0^{1-\sigma} C_1^\sigma \|\phi\|_0 \|\psi\|_0, \qquad 0 \le \sigma \le 1.$$

But $F(\sigma) = \langle \Lambda^{t(\sigma)} T \Lambda^{-s(\sigma)}\phi \mid \psi \rangle$, and $\mathcal{S}$ is dense in H_0. Hence this estimate plus the self-duality of H_0 implies that $\Lambda^{t(\sigma)} T \Lambda^{-s(\sigma)}$ is bounded on H_0, and this in turn means that T is bounded from $H_{s(\sigma)}$ to $H_{t(\sigma)}$:

$$\|Tf\|_{t(\sigma)} = \|\Lambda^{t(\sigma)} T \Lambda^{-s(\sigma)} \Lambda^{s(\sigma)} f\|_0$$
$$\le C_0^{1-\sigma} C_1^\sigma \|\Lambda^{s(\sigma)} f\|_0 = C_0^{1-\sigma} C_1^\sigma \|f\|_{s(\sigma)}. \qquad \blacksquare$$

As an application of Theorem (6.23), we obtain an easy proof of the local invariance of Sobolev spaces under smooth coordinate changes.

(6.24) Theorem.
Suppose Ω and Ω' are open sets in $\mathbb{R}^n$ and $\Theta : \Omega \to \Omega'$ a C^∞ map with C^∞ inverse. For any open set Ω_0' with compact closure in Ω', the map $f \to f \circ \Theta$ is bounded from $H_s^0(\Omega_0')$ to $H_s^0(\Theta^{-1}(\Omega_0'))$ for all $s \in \mathbb{R}$.

Proof: Let $J : \Omega \to \mathbb{R}$ and $\tilde{J} : \Omega' \to \mathbb{R}$ be the Jacobian determinants of Θ and Θ^{-1}; thus J and $\tilde{J}$ are C^∞ and nonvanishing, and $\tilde{J}(y) = [J(\Theta^{-1}y)]^{-1}$. Since J is bounded away from zero on $\Theta^{-1}(\Omega_0')$ whenever Ω_0' has compact closure in Ω', for any $f \in H_0^0(\Omega_0')$ we have

$$\int |f(y)|^2 \, dy = \int |(f \circ \Theta)(x)|^2 J(x) \, dx \ge C \int |(f \circ \Theta)(x)|^2 \, dx,$$

which proves the theorem in the case $s = 0$. In view of Theorem (6.1) and the chain rule, a similar calculation proves the theorem in the case where s is a positive integer. The case $s \geq 0$ then follows from Theorem (6.23).

From this we obtain the case $s < 0$ by a duality argument. Let Ω_1' be a neighborhood of $\overline{\Omega_0'}$ with compact closure in Ω'; then for all $f \in C_c^\infty(\Omega_0')$,

$$\|f \circ \Theta\|_s = \sup\left\{ \left| \int (f \circ \Theta)(x)\overline{g(x)}\,dx \right| : g \in C_c^\infty(\Theta^{-1}(\Omega_1')), \ \|g\|_{-s} \leq 1 \right\}.$$

But for such f and g, we have

$$\int (f \circ \Theta)(x)\overline{g(x)}\,dx = \int f(y)\overline{(g \circ \Theta^{-1})(y)}\,\tilde{J}(y)\,dy.$$

That is, the adjoint of the map $f \to f \circ \Theta$ is $g \to \tilde{J}(g \circ \Theta^{-1})$. The map $g \to g \circ \Theta^{-1}$ is bounded from $H_t^0(\Theta^{-1}(\Omega_1'))$ to $H_t^0(\Omega_1')$ for $t \geq 0$ by the preceding argument, and the map $h \to \tilde{J}h$ is bounded on $H_t^0(\Omega_1')$ by Proposition (6.12) since $\tilde{J}$ agrees with a Schwartz function on $\overline{\Omega_1'}$. Hence $g \to \tilde{J}(g \circ \Theta^{-1})$ is bounded from $H_t^0(\Theta^{-1}(\Omega_1'))$ to $H_t^0(\Omega_1')$ for $t \geq 0$. Therefore, for $s < 0$ and $f \in C_c^\infty(\Omega_0')$,

$$\|f \circ \Theta\|_s \leq \sup\left\{ \|f\|_s \|\tilde{J}(g \circ \Theta^{-1})\|_{-s} : g \in C_c^\infty(\Theta^{-1}(\Omega_1')), \ \|g\|_{-s} \leq 1 \right\}$$

$$\leq C_s \|f\|_s,$$

which implies that $f \to f \circ \Theta$ is bounded from $H_s^0(\Omega'')$ to $H_s^0(\Theta^{-1}(\Omega''))$. ∎

(6.25) Corollary.
The map $f \to f \circ \Theta$ is a bijection from $H_s^{\mathrm{loc}}(\Omega')$ to $H_s^{\mathrm{loc}}(\Omega)$.

Proof: We use Proposition (6.13). Suppose $f \in H_s^{\mathrm{loc}}(\Omega')$. If $\phi \in C_c^\infty(\Omega')$, then $\phi f \in H_s^0(\Omega'')$ for any precompact neighborhood Ω'' of supp ϕ in Ω', so $(\phi \circ \Theta)(f \circ \Theta) = (\phi f) \circ \Theta \in H_s$. But the map $\phi \to \phi \circ \Theta$ is clearly a bijection from $C_c^\infty(\Omega')$ to $C_c^\infty(\Omega)$, so this means that $f \circ \Theta \in H_s^{\mathrm{loc}}(\Omega)$. The same argument, with Θ replaced by Θ^{-1}, establishes the converse. ∎

As a consequence of Corollary (6.25), the space $H_s^{\mathrm{loc}}(M)$ can be intrinsically defined on ony C^∞ manifold M, and the space $H_s(M)$ can be intrinsically defined whenever M is compact. Let us explain this in detail for compact hypersurfaces in $\mathbb{R}^n$, the only case we shall need. (For those who know about manifolds, this should be a sufficient hint for the general case.)

Suppose S is a compact C^∞ hypersurface in $\mathbb{R}^n$. Then S can be covered by finitely many open sets $U_1, \ldots, U_M$ in $\mathbb{R}^n$ such that for $m = 1, \ldots, M$ there is a C^∞ map (with C^∞ inverse) Θ_m from U_m to a neighborhood of the origin in $\mathbb{R}^n$ such that $\Theta_m(S \cap U_m)$ is the intersection of $\Theta_m(U_m)$ with the hyperplane $x_n = 0$; we identify this hyperplane with $\mathbb{R}^{n-1}$. Let $\zeta_1, \ldots, \zeta_M$ be a partition of unity on S subordinate to the covering $\{U_m\}$ (Theorem (0.19)). Suppose f is a function on S, or more generally a distribution on S (= a continuous linear functional on $C^\infty(S)$). We say that $f \in H_s(S)$ if $(\zeta_m f) \circ \Theta_m^{-1} \in H_s(\mathbb{R}^{n-1})$ for each m. We can define a norm on $H_s(S)$ that makes $H_s(S)$ into a Hilbert space by

$$\|f\|_s^2 = \sum_1^m \|(\zeta_m f) \circ \Theta_m^{-1}\|_s^2 .$$

If we choose a different covering by coordinate patches or a different partition of unity, the norm we obtain will be different from but equivalent to this one, by Theorem (6.24) and Proposition (6.12); hence the space $H_s(S)$ is well-defined, independent of these choices.

If we combine this construction with Theorem (6.9), we immediately obtain:

(6.26) Corollary.
Let S be a compact C^∞ hypersurface in $\mathbb{R}^n$. If $s > \frac{1}{2}$, the restriction map $f \to f|S$ from $\mathcal{S}(\mathbb{R}^n)$ to $C^\infty(S)$ extends to a bounded linear map from $H_s(\mathbb{R}^n)$ to $H_{s-(1/2)}(S)$.

EXERCISES

1. Suppose $0 \neq \phi \in C_c^\infty(\mathbb{R}^n)$ and $\{a_j\}$ is a sequence in $\mathbb{R}^n$ with $|a_j| \to \infty$, and let $\phi_j(x) = \phi(x - a_j)$. Show that $\{\phi_j\}$ is bounded in H_s for every $s \in \mathbb{R}$ but has no convergent subsequence in H_t for any $t \in \mathbb{R}$.

2. In Exercise 1, replace ϕ by the point mass δ. Conclude that if $s < -\frac{1}{2}n$, the inclusion map $H_s^0(\Omega) \to H_t^0(\Omega)$ ($t < s$) is never compact unless Ω is bounded.

3. Fill in the details of the argument preceding Corollary (6.26) to show that the space $H_s(S)$ (S a compact C^∞ hypersurface in $\mathbb{R}^n$) is well-defined.

4. Deduce from the result of Exercise 4 in §6A that if $s > \frac{1}{2}n$ and $\phi \in H_s$, the map $f \to \phi f$ is bounded on H_t for $|t| \leq s$.

C. Local Regularity of Elliptic Operators

As the first application of the Sobolev machine, we shall derive the basic L^2 regularity properties of an elliptic operator L with C^∞ coefficients. The method is first to prove estimates for the derivatives of a function u in terms of derivatives of Lu, assuming that these derivatives exist; such estimates are known as **a priori estimates**. One then uses these estimates to show that smoothness of Lu implies smoothness of u.

Let

$$L = \sum_{|\alpha| \le k} a_\alpha(x) \partial^\alpha$$

be a differential operator of order k with C^∞ coefficients a_α (a qualification that will be assumed implicitly hereafter). We recall that L is **elliptic** at x_0 if $\sum_{|\alpha|=k} a_\alpha(x_0)\xi^\alpha \ne 0$ for all nonzero $\xi \in \mathbb{R}^n$. In this case we clearly have

(6.27)
$$\left| \sum_{|\alpha| \le k} a_\alpha(x_0)\xi^\alpha \right| \ge A|\xi|^k$$

for some $A > 0$, since both sides are nonzero and homogeneous of degree k. (A can be taken to be the minimum value of the left side of (6.27) on the unit sphere $|\xi| = 1$.) Moreover, since the a_α's are smooth, if L is elliptic on some compact set V, the constant A can be chosen to be independent of $x_0 \in V$.

Our first major result is the following a *priori* estimate.

(6.28) Theorem.
Suppose Ω is a bounded open set in $\mathbb{R}^n$ and $L = \sum_{|\alpha| \le k} a_\alpha \partial^\alpha$ is elliptic on a neighborhood of $\overline{\Omega}$. Then for any $s \in \mathbb{R}$ there is a constant $C > 0$ such that for all $u \in H^0_s(\Omega)$,

(6.29)
$$\|u\|_s \le C \left(\|Lu\|_{s-k} + \|u\|_{s-1} \right).$$

Proof: The argument proceeds in three steps: first we do the case where the a_α's are constant and $a_\alpha = 0$ for $|\alpha| < k$, then we remove the restriction that the a_α's be constant for $|\alpha| = k$, and finally we do the general case.

Step 1. Suppose $a_\alpha = 0$ for $|\alpha| < k$ and a_α is constant for $|\alpha| = k$. Then for $u \in H_s$ we can express Lu by the Fourier transform:

$$\widehat{Lu}(\xi) = (2\pi i)^k \sum_{|\alpha|=k} a_\alpha \xi^\alpha \widehat{u}(\xi).$$

Therefore, by (6.27),

$$(1+|\xi|^2)^s|\widehat{u}(\xi)|^2 \leq 2^k(1+|\xi|^2)^{s-k}(1+|\xi|^2)^k|\widehat{u}(\xi)|^2$$
$$\leq 2^k A^{-1}(1+|\xi|^2)^{s-k}|\widehat{Lu}(\xi)|^2 + 2^k(1+|\xi|^2)^{s-k}|\widehat{u}(\xi)|^2.$$

Integrating both sides and using the inequality $\|u\|_{s-k} \leq \|u\|_{s-1}$, we obtain

$$\|u\|_s^2 \leq 2^k A^{-1}\|Lu\|_{s-k}^2 + 2^k\|u\|_{s-1}^2,$$

which gives

$$\|u\|_s \leq C_0(\|Lu\|_{s-k} + \|u\|_{s-1})$$

with $C_0 = 2^{k/2}\max(A^{-1/2}, 1)$.

   Step 2. Assume again that $a_\alpha = 0$ for $|\alpha| < k$, but allow a_α to be variable for $|\alpha| = k$. For each $x_0 \in \Omega$, let

$$L_{x_0} = \sum_{|\alpha|=k} a_\alpha(x_0)\partial^\alpha.$$

Since $\overline{\Omega}$ is compact, we can take the constant A in (6.27) to be independent of $x_0 \in \Omega$, so by the result of Step 1,

(6.30) $$\|u\|_s \leq C_0(\|L_{x_0}u\|_{s-k} + \|u\|_{s-1})$$

for all $u \in H_s$, where C_0 is independent of x_0. Our plan will be to estimate

$$\|Lu - L_{x_0}u\|_{s-k} = \left\|\sum_{|\alpha|=k}[a_\alpha(\cdot) - a_\alpha(x_0)]\partial^\alpha u\right\|_{s-k}$$

for u supported in a small neighborhood of x_0, and then to write an arbitrary $u \in H_s^0(\Omega)$ as a sum of functions supported in small sets.

   There is no harm in assuming that $a_\alpha \in C_c^\infty(\mathbb{R}^n)$, as we can multiply the a_α's by a smooth cutoff function that equals 1 on $\overline{\Omega}$ without affecting Lu for $u \in H_s^0(\Omega)$. Thus there is a constant $C_1 > 0$ such that

$$|a_\alpha(x) - a_\alpha(x_0)| \leq C_1|x - x_0| \qquad (|\alpha| = k,\ x \in \mathbb{R}^n,\ x_0 \in \Omega).$$

Set $\delta = (4(2\pi n)^k C_0 C_1)^{-1}$, and fix $\phi \in C_c^\infty(B_{2\delta}(0))$ with $0 \leq \phi \leq 1$ and $\phi = 1$ on $B_\delta(0)$.

   Suppose $u \in H_s^0(\Omega)$ is supported in $B_\delta(x_0)$ for some $x_0 \in \Omega$. Then

$$[a_\alpha(x) - a_\alpha(x_0)]\partial^\alpha u(x) = \psi_{x_0,\alpha}(x)\partial^\alpha u(x)$$

where
$$\psi_{x_0,\alpha}(x) = \phi(x - x_0)[a_\alpha(x) - a_\alpha(x_0)].$$

We have $\sup_x |\psi_{x_0,\alpha}(x)| \leq C_1(2\delta) = (2(2\pi n)^k C_0)^{-1}$, so by Proposition (6.18),

$$\|[a_\alpha(\cdot) - a_\alpha(x_0)]\partial^\alpha u\|_{s-k} \leq (2(2\pi n)^k C_0)^{-1}\|\partial^\alpha u\|_{s-k} + C_2\|\partial^\alpha u\|_{s-k-1}$$
$$\leq (2n^k C_0)^{-1}\|u\|_s + (2\pi)^k C_2\|u\|_{s-1},$$

where C_2 depends only on $\|\psi_{x_0,\alpha}(x)\|_{|s-k-1|+n+1}$ and so can be taken independent of x_0. Thus, since there are fewer than n^k multi-indices α with $|\alpha| \leq k$, we have

$$\|Lu - L_{x_0}u\|_{s-k} \leq \sum_{|\alpha| \leq k} \|[a_\alpha(\cdot) - a_\alpha(x_0)]\partial^\alpha u\|_{s-k}$$
$$\leq (2C_0)^{-1}\|u\|_s + (2\pi n)^k C_2\|u\|_{s-1}.$$

Combining this with (6.30),

$$\|u\|_s \leq C_0(\|Lu\|_{s-k} + \|L_{x_0}u - Lu\|_{s-k} + \|u\|_{s-1})$$
$$\leq C_0\|Lu\|_{s-k} + \tfrac{1}{2}\|u\|_s + [(2\pi n)^k C_2 + 1]C_0\|u\|_{s-1},$$

so that
$$\|u\|_s \leq C_3(\|Lu\|_{s-k} + \|u\|_{s-1})$$

where $C_3 = 2[(2\pi n)^k C_2 + 1]C_0$ is independent of x_0. Thus the estimate is established for u supported in $B_\delta(x_0)$ and is uniform in x_0.

Since Ω is bounded, we can cover $\overline{\Omega}$ by a finite collection of balls $B_\delta(x_1), \ldots, B_\delta(x_N)$ with $x_j \in \Omega$. Let $\{\zeta_j\}_1^N$ be a partition of unity on $\overline{\Omega}$ subordinate to this covering (Theorem (0.19)). Then for any $u \in H_s^0(\Omega)$, $\zeta_j u$ is in $H_s^0(\Omega)$ and is supported in a ball of radius δ for each j, so

$$\|u\|_s = \left\|\sum_1^N \zeta_j u\right\|_s \leq \sum_1^N \|\zeta_j u\|_s$$
$$\leq C_3 \sum_1^N (\|L(\zeta_j u)\|_{s-k} + \|\zeta_j u\|_{s-1}).$$

But $L(\zeta_j u) = \zeta_j Lu + [L, \zeta_j]u$, and $[L, \zeta_j]$ is a differential operator of order $k - 1$ with coefficients in C_c^∞ (check it!), so by Theorem (6.3) and Proposition (6.12) we obtain the desired estimate:

$$\|u\|_s \leq C_3 \sum_1^N (\|\zeta_j Lu\|_{s-k} + \|[L, \zeta_j]u\|_{s-k} + \|\zeta_j u\|_{s-1})$$
$$\leq C_4(\|Lu\|_{s-k} + \|u\|_{s-1}).$$

Step 3. Now consider a general $L = L^0 + L^1$, where

$$L^0 = \sum_{|\alpha|=k} a_\alpha \partial^\alpha, \qquad L^1 = \sum_{|\alpha|<k} a_\alpha \partial^\alpha.$$

By the result of Step 2, we know that for all $u \in H_s^0(\Omega)$,

$$\|u\|_s \le C_4 \big(\|L^0 u\|_{s-k} + \|u\|_{s-1} \big).$$

On the other hand, as in Step 2 we may assume that the coefficients of L^1 are in C_c^∞, so by (6.3) and (6.12) again we have $\|L^1 u\|_{s-k} \le C_5 \|u\|_{s-1}$. Therefore,

$$\begin{aligned} \|u\|_s &\le C_4 \big(\|Lu\|_{s-k} + \|L^1 u\|_{s-k} + \|u\|_{s-1} \big) \\ &\le C \big(\|Lu\|_{s-k} + \|u\|_{s-1} \big) \end{aligned}$$

where $C = C_4(C_5 + 1)$. The proof is complete. ∎

(6.31) Corollary.
For any $t \le s - 1$ there is a constant $C_t > 0$ such that for all $u \in H_s^0(\Omega)$,

$$\|u\|_s \le C_t \big(\|Lu\|_{s-k} + \|u\|_t \big).$$

Proof: By Lemma (6.11), for some $C_t' > 0$ we have

$$\|u\|_{s-1} \le (2C)^{-1} \|u\|_s + C_t' \|u\|_t,$$

where C is the constant in (6.29). Substituting this into the (6.29) yields

$$\|u\|_s \le 2C \big(\|Lu\|_{s-k} + C_t' \|u\|_t \big). \quad ∎$$

We now use the Theorem (6.28) to prove the local L^2 regularity theorem for elliptic operators.

(6.32) Lemma.
Suppose Ω is an open set in $\mathbb{R}^n$ and L is an elliptic operator of order k with C^∞ coefficients on Ω. If $u \in H_s^{\mathrm{loc}}(\Omega)$ and $Lu \in H_{s-k+1}^{\mathrm{loc}}(\Omega)$, then $u \in H_{s+1}^{\mathrm{loc}}(\Omega)$.

Proof: According to Proposition (6.13), we must show that $\psi u \in H_{s+1}$ for all $\psi \in C_c^\infty(\Omega)$. By hypothesis, $\psi u \in H_s$ and $\psi L u \in H_{s-k+1}$.

Also, $[L, \psi]$ [this notation was introduced before Lemma (6.16)] is a differential operator of order $k - 1$ with coefficients in $C_c^\infty(\Omega)$, so $[L, \psi]u \in H_{s-k+1}$ by Theorem (6.3) and Proposition (6.12). Hence,

$$L(\psi u) = \psi L u + [L, \psi]u \in H_{s-k+1}.$$

We shall apply the method of difference quotients. If $1 \le j \le n$ and $h \ne 0$ is sufficiently small, the distributions $\Delta_h^j(\psi u)$ are supported in a common compact subset Ω' of Ω, so we can apply Theorem (6.28) (on Ω') to them. Combining this with Corollary (6.21), we obtain

$$\begin{aligned}
\|\Delta_h^j(\psi u)\|_s &\le C\left(\|L\Delta_h^j(\psi u)\|_{s-k} + \|\Delta_h^j(\psi u)\|_{s-1}\right) \\
&\le C\left(\|\Delta_h^j L(\psi u)\|_{s-k} + \|[L, \Delta_h^j](\psi u)\|_{s-k} + \|\Delta_h^j(\psi u)\|_{s-1}\right) \\
&\le C\left(\|\Delta_h^j L(\psi u)\|_{s-k} + C'\|\psi u\|_s + \|\Delta_h^j(\psi u)\|_{s-1}\right).
\end{aligned}$$

Now apply Theorems (6.3) and (6.19), first to deduce that the right side remains bounded as $h \to 0$, and then to conclude that $\partial_j(\psi u) \in H_s$ for all j and hence that $\psi u \in H_{s+1}$. ∎

(6.33) The Elliptic Regularity Theorem.

Suppose Ω is an open set in $\mathbb{R}^n$ and L is an elliptic operator of order k with C^∞ coefficients on Ω. Let u and f be distributions on Ω satisfying $Lu = f$. If $f \in H_s^{\mathrm{loc}}(\Omega)$ for some $s \in \mathbb{R}$, then $u \in H_{s+k}^{\mathrm{loc}}(\Omega)$.

Proof: Given $\phi \in C_c^\infty(\Omega)$, we wish to show that $\phi u \in H_{s+k}$. Choose $\psi_0 \in C_c^\infty(\Omega)$ such that $\psi_0 = 1$ on a neighborhood of supp ϕ. By Corollary (6.8), $\psi_0 u \in H_t$ for some $t \in \mathbb{R}$. By decreasing t if necessary, we may assume that $N = s + k - t$ is a positive integer. Proceeding inductively, choose C^∞ functions $\psi_1, \ldots, \psi_{N-1}$ such that ψ_j is supported in the set where $\psi_{j-1} = 1$ and $\psi_j = 1$ on a neighborhood of supp ϕ. Finally, set $\psi_N = \phi$.

We shall prove by induction that $\psi_j u \in H_{t+j}$, and the Nth step will establish the theorem. The initial case $j = 0$ is true by assumption. Suppose then that $\psi_j u \in H_{t+j}$ where $0 \le j < N$. Then $\psi_{j+1}u = \psi_{j+1}\psi_j u \in H_{t+j}$ and $L(\psi_j u) = Lu = f$ on supp ψ_{j+1}. Moreover,

$$\begin{aligned}
L(\psi_{j+1}u) = L(\psi_{j+1}\psi_j u) &= \psi_{j+1}L(\psi_j u) + [L, \psi_{j+1}](\psi_j u) \\
&= \psi_{j+1}f + [L, \psi_{j+1}](\psi_j u) \\
&\in H_s + H_{t+j-k+1} = H_{t+j-k+1}.
\end{aligned}$$

By Lemma (6.32), $\psi_{j+1}u$ is in $H_{t+j+1}^{\mathrm{loc}}(\Omega)$, hence in H_{t+j+1} since it is compactly supported. ∎

(6.34) Corollary.
Every elliptic operator with C^∞ coefficients is hypoelliptic.

Proof: If $Lu = f \in C^\infty(\Omega)$, then $f \in H_s^{loc}(\Omega)$ for all s, so $u \in H_s^{loc}(\Omega)$ for all s. By Corollary (6.7), $u \in C^\infty(\Omega)$. ∎

If L is elliptic and has analytic coefficients, then u is analytic on any open set where Lu is analytic. This can be proved by using Theorem (6.28) and keeping an excruciatingly careful count of the constants to show that the Taylor series of u converges to u; see Friedman [21]. A more illuminating method is to show that elliptic operators with analytic coefficients have analytic fundamental solutions — that is, if L is elliptic with analytic coefficients on Ω, there is a distribution $K(x, y)$ on $\Omega \times \Omega$ that is an analytic function away from $\{(x, y) : x = y\}$ and satisfies $L_x K(x, y) = \delta(x - y)$. One can then argue as in the proof of Theorem (2.27). For the construction of the fundamental solution, see John [29].

The analogues of Theorems (6.28) and (6.33) for the L^p Sobolev spaces are valid for $1 < p < \infty$, but the proofs are deeper and require the Calderón-Zygmund theory of singular integrals. The corresponding L^1 and L^∞ estimates are generally false; in particular, it is not true that if $Lu \in C^m(\Omega)$ then $u \in C^{m+k}(\Omega)$. However, if $0 < \alpha < 1$ and $Lu \in C^{m+\alpha}(\Omega)$ then $u \in C^{m+k+\alpha}(\Omega)$. (We proved this for $L = \Delta$ in §2C, and the proof in general is in the same spirit, relying on a generalized form of Theorem (2.29).) These results follow from the construction of an approximate inverse for elliptic operators that we shall present in Chapter 8 together with estimates for pseudodifferential operators that can be found in Stein [46, §VI.5] or Taylor [48, §XI.2].

D. Constant-Coefficient Hypoelliptic Operators

The ideas in the preceding section can be extended to prove hypoellipticity for operators other than elliptic ones. Indeed, in combination with some other techniques from algebra and functional analysis, they enabled Hörmander to obtain a complete algebraic characterization of those operators with constant coefficients that are hypoelliptic. In this section we present Hörmander's theorem, not only as an elegant result in its own right but as a beautiful example of the interplay of different areas of mathematics and as an application of Sobolev spaces in which the use of spaces of fractional order is crucial. (However, we omit the proof of one part of the theorem that is purely algebraic.)

We begin with some notation. It will be convenient to dispose of certain factors of $2\pi i$ that arise in Fourier analysis by setting

$$D = \frac{1}{2\pi i}\partial, \quad \text{i.e.,} \quad D_j = \frac{1}{2\pi i}\partial_j \text{ and } D^\alpha = \frac{1}{(2\pi i)^{|\alpha|}}\partial^\alpha.$$

Every polynomial P in n variables with complex coefficients then defines a differential operator $P(D)$:

$$P(\xi) = \sum_{|\alpha|\le k} c_\alpha \xi^\alpha, \qquad P(D) = \sum_{|\alpha|\le k} c_\alpha D^\alpha,$$

and every constant-coefficient operator is of this form. If P is a polynomial and α is a multi-index, we set

$$P^{(\alpha)} = \partial^\alpha P,$$

and we then have the following form of the product rule:

(6.35) $$P(D)[fg] = \sum_{|\alpha|\ge 0} \frac{1}{\alpha!}[D^\alpha f][P^{(\alpha)}(D)g].$$

(The proof is left to the reader; see Exercise 1.)

We shall need to consider complex zeros of polynomials. In what follows, ξ and η will denote elements of $\mathbb{R}^n$ and ζ will denote an element of $\mathbb{C}^n$. For any polynomial P we define

$$\mathcal{Z}(P) = \{\zeta \in \mathbb{C}^n : P(\zeta) = 0\}.$$

$\mathcal{Z}(P)$ is always unbounded when $n > 1$ (unless P is constant), because for any $\zeta_1,\ldots,\zeta_{n-1} \in \mathbb{C}$, no matter how large, there exist values of ζ_n for which $P(\zeta_1,\ldots,\zeta_n) = 0$. For $\xi \in \mathbb{R}^n$, we define $d_P(\xi)$ to be the distance from ξ to $\mathcal{Z}(P)$:

$$d_P(\xi) = \inf\{|\zeta - \xi| : \zeta \in \mathcal{Z}(P)\}.$$

Here, then, is Hörmander's theorem.

(6.36) Theorem.
If P is a polynomial of degree $k > 0$, the following are equivalent:
(H1) *$|\operatorname{Im}\zeta| \to \infty$ as $\zeta \to \infty$ in the set $\mathcal{Z}(P)$.*
(H2) *$d_P(\xi) \to \infty$ as $\xi \to \infty$ in $\mathbb{R}^n$.*
(H3) *There exist $\delta, C, R > 0$ such that $d_P(\xi) \ge C|\xi|^\delta$ when $\xi \in \mathbb{R}^n$, $|\xi| > R$.*
(H4) *There exist $\delta, C, R > 0$ such that $|P^{(\alpha)}(\xi)| \le C|\xi|^{-\delta|\alpha|}|P(\xi)|$ for all α when $\xi \in \mathbb{R}^n$, $|\xi| > R$.*
(H5) *There exists $\delta > 0$ such that if $f \in H_s^{\text{loc}}(\Omega)$ for some open $\Omega \subset \mathbb{R}^n$, every solution u of $Lu = f$ is in $H_{s+k\delta}^{\text{loc}}(\Omega)$.*
(H6) *$P(D)$ is hypoelliptic.*

Proof: Before beginning the labor of the proof, we make a few remarks. First, our arguments will show that if P satisfies (H3) for a particular δ, it satisfies (H4) for the same δ; and if it satisfies (H4) for a particular δ, it satisfies (H5) for the same δ. In fact, the optimal δ's for all these conditions are equal.

Second, some of the implications in the theorem are easy. A moment's thought shows that (H1) and (H2) are equivalent, and clearly (H3) implies (H2). Moreover, (H5) implies (H6) in view of Corollary (6.7).

We refer the reader to Hörmander [26] or [27, vol. II] for the proof that (H2) implies (H3), which is purely a matter of algebra. It requires some results from semi-algebraic geometry (the theory of sets defined by real polynomial equations and inequalities), specifically, the so-called Tarski-Seidenberg theorem. (These names should suggest to the reader that mathematical logic is casting a shadow here. Indeed, Tarski and Seidenberg's main concern was the construction of a decision procedure for solving polynomial equations and inequalities.)

Taking the implication (H2) $\Longrightarrow$ (H3) for granted, then, to prove the theorem it will suffice to show that (H3) $\Longrightarrow$ (H4), (H4) $\Longrightarrow$ (H5), and (H6) $\Longrightarrow$ (H1).

Proof that (H3) $\Longrightarrow$ (H4):   We first claim that $|P(\xi + \zeta)| \le 2^k |P(\xi)|$ if $|\zeta| \le d_P(\xi)$. To show this, consider the one-variable polynomial $g(\tau) = P(\xi + \tau\zeta)$. If $|\zeta| \le d_P(\xi)$, the zeros $\tau_1, \ldots, \tau_k$ of g satisfy $|\tau_j| \ge 1$, so as in the proof of Lemma (1.53),

$$\frac{|P(\xi + \zeta)|}{|P(\xi)|} = \left|\frac{g(1)}{g(0)}\right| = \prod_1^k \left|\frac{\tau_j - 1}{\tau_j}\right| \le 2^k.$$

Now, by applying the Cauchy integral formula to P in each variable, we see that

$$P^{(\alpha)}(\xi) = \frac{\alpha!}{(2\pi i)^n} \int_{|\zeta_1| = r} \cdots \int_{|\zeta_n| = r} \frac{P(\xi + \zeta)}{\prod_1^n \zeta_j^{\alpha_j + 1}} \, d\zeta_1 \cdots d\zeta_n$$

for any $r > 0$. If we take $r = n^{-1/2} d_P(\xi)$ then $|\zeta| = d_P(\xi)$ when $|\zeta_j| = r$ for all j, so that $|P(\xi + \zeta)| \le 2^k |P(\xi)|$, and hence

$$|P^{(\alpha)}(\xi)| \le \frac{\alpha! 2^k |P(\xi)|}{[n^{-1/2} d_P(\xi)]^{|\alpha|}}.$$

Of course $P^{(\alpha)} \equiv 0$ for $|\alpha| > k$, so this estimate together with (H3) gives (H4).

Proof that (H4) $\Longrightarrow$ (H5): This argument is similar to the proof of Theorem (6.30), with an extra twist. Suppose $f \in H_s^{\text{loc}}(\Omega)$ and $Lu = f$. Given $\phi \in C_c^\infty(\Omega)$, we wish to prove that $\phi u \in H_{s+k\delta}$. Choose $\psi_0 \in C_c^\infty(\Omega)$ with $\psi_0 = 1$ on a neighborhood of supp ϕ. By Corollary (6.8), $\psi_0 u \in H_t$ for some $t \in \mathbb{R}$, and by decreasing t if necessary we may assume that $t = s + k - 1 - M\delta$ for some positive integer M. Choose C^∞ functions $\psi_1, \ldots, \psi_M$ inductively so that ψ_m is supported in the set where $\psi_{m-1} = 1$ and $\psi_m = 1$ on a neighborhood of supp ϕ, and let $\psi_{M+1} = \phi$.

Since $\psi_0 = 1$ on supp ψ_1, by (6.35) we have

$$P(D)[\psi_1 u] = \psi_1 P(D)u + \sum_{\alpha \neq 0} \frac{1}{\alpha!}(D^\alpha \psi_1)P^{(\alpha)}(D)[\psi_0 u]$$

$$\in H_s + \sum_{\alpha \neq 0} H_{t-k+|\alpha|} = H_{t-k+1}.$$

That is,

$$\int (1 + |\xi|^2)^{t-k+1}|P(\xi)(\psi_1 u)\widehat{}(\xi)|^2 \, d\xi < \infty.$$

Thus by condition (H4),

$$\int (1 + |\xi|^2)^{t-k+1+\delta|\alpha|}|P^{(\alpha)}(\xi)(\psi_1 u)\widehat{}(\xi)|^2 \, d\xi < \infty,$$

which means that $P^{(\alpha)}(D)[\psi_1 u] \in H_{t-k+1+\delta|\alpha|}$. Next, since $\psi_1 = 1$ on supp ψ_2, by (6.35) we have

$$P(D)[\psi_2 u] = \psi_2 P(D)u + \sum_{\alpha \neq 0} \frac{1}{\alpha!}(D^\alpha \psi_2)P^{(\alpha)}(D)[\psi_1 u]$$

$$\in H_s + \sum_{\alpha \neq 0} H_{t-k+1+\delta|\alpha|} = H_{t-k+1+\delta},$$

so the argument above shows that $P^{(\alpha)}(D)[\psi_2 u] \in H_{t-k+1+\delta(1+|\alpha|)}$. Continuing inductively, we find that

$$P^{(\alpha)}(D)[\psi_m u] \in H_{t-k+1+\delta(m-1+|\alpha|)},$$

so finally, for $m = M + 1$,

$$P^{(\alpha)}(D)[\phi u] = P^{(\alpha)}(D)[\psi_{M+1} u] \in H_{t-k+1+\delta(M+|\alpha|)} = H_{s+\delta|\alpha|}.$$

Now, if $P(\xi) = \sum_{|\alpha| \leq k} c_\alpha \xi^\alpha$, pick α such that $|\alpha| = k$ and $c_\alpha \neq 0$. Then $P^{(\alpha)}(\xi) = \alpha! c_\alpha \neq 0$, so $\phi u = (\alpha! c_\alpha)^{-1} P^{(\alpha)}(D)[\phi u] \in H_{s+k\delta}$, and we are done.

Proof that (H6) $\Longrightarrow$ (H1): Suppose $P(D)$ is hypoelliptic. Let $B = B_1(0)$, and set

$$\mathcal{N} = \{f \in L^2(\mathbb{R}^n) : f = 0 \text{ outside } B \text{ and } P(D)f = 0 \text{ in } B\}.$$

(The elements of $\mathcal{N}$ are C^∞ functions on the complement of ∂B, but they are in general discontinuous on ∂B. For example, if g is any solution of $P(D)g = 0$ on a neighborhood of $\overline{B}$ then $g\chi_B \in \mathcal{N}$.) If $\{f_m\} \subset \mathcal{N}$ and $f_m \to f$ in L^2, then $f = 0$ outside B and $P(D)f$ is a distribution supported on ∂B since the same is true of $P(D)f_m$ for all m; hence $f \in \mathcal{N}$. Thus $\mathcal{N}$ is a closed subspace of L^2.

Next, let $B' = B_{1/2}(0)$. The elements of $\mathcal{N}$ are C^∞ on $\overline{B'}$; in particular, if $f \in \mathcal{N}$ then $D_j f \in L^2(B')$ for $1 \leq j \leq n$. We claim that $f \to D_j f|B'$ is a bounded linear map from the Hilbert space $\mathcal{N}$ to $L^2(B')$, so that

$$(6.37) \qquad \sum_1^n \int_{B'} |D_j f(x)|^2 \, dx \leq C \int_B |f(x)|^2 \, dx \qquad (f \in \mathcal{N}).$$

By the closed graph theorem, it suffices to show that if $f_m \to f$ in $\mathcal{N}$ and $D_j f_m \to g$ in $L^2(B')$ then $g = D_j f|B'$. But this is obvious: the hypotheses imply that $D_j f_m \to D_j f$ and $D_j f_m \to g$ in the weak topology of distributions on B', so $g = D_j f$ on B'.

Now, given $\zeta \in \mathcal{Z}(P)$, let $f(x) = e^{2\pi i \zeta \cdot x}\chi_B(x)$. Then $P(D)f(x) = P(\zeta)e^{2\pi i \zeta \cdot x} = 0$ on B, so $f \in \mathcal{N}$. Moreover, $D_j f(x) = \zeta_j e^{2\pi i \zeta \cdot x}$ for $x \in B$, so by (6.37),

$$\sum_1^n |\zeta_j|^2 \int_{B'} e^{-4\pi(\mathrm{Im}\,\zeta)\cdot x} \, dx \leq C \int_B e^{-4\pi(\mathrm{Im}\,\zeta)\cdot x} \, dx \qquad (\zeta \in \mathcal{Z}(P)).$$

If $|\mathrm{Im}\,\zeta| \leq R$, this gives

$$|\zeta|^2 \leq \frac{\int_B e^{4\pi R|x|} \, dx}{\int_{B'} e^{-4\pi R|x|} \, dx}.$$

In other words, if ζ varies in $\mathcal{Z}(P)$ but $\mathrm{Im}\,\zeta$ remains bounded then ζ remains bounded, which is precisely condition (H1). ∎

It may easily be imagined that the methods of the preceding section can be generalized to yield a hypoellipticity theorem for non-elliptic operators with variable coefficients, and this is indeed the case. Indeed, suppose $L = \sum a_\alpha(x)D^\alpha$ is an operator with C^∞ coefficients on $\Omega \subset \mathbb{R}^n$. If the polynomials $P_{x_0}(\xi) = \sum a_\alpha(x_0)\xi^\alpha$ $(x_0 \in \Omega)$ all satisfy the conditions of Theorem (6.36), and if these polynomials are all of comparable size in a certain precise sense, then L is hypoelliptic on Ω. See Hörmander [27, vol. II, §13.4]. However, this sufficient condition for hypoellipticity is far from being necessary in the variable-coefficient case. For example, the operator $L = \partial_x^2 + (\partial_y + x\partial_z)^2$ is known to be hypoelliptic on $\mathbb{R}^3$, but the operator $L_0 = \partial_x^2 + (\partial_y + a\partial_z)^2$ obtained by freezing the coefficients of L at a point (a,b,c) is not, because arbitrary functions of the form $f(x,y,z) = g(z - ay)$ satisfy $L_0 f = 0$. The hypoellipticity of this L follows from a beautiful theorem of Hörmander concerning hypoellipticity of operators of the form $L = \sum_1^k (A_j \cdot \nabla)^2 + (A_0 \cdot \nabla)$ where the A_j's are real C^∞ vector fields; see Hörmander [27, vol. III, §22.2].

EXERCISES

1. Prove (6.35). (Hint: It suffices to take P to be a monomial.)

2. Show that $P(D)$ is elliptic if and only if P satisfies (H4) with $\delta = 1$, and that no P of positive degree satisfies (H4) with $\delta > 1$.

3. Let P be a polynomial of degree k with real coefficients such that $P(D)$ is elliptic on $\mathbb{R}^n$. Let $Q(\xi, \tau) = 2\pi i\tau - P(\xi)$, so that $Q(D_x, D_t) = \partial_t - P(D)$. Show that Q satisfies condition (H4) (on $\mathbb{R}^{n+1}$) with $\delta = k^{-1}$ but not for any $\delta > k^{-1}$. (Hint: Consider the regions $|\xi|^k \leq |\tau|$ and $|\xi|^k \geq |\tau|$ separately.)

E. Sobolev Spaces on Bounded Domains

If Ω is a bounded open set in $\mathbb{R}^n$, we define the Sobolev space $H_k(\Omega)$ for k a nonnegative integer to be the completion of $C^\infty(\overline{\Omega})$ with respect to the norm

$$(6.38) \qquad \|f\|_{k,\Omega} = \left[\sum_{|\alpha| \leq k} \int_\Omega |\partial^\alpha f|^2 \right]^{1/2}.$$

By Theorem (6.1), $H_k^0(\Omega)$ is the completion of $C_c^\infty(\Omega)$ with respect to the same norm, so we have $H_k^0(\Omega) \subset H_k(\Omega) \subset H_k^{loc}(\Omega)$.

(If Ω is a bounded domain with smooth boundary, it is possible to develop a theory of Sobolev spaces $H_s(\Omega)$ of arbitrary real order. One defines $H_{-k}(\Omega)$ to be the dual of $H_k^0(\Omega)$ when k is a positive integer, and then defines $H_s(\Omega)$ for nonintegral s by an interpolation process; see Lions and Magenes [34]. However, we shall have no need of this refinement.)

The following basic properties of $H_k(\Omega)$ are obvious from the definitions:

i. If $j \leq k$, then $\|\cdot\|_{j,\Omega} \leq \|\cdot\|_{k,\Omega}$ and $H_k(\Omega)$ is included in $H_j(\Omega)$ as a dense subspace.

ii. If $|\alpha| \leq k$, ∂^α is a bounded map from $H_k(\Omega)$ to $H_{k-|\alpha|}(\Omega)$.

iii. If $\phi \in C^\infty(\overline{\Omega})$, the map $f \to \phi f$ is bounded on $H_k(\Omega)$ for all k. (This follows from the product rule for derivatives.)

iv. $H_k(\Omega)$ is invariant under C^∞ coordinate transformations on any neighborhood of $\overline{\Omega}$. (This follows from the chain rule.)

v. The restriction map $f \to f|\Omega$ is bounded from $H_k(\mathbb{R}^n)$ to $H_k(\Omega)$. (This follows via Theorem (6.1) from the estimate $\int_\Omega |f|^2 \leq \int_{\mathbb{R}^n} |f|^2$.)

Our definition of $H_k(\Omega)$ is designed to trivialize the problem of approximation by smooth functions. However, it would also be reasonable to consider the space of all functions on Ω whose distribution derivatives of order $\leq k$ are in $L^2(\Omega)$, which we denote by $W_k(\Omega)$:

$$W_k(\Omega) = \left\{ f \in L^2(\Omega) : \partial^\alpha f \in L^2(\Omega) \text{ for } |\alpha| \leq k \right\}.$$

$W_k(\Omega)$ is a Hilbert space with the norm (6.38). In general, $H_k(\Omega)$ is a proper subspace of $W_k(\Omega)$; see Exercise 1. However, if $\partial\Omega$ satisfies a mild smoothness condition, the two spaces coincide.

Specifically, we shall say that a bounded open set Ω has the **segment property** if there is an open covering $U_0, \ldots, U_N$ of $\overline{\Omega}$ with the following properties:

a. $U_0 \subset \Omega$;

b. $U_j \cap \partial\Omega \neq \varnothing$ for $j \geq 1$;

c. for each $j \geq 1$ there is a vector $y^j \in \mathbb{R}^n$ such that $x + \delta y^j \notin \overline{\Omega}$ for all $x \in U_j \setminus \Omega$ and $0 < \delta \leq 1$.

See Figure 6.1. In particular, if Ω is a domain with a C^1 boundary, then Ω has the segment property; see Exercise 2.

(6.39) Theorem.
If Ω has the segment property, then $H_k(\Omega) = W_k(\Omega)$.

Proof: We need only show that $W_k(\Omega) \subset H_k(\Omega)$. Let $U_0, \ldots, U_N$ be an open covering of $\overline{\Omega}$ as in the definition of the segment property. Choose

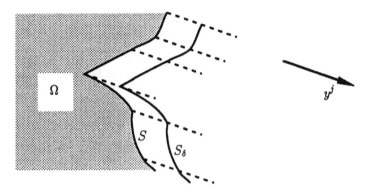

Figure 6.1. The segment property. The dotted lines represent the segments $x + ty^j$, $0 < t \le 1$, for various $x \in S = \partial\Omega$, and $S_\delta = \{x + \delta y^j : x \in S\}$.

another open covering $V_0, \ldots, V_N$ of $\overline{\Omega}$ such that $\overline{V}_j \subset U_j$ for all j, and let $\{\zeta_j\}_0^N$ be a partition of unity subordinate to the covering $\{V_j\}_0^N$. If $f \in W_k(\Omega)$, then clearly $\zeta_j f \in W_k(\Omega)$, and it is enough to show that $\zeta_j f \in H_k(\Omega)$ for all j.

For $j = 0$ this is easy. Choose $\phi \in C_c^\infty(B_1(0))$ with $\int \phi = 1$, and set $\phi_\epsilon(x) = \epsilon^{-n}\phi(\epsilon^{-1}x)$. Then $(\zeta_0 f) * \phi_\epsilon$ is C^∞ and supported in Ω for ϵ sufficiently small, and $\partial^\alpha[(\zeta_0 f) * \phi_\epsilon] = \partial^\alpha(\zeta_0 f) * \phi_\epsilon \to \partial^\alpha(\zeta_0 f)$ in $L^2(\Omega)$ as $\epsilon \to 0$ for $|\alpha| \le k$, by Theorem (0.13). Thus $\zeta_0 f \in H_k^0(\Omega) \subset H_k(\Omega)$.

Writing f instead of $\zeta_j f$, then, it suffices to assume that f is supported in some V_j, $j \ge 1$, where we extend f to be zero outside Ω. Set $S = \partial\Omega \cap V_j$. Then f and its distribution derivatives of order $\le k$ agree with L^2 functions on $\mathbb{R}^n \setminus S$. For $0 < \delta \le 1$, define $f^\delta(x) = f(x - \delta y^j)$ and $S^\delta = \{x + \delta y^j : x \in S\}$, where y^j is as in the definition of the segment property. Then f^δ and $\partial^\alpha f^\delta$ $(|\alpha| \le k)$ are in L^2 on $\mathbb{R}^n \setminus S^\delta$, f^δ is supported in U_j for δ sufficiently small, and $\overline{\Omega} \cap S_\delta = \varnothing$. It follows easily from Lemma (0.12) that

$$\sum_{|\alpha| \le k} \int_\Omega |\partial^\alpha f^\delta - \partial^\alpha f|^2 \to 0 \text{ as } \delta \to 0,$$

so it is enough to show that $f^\delta|\Omega \in H_k(\Omega)$. Given $\delta > 0$, choose $\phi \in C_c^\infty$ such that $\phi = 1$ on Ω and $\phi = 0$ near S^δ. Then $\phi f^\delta \in H_k(\mathbb{R}^n)$, so ϕf^δ is the limit in the H_k norm of functions in $\mathcal{S}(\mathbb{R}^n)$. It follows that $f^\delta|\Omega = \phi f^\delta|\Omega$ is the limit in the norm (6.38) of functions in $C^\infty(\overline{\Omega})$, i.e., $f \in H_k(\Omega)$. ∎

(6.40) Corollary.
If Ω has the segment property, then $f \in H_{k+j}(\Omega)$ if and only if $\partial^\alpha f \in H_k(\Omega)$ for $|\alpha| \le j$.

Proof: The assertion is obvious if H_{k+j} and H_k are replaced by W_{k+j} and W_k. ∎

We now derive a useful construction for extending elements of $H_k(\Omega)$ to a neighborhood of $\overline{\Omega}$ when Ω is a domain with smooth boundary. The starting point is the following gem of linear algebra. If $a_1, \ldots, a_m$ are complex numbers (or elements of an arbitrary field), the **Vandermonde matrix** $V(a_1, \ldots, a_m)$ is the $m \times m$ matrix whose jkth entry is $(a_j)^{k-1}$, $1 \le j, k \le m$.

(6.41) Lemma.
$\det V(a_1, \ldots, a_m) = \prod_{1 \le j < k \le m}(a_k - a_j)$. In particular, $V(a_1, \ldots, a_m)$ is nonsingular if and only if the a_j's are all distinct.

Proof: Clearly $\det V(a_1, \ldots, a_m) = 0$ if any two a_j's are equal, since then two rows of $V(a_1, \ldots, a_m)$ are equal. So suppose $a_1, \ldots, a_{m-1}$ are distinct, and regard $\det V(a_1, \ldots, a_m)$ as a function of a_m. It is a polynomial of degree $m - 1$ that vanishes when $a_m = a_j$ for some $j < m$, hence equals $c \prod_{j=1}^{m-1}(a_m - a_j)$. Here c is the coefficient of $(a_m)^{m-1}$, which is nothing but $\det V(a_1, \ldots, a_{m-1})$. The proof is therefore completed by induction. ∎

Returning now to Sobolev spaces, we introduce a notation that will be used repeatedly in the sequel:

$$(6.42) \qquad N(r) = \{x \in \mathbb{R}^n : |x| < r \text{ and } x_n < 0\}.$$

(6.43) Lemma.
Let $B_r = B_r(0)$, let $C_0^k(B_r)$ be the space of C^k functions supported in B_r, and let $C_-^k(N(r))$ be the space of C^k functions on $\overline{N(r)}$ that vanish near $|x| = r$. For each positive integer k there is a linear map $E_k : C_-^k(N(r)) \to C_0^k(B_r)$ such that $E_k f | N(r) = f$ and $\|E_k f\|_j \le C_j \|f\|_{j,N(r)}$ for $0 \le j \le k$, where C_j is independent of f and r.

Proof: By Lemma (6.41) the matrix $V(-1, \ldots, -k-1)$ is nonsingular, so there exist numbers $c_1, \ldots, c_{k+1}$ such that

$$\sum_{l=1}^{k+1} (-l)^m c_l = 1, \qquad (m = 0, \ldots, k).$$

We define the map E_k on $C_-^k(N(r))$ by

$$E_k f(x) = \begin{cases} f(x) & (x_n \leq 0), \\ \sum_{l=1}^{k+1} c_l f(x_1, \ldots, x_{n-1}, -lx_n) & (x_n > 0). \end{cases}$$

Then for $x_n > 0$ and $|\alpha| \leq k$ we have

$$\partial^\alpha E_k f(x) = \sum_{l=1}^{k+1} (-l)^{\alpha_n} c_l (\partial^\alpha f)(x_1, \ldots, x_{n-1}, -lx_n) \qquad (x_n > 0),$$

so the limits of $\partial^\alpha E_k f$ as x_n approaches zero from above or below are equal. Also, it is clear that $E_k f$ is supported in B_r, and hence $E_k f \in C_0^k(B_r)$. Finally,

$$\int_{B_r} |\partial^\alpha E_k f|^2 \leq \left[1 + \sum_{l=1}^{k+1} l^{2\alpha_n - 1} |c_l|^2 \right] \int_N (r) |\partial^\alpha f|^2,$$

whence $\|E_k f\|_j \leq C_j \|f\|_{j,N(r)}$ with C_j independent of f and r.   ∎

(6.44) Theorem.
Let Ω be a bounded domain on $\mathbb{R}^n$ with C^∞ boundary, and let $\widetilde{\Omega}$ be a bounded neighborhood of $\overline{\Omega}$. For each positive integer k there is a bounded linear map $E_k : H_k(\Omega) \to H_k^0(\widetilde{\Omega})$ such that $E_k f|\Omega = f$. E_k is also bounded from $H_j(\Omega)$ to $H_j^0(\widetilde{\Omega})$ for $0 \leq j \leq k$.

Proof: Let $V_0, \ldots, V_M$ be an open covering of $\overline{\Omega}$ with the following properties: (i) $V_m \subset \widetilde{\Omega}$ for all m, (ii) $\overline{V}_0 \subset \Omega$, (iii) for $m \geq 1$, V_m can be mapped to a ball $B_r(0)$ by a C^∞ map ψ_m with C^∞ inverse such that $\psi_m(V_m \cap \Omega) = N(r)$. Choose a partition of unity $\{\zeta_m\}_0^N$ on $\overline{\Omega}$ subordinate to this covering, and define $E_k f$ for $f \in C^k(\overline{\Omega})$ by

$$E_k f = \zeta_0 f + \sum_{m=1}^{N} [E_k((\zeta_m f) \circ \psi_m^{-1})] \circ \psi_m,$$

where the E_k on the right is given by Lemma (6.43). Then $E_k f$ is C^k and is supported in $\widetilde{\Omega}$, hence is in $H_k^0(\widetilde{\Omega})$. From Lemma (6.43) and the product and chain rules it follows that $\|E_k f\|_j \leq C_j \|f\|_{j,\Omega}$ for $j \leq k$, so E_k extends uniquely to a bounded map from $H_j(\Omega)$ to $H_j^0(\widetilde{\Omega})$.   ∎

Remark: By refining the argument in Lemma (6.43) one can construct an extension operator E_∞ that works for all k simultaneously; see Seeley [44].

With the aid if the extension map E_k we can easily obtain analogues of some of the major results of §6A and §6B for the spaces $H_k(\Omega)$.

(6.45) The Sobolev Lemma.
If Ω is a bounded domain with C^∞ boundary and $k > j + \frac{1}{2}n$, then $H_k(\Omega) \subset C^j(\overline{\Omega})$.

Proof: If $f \in H_k(\Omega)$ then $E_k f \in H_k^0(\widetilde{\Omega}) \subset H_k(\mathbb{R}^n) \subset C^j(\mathbb{R}^n)$, so $f \in C^j(\overline{\Omega})$. ∎

(6.46) Rellich's Theorem.
If Ω is a bounded domain with C^∞ boundary and $0 \le j < k$, the inclusion map $H_k(\Omega) \to H_j(\Omega)$ is compact.

Proof: If $\{f_l\}$ is a bounded sequence in $H_k(\Omega)$, $\{E_k f_l\}$ is bounded in $H_k^0(\widetilde{\Omega})$, so a subsequence $\{E_k f_{l_m}\}$ converges in $H_j^0(\widetilde{\Omega})$, and then $\{f_{l_m}\}$ converges in $H_j(\Omega)$. ∎

(6.47) Theorem.
If Ω is a bounded domain with C^∞ boundary and $k \ge 1$, the restriction map $f \to f|S$ from $C^k(\overline{\Omega})$ to $C^k(S)$ extends to a bounded map from $H_k(\Omega)$ to $H_{k-(1/2)}(\partial\Omega)$.

Proof: This is immediate from Corollary (6.26), since $E_k f|\partial\Omega = f|\partial\Omega$ for $f \in C^k(\overline{\Omega})$. ∎

(6.48) Corollary.
If $f \in H_k(\Omega)$, then $\partial^\alpha f|\partial\Omega$ is well defined as an element of $L^2(\partial\Omega)$ for $|\alpha| \le k - 1$.

(This is also an easy consequence of Corollary (2.40).)

(6.49) Corollary.
If $f \in H_k^0(\Omega)$ then $\partial^\alpha f|\partial\Omega = 0$ for $|\alpha| \le k - 1$.

Proof: This is true for $f \in C_c^\infty(\Omega)$, hence in general since $C_c^\infty(\Omega)$ is dense in $H_k^0(\Omega)$. ∎

The converse of Corollary (6.49) — that if $f \in H_k(\Omega)$ and $\partial^\alpha f|\partial\Omega = 0$ for $|\alpha| \le k - 1$ then $f \in H_k^0(\Omega)$ — is also true. The full proof is a bit technical; it can be found in Adams [1, Theorem 7.55] or Lions and Magenes [34, Theorem 11.5]. We shall content ourselves with proving the following slightly weaker result.

(6.50) Proposition.
Suppose Ω is a bounded domain with C^∞ boundary. If $f \in C^k(\overline{\Omega})$ and $\partial^\alpha f = 0$ on $\partial\Omega$ for $0 \leq |\alpha| \leq k - 1$, then $f \in H_k^0(\Omega)$.

Proof: We use the same construction as in the proof of Theorem (6.39), noting that Ω has the segment property (Exercise 2). By using a suitable partition of unity, it suffices to assume either that f is supported in Ω or that f is supported in a set V for which there exists $y \in \mathbb{R}^n$ such that $x + \delta y \notin \overline{\Omega}$ for $x \in V \setminus \Omega$ and $0 < \delta \leq 1$. The first case is obvious: just convolve f with a suitable approximation to the identity to obtain it as the limit of elements of $C_c^\infty(\Omega)$. For the second, we extend f to be zero outside Ω; then $f \in C^{k-1}(\mathbb{R}^n)$ and the k-th derivatives of f have only jump discontinuities along $\partial\Omega$, so $f \in H_k(\mathbb{R}^n)$. Define $f^\delta(x) = f(x + \delta y)$ for $0 < \delta \leq 1$. Then f^δ is supported in Ω, so $f^\delta \in H_k^0(\Omega)$ as above. But clearly $f^\delta \to f$ in the H_k norm as $\delta \to 0$, so $f \in H_k^0(\Omega)$.   ∎

Finally, we shall need the analogues of Theorem (6.19) and Proposition (6.20) for $H_k(N(r))$, where $N(r)$ is defined by (6.42). We note that if f is a function on $N(r)$ that vanishes near $|x| = r$, then the same is true of the difference quotient $\Delta_\mathbf{h}^\alpha f$ for $\mathbf{h}$ sufficiently small, provided $\alpha_n = 0$.

(6.51) Theorem.
Suppose $f \in H_k(N(r))$ and f vanishes near $|x| = r$, and let α be a multi-index with $\alpha_n = 0$. Then the distribution derivative $\partial^\alpha u$ is in $H_k(N(r))$ if and only if

$$A(f, \alpha) = \limsup_{|\mathbf{h}| \to 0} \|\Delta_\mathbf{h}^\alpha f\|_{k,N(r)} < \infty,$$

in which case $\|\partial^\alpha f\|_{k,N(r)} \leq C_k A(f, \alpha)$ where C_k is independent of f and r.

Proof: Let E_k be the extension map of Lemma (6.43). If $\partial^\alpha f \in H_k(N(r))$, then $E_k \partial^\alpha f \in H_k$. But $\partial^\alpha E_k f = E_k \partial^\alpha f$ and $\Delta_\mathbf{h}^\alpha E_k f = E_k \Delta_\mathbf{h}^\alpha f$ when $\alpha_n = 0$, so by Theorem (6.19),

$$A(f, \alpha) \leq \limsup_{|\mathbf{h}| \to 0} \|\Delta_\mathbf{h}^\alpha E_k f\|_k = \|\partial^\alpha E_k f\|_k < \infty.$$

Conversely, if $A(f, \alpha) < \infty$, then

$$\limsup_{|\mathbf{h}| \to 0} \|\Delta_{\mathbf{h}}^{\alpha} E_k f\|_k = \limsup_{|\mathbf{h}| \to 0} \|E_k \Delta_{\mathbf{h}}^{\alpha} f\|_k \leq C_k A(f, \alpha) < \infty,$$

where C_k is the constant in Lemma (6.43). Hence $\partial^{\alpha} E_k f \in H_k$, so $\partial^{\alpha} f \in H_k(N(r))$ and

$$\|\partial^{\alpha} f\|_{k, N(r)} \leq \|\partial^{\alpha} E_k f\|_k \leq C_k A(f, \alpha). \qquad \blacksquare$$

(6.52) Proposition.
Suppose $\phi \in C^{\infty}(\overline{N(r)})$ and α is a multi-index with $\alpha_n = 0$. There is a constant C such that for all $f \in H_k(N(r))$ which vanish near $|x| = r$ and all sufficiently small $\mathbf{h}$,

$$\|[\Delta_{\mathbf{h}}^{\alpha}, \phi] f\|_{k, N(r)} \leq C \sum_{\beta < \alpha} \|\partial^{\beta} f\|_{k, N(r)}.$$

(Here "$\beta < \alpha$" means that $\beta_j \leq \alpha_j$ for all j with strict inequality for at least one j.)

Proof: Same as the proof of Proposition (6.21). $\qquad \blacksquare$

EXERCISES

1. Let $\Omega = \{re^{i\theta} \in \mathbb{C} \cong \mathbb{R}^2 : -\pi < \theta < \pi\}$ (the unit disc with a line segment removed), and define the function f on Ω by $f(re^{i\theta}) = \theta \in (-\pi, \pi)$. Show that $f \in W_1(\Omega) \setminus H_1(\Omega)$.

2. Show that if Ω is a bounded domain with C^1 boundary, then Ω has the segment property. (Hint: Take the sets U_j for $j \geq 1$ to be small balls centered at suitably chosen points $x^j \in \partial\Omega$, and y^j to be a small positive multiple of the unit outward normal at x^j.)

Chapter 7
ELLIPTIC BOUNDARY VALUE PROBLEMS

Let L be an elliptic differential operator with C^∞ coefficients and let Ω be a bounded domain in $\mathbb{R}^n$ with C^∞ boundary S. In this chapter we shall study the equation $Lu = f$ on Ω, where u is to satisfy certain boundary conditions on S, the object being to prove existence, uniqueness, and regularity theorems. Our approach will be to formulate the problems in terms of sesquilinear forms that generalize the Dirichlet integral of §2F and then to apply some Hilbert space theory. For many of these problems a version of Dirichlet's principle holds — that is, solution of the problem is equivalent to minimizing the generalized Dirichlet integral in a suitable class of functions — and for this reason, boundary value problems set up in terms of Dirichlet forms are often said to be in "variational form." However, the calculus of variations will play no direct role in our work.

A. Strong Ellipticity

In Chapter 2 we saw that we obtain a "good" boundary value problem for the equation $\Delta u = f$ by specifying either u or $\partial_\nu u$ on the boundary. In general, for an elliptic operator of order k it turns out to be appropriate to impose $\frac{1}{2}k$ independent conditions on the Cauchy data of u on the boundary.

This causes problems if k is odd. For example, consider the Cauchy-Riemann operator

$$\partial_{\bar{z}} = \tfrac{1}{2}(\partial_x + i\partial_y)$$

on $\Omega \subset \mathbb{R}^2$. Solutions of $\partial_{\bar{z}}u = 0$ are holomorphic functions of $z = x + iy$ on Ω, so the Dirichlet data ($u = f$ on $S = \partial\Omega$) are overdetermined: not every function f on S extends holomorphically to Ω. For example, if Ω is the unit disc, a function $f(e^{i\theta}) = \sum_{-\infty}^{\infty} a_m e^{im\theta}$ on the unit circle extends holomorphically if and only if $a_m = 0$ for $m < 0$, in which case the extension is $u(z) = \sum_0^{\infty} a_m z^m$. Thus in some sense we can prescribe "half of the Dirichlet data" of u — namely, the Fourier coefficients a_m with $m \geq 0$.

Fortunately, elliptic operators of odd order are not common:

(7.1) Proposition.
Let $L = \sum_{|\alpha| \leq k} a_\alpha \partial^\alpha$ be elliptic at $x_0 \in \mathbb{R}^n$. If $n \geq 3$, or if $n = 2$ and the numbers $a_\alpha(x_0)$ are real for $|\alpha| = k$, then k is even.

Proof: Ellipticity means that the function

$$\sigma(\xi) = \sum_{|\alpha|=k} a_\alpha(x_0)\xi^\alpha$$

is nonvanishing on $\mathbb{R}^n \setminus \{0\}$. For each nonzero $\xi' = (\xi_1, \ldots, \xi_{n-1}) \in \mathbb{R}^{n-1}$ let us consider the polynomial of degree k

$$P_{\xi'}(z) = \sigma(\xi_1, \ldots, \xi_{n-1}, z) \qquad (z \in \mathbb{C}).$$

Let $N_+(\xi')$ [resp. $N_-(\xi')$] be the number of zeros of $P_{\xi'}$ with positive [resp. negative] imaginary part. Since $P_{\xi'}(x) \neq 0$ for z real, $N_+(\xi') + N_-(\xi') = k$ for all ξ'. Also, it follows from Rouché's theorem that N_+ and N_- are locally constant on $\mathbb{R}^{n-1} \setminus \{0\}$. For $n \geq 3$ this set is connected, so N_+ and N_- are constant. But $N_+(-\xi') = N_-(\xi')$ since $\sigma(-\xi) = (-1)^k \sigma(\xi)$, so $N_+ = N_-$, and hence $k = N_+ + N_- = 2N_+$ is even.

On the other hand, suppose $n \geq 2$ and the numbers $a_\alpha(x_0)$ are real. Since $\sigma(\xi)$ is nonzero for $\xi \in \mathbb{R}^n \setminus \{0\}$ and the latter set is connected, σ must be everywhere positive or everywhere negative. Since $\sigma(-\xi) = (-1)^k \sigma(\xi)$, this forces k to be even. ∎

It is also possible for strange things to happen with operators of even order. Consider the following example, due to A. Bitsadze: The operator

$$\partial_{\bar{z}}^2 = \tfrac{1}{4}(\partial_x^2 + 2i\partial_x\partial_y - \partial_y^2)$$

is elliptic on $\mathbb{R}^2$, and the general solution of $\partial_{\bar{z}}^2 u = 0$ is

$$u(x, y) = f(z) + \bar{z}g(z) \qquad (z = x + iy)$$

where f and g are holomorphic functions. In particular, if we choose f holomorphic on the unit disc B and continuous on $\overline{B}$, and set $g(z) = -zf(z)$, we see that $u = (1 - |z|^2)f(z)$ solves the Dirichlet problem

$$\partial_{\overline{z}}^2 u = 0 \text{ on } B, \qquad u = 0 \text{ on } \partial B.$$

Since there are many such f's, the solution of this Dirichlet problem is far from unique, and we can say virtually nothing about the smoothness of the solution along the boundary.

We therefore introduce a restricted class of elliptic operators which will exclude such pathological cases. We say that the operator

$$L = \sum_{|\alpha| \le k} a_\alpha \partial^\alpha$$

(with C^∞ coefficients, as usual) is **strongly elliptic** on $\overline{\Omega}$ if there is a C^∞ complex-valued function γ of absolute value 1 on $\overline{\Omega}$ and a constant $C > 0$ such that

$$\mathrm{Re}\left[\gamma(x) \sum_{|\alpha|=k} a_\alpha(x)\xi^\alpha \right] \ge C|\xi|^k \qquad (\xi \in \mathbb{R}^n, \ x \in \overline{\Omega}).$$

It follows that any elliptic operator with real coefficients on $\overline{\Omega}$ is strongly elliptic on $\overline{\Omega}$ (take $\gamma = \pm 1$), and that a strongly elliptic operator has even order $k = 2m$ (because $(-\xi)^\alpha = (-1)^{|\alpha|}\xi^\alpha$). Since the equation $Lu = f$ is equivalent to the equation $(-1)^m \gamma Lu = (-1)^m \gamma f$, by replacing L by $(-1)^m \gamma L$ we may assume that

(7.2) $$(-1)^m \, \mathrm{Re} \sum_{|\alpha|=2m} a_\alpha(x)\xi^\alpha \ge C|\xi|^{2m} \qquad (\xi \in \mathbb{R}^n, \ x \in \overline{\Omega}).$$

Henceforth, when we speak of a strongly elliptic operator we shall assume that this normalization has been made. (In particular, this means replacing the Laplacian Δ by $-\Delta$. The significance of the factor of $(-1)^m$ will appear in the next section.)

EXERCISES

1. Show that the Bitsadze operator $\partial_{\overline{z}}^2$ is not strongly elliptic.

2. Let $(x, y, w_1, \dots, w_n)$ be coordinates on $\mathbb{R}^{n+2}$, and let L be a second order operator in the variables $w_1, \dots, w_n$.
   a. Show that the operator $\partial_{\overline{z}}^2 + L$ ($z = x + iy$) cannot be elliptic at any point of $\mathbb{R}^{n+2}$.
   b. For which L is $\partial_x^2 + \partial_y^2 + L$ elliptic on $\mathbb{R}^{n+2}$?

B. On Integration by Parts

Henceforth Ω will denote a bounded domain in $\mathbb{R}^n$ with C^∞ boundary S, and

$$L = \sum_{|\alpha| \leq 2m} a_\alpha \partial^\alpha$$

will denote a strongly elliptic operator on $\overline{\Omega}$ satisfying (7.2). Also, $\langle v \,|\, u \rangle$ will denote the inner product of v and u in $L^2(\Omega)$, i.e., $\langle v \,|\, u \rangle = \int_\Omega v\overline{u}$.

The **formal adjoint** of L is the differential operator L^* on Ω determined by the formula

$$\langle L^* v \,|\, u \rangle = \langle v \,|\, Lu \rangle \text{ for all } u, v \in C_c^\infty(\Omega).$$

(This differs from the dual of L that we used in defining the action of L on distributions in that we are using Hermitian inner products.) Working out the integration by parts implicit in this formula, we see that

$$L^* v = \sum_{|\alpha| \leq 2m} (-1)^{|\alpha|} \partial^\alpha (\overline{a}_\alpha v).$$

In particular, the $2m$-th order part of L^* is $\sum_{|\alpha|=2m} \overline{a}_\alpha \partial^\alpha$, so L^* is strongly elliptic whenever L is.

Frequently it will be convenient to integrate by parts only m times to obtain an expression

$$\langle v \,|\, Lu \rangle = \sum_{|\alpha|,|\beta| \leq m} \langle \partial^\alpha v \,|\, a_{\alpha\beta} \partial^\alpha u \rangle.$$

In general, a sesquilinear form (linear in the first variable, conjugate-linear in the second) of the type

$$(7.3) \qquad D(v, u) = \sum_{|\alpha|,|\beta| \leq m} \langle \partial^\alpha v \,|\, a_{\alpha\beta} \partial^\beta u \rangle \qquad (a_{\alpha\beta} \in C^\infty(\overline{\Omega}))$$

is called a **Dirichlet form** of order m. D is called a **Dirichlet form for the operator L** if

$$D(v, u) = \langle v \,|\, Lu \rangle \text{ for all } u, v \in C_c^\infty(\Omega).$$

We have already met in §2F the special case of this idea which historically provided the inspiration for the general construction, namely the Dirichlet form

$$D(v, u) = \sum_1^n \langle \partial_j v \,|\, \partial_j u \rangle = \int_\Omega \nabla v \cdot \overline{\nabla u}$$

for $-\Delta$.

Any operator L has many different Dirichlet forms. For example, in dimension $n = 2$, another Dirichlet form for $-\Delta$ is

$$D'(v, u) = \langle (\partial_x + i\partial_y)v \,|\, (\partial_x + i\partial_y)u \rangle = 4\langle \partial_{\bar{z}} v \,|\, \partial_{\bar{z}} u \rangle.$$

For another example, two Dirichlet forms for Δ^2 on $\mathbb{R}^2$ are

$$D_1(v, u) = \langle \Delta v \,|\, \Delta u \rangle,$$
$$D_2(v, u) = \langle (\partial_x^2 - \partial_y^2)v \,|\, (\partial_x^2 - \partial_y^2)u \rangle + 4\langle \partial_x \partial_y v \,|\, \partial_x \partial_y u \rangle.$$

The choice of Dirichlet form will be partly determined by the boundary conditions we wish to consider.

The Dirichlet form (7.3) is said to be **strongly elliptic** on $\overline{\Omega}$ if for some constant $C > 0$,

$$\operatorname{Re} \sum_{|\alpha|=|\beta|=m} a_{\alpha\beta}(x)\xi^{\alpha+\beta} \geq C|\xi|^{2m} \qquad (\xi \in \mathbb{R}^n,\ x \in \overline{\Omega}).$$

If D is a Dirichlet form for $L = \sum a_\alpha \partial^\alpha$, it is easy to see that

$$a_\gamma = (-1)^m \sum_{\alpha+\beta=\gamma} a_{\alpha\beta} \qquad (|\gamma| = 2m),$$

and it follows immediately that L is strongly elliptic and satisfies (7.2) if and only if every Dirichlet form for L is strongly elliptic. (This is the reason for the factor of $(-1)^m$ in (7.2).)

We now have to face the task of determining what happens to the formula $\langle v \,|\, Lu \rangle = D(v, u)$ when v and u fail to vanish near the boundary. For this we need some terminology.

Let J be a finite set of nonnegative integers. A set $\{M_j\}_{j \in J}$ of differential operators defined near $S = \partial\Omega$ will be called a **normal J-system** on S if the order of M_j is j and S is non-characteristic for each M_j. (The most obvious example is the system $\{\partial_\nu^j\}_{j \in J}$ of powers of the normal derivative defined on a tubular neighborhood of S by (0.3).) If k and l are nonnegative integers with $l \geq k$, we shall denote the set $\{k, k+1, \ldots, l\}$ by $[k, l]$.

What we wish to prove is the following:

(7.4) Theorem.
Let D be a Dirichlet form of order m associated to the operator L on Ω. Given a normal $[0, m-1]$-system $\{M_j\}$ on S, there exist differential operators N_j of order j, $m \leq j \leq 2m-1$, defined near S such that

$$D(v, u) - \langle v \,|\, Lu \rangle = \sum_0^{m-1} \int_S (M_j v)(\overline{N_{2m-1-j} u}) \, d\sigma.$$

The proof is straightforward but tedious, and we shall leave some of the details of the calculations to the compulsive reader. It proceeds in three steps, working up from the simplest special case to the general one. To begin with, we set

$$X^- = \{x \in \mathbb{R}^n : x_n < 0\}, \qquad X^0 = \{x \in \mathbb{R}^n : x_n = 0\}.$$

(7.5) Lemma.
Let D be a Dirichlet form for L on X^-. There exist differential operators N_j of order j, $m \le j \le 2m-1$, defined near X^0 such that

$$D(v,u) - \int_{X^-} v(\overline{Lu}) = \sum_0^{m-1} \int_{X^0} (\partial_n^j v)(\overline{N_{2m-1-j}u})$$

for all $v, u \in C^\infty(\overline{X^-})$ of which at least one vanishes for $|x|$ large.

Proof: Integration by parts shows that

$$\int_{X^-} (\partial_j v)\overline{u} + \int_{X^-} v(\partial_j \overline{u}) = 0 \text{ for } j < n,$$

$$\int_{X^-} (\partial_n v)\overline{u} + \int_{X^-} v(\partial_n \overline{u}) = \int_{X^0} v\overline{u},$$

and repeated application of these formulas yields

$$\int_{X^-} (\partial^\alpha v)\overline{u} - (-1)^{|\alpha|} \int_{X^-} v(\partial^\alpha \overline{u})$$
$$= \sum_1^{\alpha_n} \int_{X^0} (-1)^{|\alpha'|+i-1}(\partial_n^{\alpha_n - i}v)(\partial^{(\alpha',i-1)}\overline{u}),$$

where $\alpha' = (\alpha_1, \ldots, \alpha_{n-1})$. Now let D be given by (7.3). Replacing u by $a_{\alpha\beta}\partial^\beta u$ and summing over α and β, we obtain

$$D(v,u) - \int_{X^-} v(\overline{Lu}) = \sum_{j=0}^{m-1} \int_{X^0} (\partial_n^j v)(\overline{N_{2m-1-j}u})$$

where

$$N_{2m-1-j}u = \sum_{|\alpha|,|\beta| \le m; \ \alpha_n \le j+1} (-1)^{|\alpha|-j-1}\partial^{(\alpha',\alpha_n-j-1)}[a_{\alpha\beta}\partial^\beta u]. \qquad \blacksquare$$

(7.6) Lemma.

let D, L, X^-, X^0 be as above. If $\{M_j\}$ is a normal $[0, m-1]$-system on X^0, there exist differential operators N'_j of order j, $m \le j \le 2m-1$, such that

$$D(v, u) - \int_{X^-} v(\overline{Lu}) = \sum_{0}^{m-1} \int_{X^0} (M_j v)(\overline{N'_{2m-1-j} u})$$

for all $v, u \in C^\infty(\overline{X^-})$ of which at least one vanishes for $|x|$ large.

Proof: The assumption on the M_j means that we can write

$$\partial_n^j = b_j M_j + \sum_{|\alpha| \le j,\ \alpha_n < j} b_{\alpha j} \partial^\alpha \qquad (b_j, b_{\alpha j} \in C^\infty).$$

If $\{N_j\}$ are the operators of Lemma (7.5), by integration by parts on X^0 we find that

$$\int_{X^0} (\partial_n^j v)(\overline{N_{2m-1-j} u}) = \int_{X^0} (M_j v)(\overline{\overline{b}_j N_{2m-1-j} u}) + \sum_{i=0}^{j-1} (\partial_n^i v)(\overline{A^j_{2m-1-i} u}),$$

where

$$A^j_{2m-1-i} u = \sum_{|\alpha'| \le j-1} (-1)^{|\alpha'|} \partial^{\alpha'} [\overline{b}_{(\alpha', i)j} N_{2m-1-j} u]$$

and again, $\alpha' = (\alpha_1, \ldots, \alpha_{n-1})$. Clearly A^j_{2m-1-i} is a differential operator of order $2m-1-i$. Applying this formula with $j = m-1$ to the result of Lemma (7.5), we obtain

$$D(v, u) - \int_{X^-} v(\overline{Lu}) = \int_{X^0} (M_{m-1} v)(\overline{N_m^1 u}) + \sum_{i=0}^{m-2} \int_{X^0} (\partial_n^i v)(\overline{N^1_{2m-1-i} u}),$$

where $N_m^1 = \overline{b}_m N_m$ and $N^1_{2m-1-i} = A^{m-1}_{2m-1-i} + N_{2m-1-i}$ for $i < m-1$.

Now repeat this argument with m replaced by $m-1$ and N_j replaced by N_j^1, obtaining a formula

$$\sum_{i=0}^{m-2} \int_{X^0} (\partial_n^i v)(\overline{N^1_{2m-1-i} u})$$

$$= \int_{X^0} (M_{m-2} v)(\overline{N^2_{m+1} u}) + \sum_{i=0}^{m-3} \int_{X^0} (\partial_n^i v)(\overline{N^2_{2m-1-i} u}),$$

whence

$$D(v, u) - \int_{X^-} v(\overline{Lu}) = \int_{X^0} (M_{m-1}v)(\overline{N_m^1 u}) + \int_{X^0} (M_{m-2}v)(\overline{N_{m+1}^2 u})$$

$$+ \sum_{i=0}^{m-3} \int_{X^0} (\partial_n^i v)(\overline{N_{2m-1-i}^2 u}).$$

Continuing this process inductively, after m steps we obtain the desired result with $N'_{m+j} = N_{m+j}^{j+1}$. ∎

Proof of Theorem (7.4): Let $V_0, \ldots, V_N$ be an open covering of $\overline{\Omega}$ such that $\overline{V}_0 \subset \Omega$, $V_i \cap S \neq \varnothing$ for $i \geq 1$, and for each $i \geq 1$,

$$V_i \cap \Omega = \{x \in V_i : x_l < f_i(x_1, \ldots, x_{l-1}, x_{l+1}, \ldots, x_n)\}$$

for some l and some smooth function f_i. Let $\zeta_0, \ldots, \zeta_N$ be a partition of unity on $\overline{\Omega}$ subordinate to this covering. We write

$$D(v, u) - \langle v \mid Lu \rangle = \sum_0^N [D(v, \zeta_i u) - \langle v \mid L(\zeta_i u)\rangle].$$

Clearly $D(v, \zeta_0 u) = \langle v \mid L(\zeta_0 u)\rangle$. For $i \geq 1$, make the coordinate transformation

$$y_j = \begin{cases} x_j & \text{for } j < l, \\ x_{j+1} & \text{for } l \leq j < n, \\ x_l - f_i(x_1, \ldots) & \text{for } j = n, \end{cases}$$

which takes $V_i \cap S$ to a portion of the hyperplane X^0 and $V_i \cap \Omega$ to a portion of the half-space X^-. The operator L assumes a new form L_i in these coordinates, and the system $\{M_j\}$ becomes a normal system $\{_iM_j\}$ on X^0. Moreover, $|\det(\partial y_j / \partial x_k)| = 1$, so L^2 inner products are preserved and D is transformed to a Dirichlet form D_i for L_i. Applying Lemma (5.6), we obtain for each i a system $\{_iN'_j\}$ $(m \leq j \leq 2m - 1)$ of differential operators defined near $y_n = 0$ such that

$$D_i(v, \zeta_i u) - \langle v \mid L_i(\zeta_i u)\rangle = \sum_{j=0}^{m-1} \int_{X^0} (_iM_j v)(\overline{_iN'_{2m-1-j}(\zeta_i u)}).$$

Transforming back to the original coordinates, $_iN'_j$ becomes an operator $_iN''_j$ defined near $V_i \cap S$, and we have

$$D(v, \zeta_i u) - \langle v \mid L(\zeta_i u)\rangle = \sum_{j=0}^{m-1} \int_S (M_j v)(\overline{_iN_{2m-1-j}(\zeta_i u)}) \, d\sigma,$$

where $_iN_j$ is $_iN_j''$ multiplied by the change-of-measure factor from Euclidean measure on X^0 to surface measure on S. Therefore, if we set

$$N_j u = \sum_1^n {}_iN_j(\zeta_i u),$$

N_j is a differential operator of order j defined near S, and it fulfills the conditions of the theorem. ∎

Remark: A close examination of the proof shows that if D is strongly elliptic, $\{N_j\}$ is a normal $[m, 2m - 1]$-system on S.

We need one further result along these lines:

(7.7) Proposition.
Let $\{M_j\}$ be a normal $[0, k]$-system on S. Given $f_0, \ldots, f_k \in C^\infty(S)$, there exists $w \in C_c^\infty(\mathbb{R}^n)$ such that $M_j w = f_j$ on S.

Proof: First suppose $M_j = \partial_\nu^j$. Let V be the tubular neighborhood of S given by (0.2), and choose $\phi \in C_c^\infty(\mathbb{R})$ with $\phi(t) = 1$ for $|t| < \frac{1}{4}\epsilon$ and $\phi(t) = 0$ for $|t| > \frac{1}{2}\epsilon$. We can then define w on V by

$$w(x + t\nu(x)) = \phi(t) \sum_0^k \frac{t^j}{j!} f_j(x)$$

and set $w = 0$ outside V.

In general, we have

$$M_j = \sum_0^j a_{ij}(x, \partial) \partial_\nu^i$$

where $a_{ij}(x, \partial)$ is a differential operator of order $j-i$ which acts tangentially to S and $a_{jj}(x, \partial) = a_{jj}(x)$ never vanishes. Hence $M_j w = f_j$ on S if and only if

$$\partial_\nu^j w(x) = \frac{1}{a_{jj}(x)} \left[f_j(x) - \sum_{i=0}^{j-1} a_{ij}(x, \partial) \partial_\nu^i w(x) \right].$$

But the quantity on the right is completely determined by the f_j's and their derivatives on S, by induction on j. Namely, for $j = 0$ it is f_0/a_{00}; for $j = 1$ it is

$$\frac{1}{a_{11}} [f_1 - a_{01}(x, \partial)w] = \frac{1}{a_{11}} \left[f_1 - a_{01}(x, \partial) \left(\frac{f_0}{a_{00}} \right) \right],$$

and so forth. Thus the problem is reduced to the case $M_j = \partial_\nu^j$, which we have solved. ∎

C. Dirichlet Forms and Boundary Conditions

We now construct a general scheme for setting up boundary value problems in terms of Dirichlet forms. To fix the ideas, let us first consider the Dirichlet problem (with homogeneous boundary conditions) for a strongly elliptic operator L of order $2m$ on Ω. This is roughly the following: *Given a reasonable function f on Ω, find a function u satisfying $Lu = f$ on Ω and $\partial_\nu^j u = 0$ on $S = \partial\Omega$ for $0 \le j < m$.* Our definition of "reasonable" will be $f \in H_0(\Omega) = L^2(\Omega)$, and we shall accordingly allow u to be a distribution solution of $Lu = f$. If f is sufficiently smooth, the Sobolev lemma and the local regularity theorem (6.33) will guarantee u to be a classical solution.

To start out, we must search for u in a class of functions for which the boundary conditions make sense, and in view of Corollary (6.48) the natural candidate is $H_m(\Omega)$. Moreover, since the condition $\partial_\nu^j u = 0$ on S for $j < m$ implies that all derivatives of u of order $< m$ vanish on S, by (6.49) and (6.50) a reasonable interpretation of the Dirichlet conditions is that u should be in $H_m^0(\Omega)$. We therefore reformulate the Dirichlet problem as follows: *Given $f \in L^2(\Omega) = H_0(\Omega)$, find $u \in H_m^0(\Omega)$ satisfying $Lu = f$ on Ω.*

This is still somewhat awkward, since if $u \in H_m^0(\Omega)$ then Lu is a priori only known to be in $H_{-m}(\mathbb{R}^n)$. It would be better to formulate the problem so that only derivatives of u up to order m occur, and this is where a Dirichlet form comes in handy.

Let D be a Dirichlet form for L on Ω (it doesn't matter which one at this point). We note that $D(v, u)$ is well defined for $v, u \in H_m(\Omega)$. Moreover, if $u \in H_m(\Omega)$ and $v \in C_c^\infty(\Omega)$ we have $D(v, u) = \langle L^* v \,|\, u \rangle$; thus u is a distribution solution of $Lu = f$ if and only if $D(v, u) = \langle v \,|\, f \rangle$ for all $v \in C_c^\infty(\Omega)$. In this case we also have $D(v, u) = \langle v \,|\, f \rangle$ for all $v \in H_m^0(\Omega)$ by passing to limits. Thus the final version of our problem is the following:

The Dirichlet Problem: *Given $f \in H_0(\Omega)$, find $u \in H_m^0(\Omega)$ such that $D(v, u) = \langle v \,|\, f \rangle$ for all $v \in H_m^0(\Omega)$.*

By the local regularity theorem (6.33), we know that any solution u will be in $H_m^0(\Omega) \cap H_{2m}^{loc}(\Omega)$. It actually turns out that $u \in H_m^0(\Omega) \cap H_{2m}(\Omega)$; more generally, if $f \in H_k(\Omega)$ then $u \in H_m^0(\Omega) \cap H_{k+2m}(\Omega)$.

The inhomogenous problem

$$Lu = f \text{ on } \Omega, \qquad \partial_\nu^j u = g_j \text{ on } S \text{ for } 0 \le j < m,$$

can be reduced to the homogeneous problem, provided the g_j's are reasonably nice. Namely, since we expect the solution to be in $H_{2m}(\Omega)$ we require that there exist a function $g \in H_{2m}(\Omega)$ such that $\partial_\nu^j g = g_j$ on S for $0 \le j < m$ in the sense of Corollary (6.48). Then, setting $w = u - g$, we are reduced to solving

$$Lw = f - Lg \text{ on } \Omega, \qquad \partial_\nu^j w = 0 \text{ on } S \text{ for } 0 \le j < m,$$

or in other words, to finding $w \in H_m^0(\Omega)$ such that $D(v, w) = \langle v \mid f - Lg \rangle$ for all $v \in H_m^0(\Omega)$.

Henceforth we shall consider only homogeneous boundary conditions, with the understanding that inhomogeneous problems can be handled as above. A much more detailed discussion of inhomogeneous problems can be found in Lions and Magenes [34].

We can set up other boundary value problems in a similar fashion. Let $\mathcal{X}$ be a closed subspace of $H_m(\Omega)$ which includes $H_m^0(\Omega)$. Any $u \in H_m(\Omega)$ agrees locally inside Ω with elements of $\mathcal{X}$, since $\zeta u \in H_m^0(\Omega) \subset \mathcal{X}$ for any $\zeta \in C_c^\infty(\Omega)$. Thus for $u \in H_m(\Omega)$ the condition $u \in \mathcal{X}$ is a condition on the behavior of u at the boundary. Therefore, we might consider boundary value problems of the following form: Given $f \in H_0(\Omega)$, find $u \in \mathcal{X}$ such that $Lu = f$ on Ω. In general, such a problem is underdetermined, as the condition $u \in \mathcal{X}$ does not give enough boundary data. Here again, the formulation in terms of a Dirichlet form for L comes to the rescue. Consider:

The $(D, \mathcal{X})$ Boundary Value Problem: *Let D be a Dirichlet form for L and let $\mathcal{X}$ be a closed subspace of $H_m(\Omega)$ containing $H_m^0(\Omega)$. Given $f \in H_0(\Omega)$, find $u \in \mathcal{X}$ such that $D(v, u) = \langle v \mid f \rangle$ for all $v \in \mathcal{X}$.*

We make several comments on the meaning of this problem:

i. Since $C_c^\infty(\Omega) \subset \mathcal{X}$, a solution of this problem will satisfy $\langle L^* v \mid u \rangle = \langle v \mid f \rangle$ for all $v \in C_c^\infty(\Omega)$ and hence will be a distribution solution of $Lu = f$.

ii. However, we require that $D(v, u) = \langle v \mid f \rangle$ not just for $v \in C_c^\infty(\Omega)$ but for all $v \in \mathcal{X}$. This automatically imposes some additional boundary conditions on u — namely, the ones that ensure that the boundary terms coming from the integration by parts from $D(v, u)$ to $\langle v \mid f \rangle$ vanish. These are the so-called **free** (or **natural** or **unstable**) **boundary conditions**. As we shall see, the requirement that u and v both belong to the same space is just what is needed to make the existence and uniqueness theory run smoothly, so this formulation automatically provides the right number of boundary conditions.

iii. If $\mathfrak{X} \neq H_m^0(\Omega)$, the choice of Dirichlet form matters, because it affects how the integration by parts proceeds.

By choosing D and $\mathfrak{X}$ properly, we can construct a wide variety of boundary value problems in this fashion. Let us look at some examples.

(7.8) Example.
$\mathfrak{X} = H_m^0(\Omega)$. This is the Dirichlet problem.

(7.9) Example.
$\mathfrak{X} = H_m(\Omega)$. Here there are no *a priori* boundary conditions. To see what the free boundary conditions are, suppose that u and f are smooth functions on $\overline{\Omega}$ satisfying $D(v, u) = \langle v \mid f \rangle$ for all $v \in C^\infty(\overline{\Omega})$. By Theorem (7.4) there are operators N_j of order j, $m \leq j < 2m$, such that

$$0 = D(v, u) - \langle v \mid f \rangle = \sum_{j=0}^{m-1} \int_S (\partial_\nu^j v)(\overline{N_{2m-1-j} u})\, d\sigma.$$

But by Proposition (7.7) we can choose v so that $\partial_\nu^j v = N_{2m-1-j} u$ for all j, so it follows that

$$N_{2m-1-j} u = 0 \text{ on } S \text{ for } 0 \leq j < m.$$

Thus we obtain m boundary conditions. Note that these boundary conditions are of orders $m, \dots, 2m - 1$, so *a priori* they do not make sense for $u \in H_m(\Omega)$. However, under suitable hypotheses the solution will turn out to be in $H_{2m}(\Omega)$, so the boundary conditions are well defined in the sense of Corollary (6.48).

(7.10) Example.
More generally, let $\{M_j\}$ be a normal $[0, m - 1]$-system on S, and let J be a subset of $[0, m - 1]$ and J' the complementary subset. We take

$$\mathfrak{X} = \text{closure in } H_m(\Omega) \text{ of } \{v \in C^\infty(\overline{\Omega}) : M_j v = 0 \text{ on } S \text{ for } j \in J\}.$$

As above, suppose $u \in \mathfrak{X}$ and f are smooth functions on $\overline{\Omega}$ satisfying $D(v, u) = \langle v \mid f \rangle$ for all $v \in \mathfrak{X}$, and let $\{N_j\}$ be the operators of Theorem (7.4). Then if $v \in C^\infty(\overline{\Omega}) \cap \mathfrak{X}$, we have

$$0 = D(v, u) - \langle v \mid f \rangle = \sum_{j \in J'} \int_S (M_j v)(\overline{N_{2m-1-j} u})\, d\sigma.$$

But by Proposition (7.7) we can find $v \in C^\infty(\overline{\Omega})$ such that $M_j v = 0$ for $j \in J$ and $M_j v = N_{2m-1-j} u$ for $j \in J'$. It follows that the boundary conditions on u are

$$M_j u = 0 \text{ on } S \ (j \in J), \quad N_{2m-1-j} u = 0 \text{ on } S \ (j \in J').$$

(7.11) Example.
Now let us be specific. Let

$$L = -\sum_{i,j=1}^{n} a_{ij}\partial_i\partial_j + \sum_{j=1}^{n} a_j\partial_j + a$$

be a second order operator with real coefficients, where (a_{ij}) is positive definite (and in particular, symmetric). Let us rewrite L as

$$L = -\sum b_{ij}\partial_i\partial_j + \sum b_j\partial_j + \sum c_j\partial_j + a$$

where b_{ij}, b_j, and c_j are real, $b_{ij} + b_{ji} = 2a_{ij}$, and $b_j + c_j = a_j$. By the divergence theorem,

$$\langle v \mid -b_{ij}\partial_i\partial_j u\rangle = \langle\partial_i(b_{ij}v)\mid\partial_j u\rangle - \int_S v b_{ij}\nu_i\partial_j\overline{u}\,d\sigma$$

$$= \langle\partial_i v\mid b_{ij}\partial_j u\rangle + \langle v\mid(\partial_i b_{ij})\partial_j u\rangle - \int_S v b_{ij}\nu_i\partial_j\overline{u}\,d\sigma,$$

and

$$\langle v\mid b_j\partial_j u\rangle = -\langle\partial_j(b_j v)\mid u\rangle + \int_S v b_j\nu_j\overline{u}\,d\sigma$$

$$= -\langle\partial_j v\mid b_j u\rangle - \langle v\mid(\partial_j b_j)u\rangle + \int_S v b_j\nu_j\overline{u}\,d\sigma.$$

Thus if we define the Dirichlet form D for L by

$$D(v,u) = \sum_{i,j}\langle\partial_i v\mid b_{ij}\partial_j u\rangle - \sum_j\langle\partial_j v\mid b_j u\rangle$$

$$+ \sum_j\langle v\mid(\textstyle\sum_i\partial_i b_{ij} + c_j)\partial_j u\rangle + \langle v\mid(a - \textstyle\sum_j\partial_j b_j)u\rangle,$$

we have

$$D(v,u) - \langle v\mid Lu\rangle = \int_S v\left[\sum_{ij} b_{ij}\nu_i\partial_j\overline{u} - \sum_j b_j\nu_j\overline{u}\right]d\sigma.$$

Hence if we take $\mathfrak{X} = H_1(\Omega)$, so that $v|S$ is arbitrary, we obtain the boundary condition

$$\sum_{ij} b_{ij}\nu_i\partial_j u - \sum_j b_j\nu_j u = 0 \text{ on } S.$$

If we set

$$\mu_j = \sum_i b_{ij}\nu_i, \quad \mu = (\mu_1,\dots,\mu_n), \quad \partial_\mu u = \mu\cdot\nabla u, \quad \beta = \sum b_j\nu_j,$$

this equation becomes

$$\partial_\mu u - \beta u = 0 \text{ on } S.$$

By choosing the quantities b_{ij} and b_j properly, we can obtain such conditions for more or less arbitrary μ and β, provided that $\mu \cdot \nu > 0$. (This condition is necessary since $\sum b_{ij} \nu_i \nu_j = \sum a_{ij} \nu_i \nu_j > 0$.) In particular, if we take $b_{ij} = a_{ij}$, μ is called the **conormal** to S with respect to L and is denoted by ν^*. (If $a_{ij} = \delta_{ij}$ as for the Laplacian, of course, then $\nu = \nu^*$.) If we also take $b_1 = \cdots = b_n = 0$, we obtain the **Neumann boundary condition**

$$\partial_\nu \cdot u = 0 \text{ on } S.$$

On the other hand, if we take $b_{ij} = a_{ij} + c_{ij}$ where $c_{ji} = -c_{ij}$, we have $\mu = \nu^* + \tau$ where $\tau = \sum_i c_{ij} \nu_i$. Since (c_{ij}) is skew-symmetric, τ is perpendicular to ν, i.e., τ is tangent to S. Thus, given $\beta \in C^\infty(S)$ and any smooth tangential vector field τ on S, to construct the boundary conditions

$$\partial_\nu \cdot u + \partial_\tau u - \beta u = 0 \text{ on } S$$

for the **oblique derivative problem**, we choose the skew-symmetric matrix $(c_{ij}) = (b_{ij} - a_{ij})$ so that $\sum_j c_{ij} \nu_i = \tau_j$ and the vector field $b = (b_1, \ldots, b_n)$ so that $b \cdot \nu = \beta$.

EXERCISES

1. Consider the Dirichlet form $D(v, u) = \langle \Delta v \,|\, \Delta u \rangle$ for Δ^2 on $\mathbb{R}^n$. What are the free boundary conditions for the $(D, \mathcal{X})$ problem if (a) $\mathcal{X} = H_2(\Omega)$, (b) $\mathcal{X} = H_2(\Omega) \cap H_1^0(\Omega)$?

2. Let Ω be a bounded domain in $\mathbb{R}^n$ and c a positive constant. Find a Dirichlet form D for $-\Delta$ on Ω such that $D(v, u) = \overline{D(u, v)}$ and the $(D, H_1(\Omega))$ problem is the problem

$$\Delta u = f \text{ on } \Omega, \quad \partial_\nu u + cu = 0 \text{ on } \partial\Omega.$$

(This problem arises in the study of heat flow where the boundary condition is given by Newton's law of cooling; see §4C.)

D. The Coercive Estimate

Our setup is still too general to produce nice results. For example, consider the Dirichlet form

$$D(v, u) = 4\langle \partial_{\bar{z}} v \mid \partial_{\bar{z}} u \rangle = \langle (\partial_x + i\partial_y)v \mid (\partial_x + i\partial_y)u \rangle$$

for $-\Delta$ on $\mathbb{R}^2$. Let Ω be the unit disc, and let $\mathcal{X} = H_1(\Omega)$. Consider the $(D, \mathcal{X})$ problem with $f = 0$, i.e., find $u \in H_1(\Omega)$ such that $D(v, u) = 0$ for all $v \in H_1(\Omega)$. Taking $v = u$, we see that $u \in H_1(\Omega)$ solves this problem precisely when $\partial_{\bar{z}} u = 0$ on Ω; that is, when u is a holomorphic function of $z = x + iy$ in Ω. Thus uniqueness and regularity at the boundary both fail miserably: the solutions form an infinite-dimensional space, and most of them are not smooth at the boundary even though $f = 0$ is C^∞ there. Moreover, even for those solutions that are smooth on $\overline{\Omega}$ we can obtain no estimates on their derivatives. For example, let $u_a(z) = \log(z + a)$ where $a - 1$ is real and positive and log is defined by cutting the plane along the negative real axis from $-\infty$ to $-a$. Then $u_a \in C^\infty(\overline{\Omega})$ and $\|u_a\|_{0,\Omega} = \int_\Omega |u_a|^2$ is a bounded function of a since the logarithmic singularity is square-integrable. However, $\partial_z^k u_a(z) = (-1)^{k-1}(k-1)!(z+a)^{-k}$ for $k > 0$, and the L^2 norm of this function on Ω blows up as $a \to 1$. (On the other hand, if we take $\mathcal{X}$ to be $H_1^0(\Omega)$ instead of $H_1(\Omega)$, there is no problem: a holomorphic function on Ω that vanishes on $\partial\Omega$ is zero.)

Another example of this phenomenon that works in higher dimensions can be found in Exercise 1.

We therefore introduce the following definition, which is designed precisely to avoid such pathologies. A Dirichlet form D of order m on Ω is said to be **coercive** over $\mathcal{X}$, where $H_m^0(\Omega) \subset \mathcal{X} \subset H_m(\Omega)$, if there exist constants $C > 0$ and $\lambda \geq 0$ such that

$$(7.12) \qquad \operatorname{Re} D(u, u) \geq C\|u\|_{m,\Omega}^2 - \lambda\|u\|_{0,\Omega}^2 \qquad (u \in \mathcal{X}).$$

D is called **strictly coercive** over $\mathcal{X}$ if we can take $\lambda = 0$ in (7.12). (Notice that if D is a Dirichlet form for L satisfying (7.12), then

$$D'(v, u) = D(v, u) + \lambda\langle v \mid u \rangle$$

is a strictly coercive Dirichlet form for $L+\lambda$.) The example discussed above is not coercive, as one sees by taking $u = u_a$ and letting $a \to 1$.

If the space $\mathcal{X}$ is defined by a normal system of boundary conditions as in (7.10), general conditions of an algebraic nature on D and the boundary operators M_j and N_j are known which ensure that D is coercive over $\mathcal{X}$; see Agmon [2]. However, the study of these conditions is beyond the scope of this book, and we shall content ourselves with a discussion of coerciveness for the more specific problems (7.8) and (7.11).

Our first result is very simple, but it covers the second order Neumann and oblique derivative problems discussed in (7.11).

(7.13) Theorem.
Let
$$D(v, u) = \sum_{i,j} \langle \partial_i v \mid b_{ij} \partial_j u \rangle + \sum_j \langle \partial_j v \mid b_j u \rangle + \sum_j \langle v \mid b'_j \partial_j u \rangle + \langle v \mid bu \rangle$$
be a strongly elliptic Dirichlet form of order 1 on Ω, and suppose the b_{ij}'s are real-valued. Then D is coercive over $H_1(\Omega)$ (and hence over any $\mathcal{X} \subset H_1(\Omega)$).

Proof: Set $a_{ij} = \frac{1}{2}(b_{ij} + b_{ji})$. Since the b_{ij}'s are real, strong ellipticity means that for some $C_0 > 0$,
$$\sum a_{ij} \xi_i \xi_j = \sum b_{ij} \xi_i \xi_j \geq C_0 |\xi|^2$$
for all $\xi \in \mathbb{R}^n$. Thus (a_{ij}) is positive definite, so if ξ is any *complex* n-vector,
$$\operatorname{Re} \sum b_{ij} \xi_i \bar{\xi}_j = \sum a_{ij} \xi_i \bar{\xi}_j \geq C_0 |\xi|^2.$$
Setting $\xi = \nabla u$, where $u \in H_1(\Omega)$, we obtain
$$\operatorname{Re} \sum b_{ij} (\partial_i u)(\partial_j \bar{u}) \geq C_0 \sum |\partial_j u|^2,$$
so an integration over Ω yields
$$\operatorname{Re} \sum \langle \partial_i u \mid b_{ij} \partial_j u \rangle \geq C_0 \sum \|\partial_j u\|_{0,\Omega}^2 = C_0 \left(\|u\|_{1,\Omega}^2 - \|u\|_{0,\Omega}^2 \right).$$
Also, for some $C_1 > 0$ (independent of u) we clearly have
$$|\langle \partial_j u \mid b_j u \rangle| \leq \|u\|_{1,\Omega} \|b_j u\|_{0,\Omega} \leq C_1 \|u\|_{1,\Omega} \|u\|_{0,\Omega},$$
$$|\langle u \mid b'_j \partial_j u \rangle| \leq \|u\|_{1,\Omega} \|b'_j \partial_j u\|_{0,\Omega} \leq C_1 \|u\|_{1,\Omega} \|u\|_{0,\Omega},$$
$$|\langle u \mid bu \rangle| \leq C_1 \|u\|_{0,\Omega}^2 \leq C_1 \|u\|_{1,\Omega} \|u\|_{0,\Omega}.$$
Therefore, setting $C_2 = (2n + 1)C_1$, we have
$$\operatorname{Re} D(u, u) \geq C_0 \left(\|u\|_{1,\Omega}^2 - \|u\|_{0,\Omega}^2 \right) - C_2 \|u\|_{1,\Omega} \|u\|_{0,\Omega}.$$
But since $\alpha\beta \leq \frac{1}{2}(\alpha^2 + \beta^2)$ for all $\alpha, \beta > 0$,
$$C_2 \|u\|_{1,\Omega} \|u\|_{0,\Omega} \leq \frac{C_0}{2} \|u\|_{1,\Omega}^2 + \frac{C_2^2}{2C_0} \|u\|_{0,\Omega}^2,$$
so
$$\operatorname{Re} D(u, u) \geq \frac{C_0}{2} \|u\|_{1,\Omega}^2 - \frac{2C_0^2 + C_2^2}{2C_0} \|u\|_{0,\Omega}^2. \qquad \blacksquare$$

Another simple and useful result is the following.

(7.14) Theorem.
Suppose

$$D(v, u) = \sum_{|\alpha|=|\beta|=m} \langle \partial^\alpha v \,|\, a_{\alpha\beta} \partial^\beta u \rangle$$

where the $a_{\alpha\beta}$'s are constants. If D is strongly elliptic, then D is strictly coercive over $H_m^0(\Omega)$.

Proof: If $u \in H_m^0(\Omega) \subset H_m(\mathbb{R}^n)$, by the Plancherel theorem and the definition of strong ellipticity we have

$$\operatorname{Re} D(u, u) = \operatorname{Re} \sum_{|\alpha|=|\beta|=m} \int (2\pi)^{2m} a_{\alpha\beta} \xi^{\alpha+\beta} |\widehat{u}(\xi)|^2 \, d\xi$$

$$\geq C \int (2\pi)^{2m} |\xi|^{2m} |\widehat{u}(\xi)|^2 \, d\xi$$

$$\geq C' \sum_{|\alpha|=m} \int (2\pi)^{2m} |\xi^\alpha|^2 |\widehat{u}(\xi)|^2 \, d\xi$$

$$= C' \sum_{|\alpha|=m} \|\partial^\alpha u\|_0^2.$$

The result therefore follows from Proposition (6.15).   ∎

This theorem can be generalized:

(7.15) Gårding's Inequality.
Every strongly elliptic Dirichlet form of order m on $\overline{\Omega}$ is coercive over $H_m^0(\Omega)$.

Remark: The inequality in question is of course (7.12). Gårding's inequality is a milestone in the theory of elliptic equations; it antedates, and is the inspiration for, the general notion of coerciveness.

Proof: The outline of the argument is much the same as the proof of Theorem (6.28). Theorem (7.14) provides the first step, the case of constant coefficients and no lower order terms.

Given $u \in H_m^0(\Omega)$, let us write

$$D(u, u) = \sum_{|\alpha|=|\beta|=m} \langle \partial^\alpha u \,|\, a_{\alpha\beta} \partial^\beta u \rangle + \sum_{\min(|\alpha|,|\beta|)<m} \langle \partial^\alpha u \,|\, a_{\alpha\beta} \partial^\beta u \rangle$$

$$= D_m(u) + D_0(u).$$

To begin with, since either $|\alpha| < m$ or $|\beta| < m$ in each term of $D_0(u)$, for some constant $C_0 > 0$ we have

(7.16) $$|D_0(u)| \leq C_0 \|u\|_{m,\Omega} \|u\|_{m-1,\Omega}.$$

Next, we shall show that $\operatorname{Re} D_m(u)$ dominates $\|u\|_{m,\Omega}^2$ when the support of u is small. For any $x_0 \in \Omega$ we can write

$$D_m(u) = \sum_{|\alpha|=|\beta|=m} \langle \partial^\alpha u \mid a_{\alpha\beta}(x_0)\partial^\beta u \rangle + \sum_{|\alpha|=|\beta|=m} \langle \partial^\alpha u \mid (a_{\alpha\beta} - a_{\alpha\beta}(x_0))\partial^\beta u \rangle.$$

By Theorem (7.14) (and its proof),

$$\operatorname{Re} \sum_{|\alpha|=|\beta|=m} \langle \partial^\alpha u \mid a_{\alpha\beta}(x_0)\partial^\beta u \rangle \geq C_1 \|u\|_{m,\Omega}^2,$$

where C_1 is independent of x_0. On the other hand, let us choose $\delta > 0$ so small that if $|x - x_0| < \delta$ then

$$\sum_{|\alpha|=|\beta|=m} |a_{\alpha\beta}(x) - a_{\alpha\beta}(x_0)| \leq \frac{C_1}{2}.$$

Then if u is supported in $B_\delta(x_0)$,

$$\sum_{|\alpha|=|\beta|=m} \langle \partial^\alpha u \mid (a_{\alpha\beta} - a_{\alpha\beta}(x_0))\partial^\beta u \rangle$$

$$\leq \sup_{|x-x_0|<\delta} \sum_{|\alpha|=|\beta|=m} |a_{\alpha\beta}(x) - a_{\alpha\beta}(x_0)| \, \|\partial^\alpha u\|_{0,\Omega} \|\partial^\beta u\|_{0,\Omega} \leq \frac{C_1}{2} \|u\|_{m,\Omega}.$$

Hence, for u supported in a ball of radius δ we have

$$\operatorname{Re} D_m(u) \geq \frac{C_1}{2} \|u\|_{m,\Omega}.$$

Now cover $\overline{\Omega}$ by finitely many balls of radius δ, say $B_\delta(x_1) \ldots, B_\delta(x_N)$, with $x_1, \ldots, x_N \in \Omega$. Choose $\phi_j \in C_c^\infty(B_\delta(x_j))$ such that $\phi_j \geq 0$ and $\sum_1^N \phi_j^2 > 0$ on $\overline{\Omega}$, and set $\zeta_j = \phi_j [\sum_1^N \phi_i^2]^{-1/2}$. Then ζ_j^2 is a partition of unity on $\overline{\Omega}$, so

$$D_m(u) = \sum_{j=1}^N \sum_{|\alpha|=|\beta|=m} \langle \zeta_j \partial^\alpha u \mid \zeta_j a_{\alpha\beta} \partial^\beta u \rangle = A_1(u) + A_2(u),$$

where

$$A_1(u) = \sum_{j=1}^{N} \sum_{|\alpha|=|\beta|=m} \langle \partial^\alpha(\zeta_j u) \,|\, a_{\alpha\beta}\partial^\beta(\zeta_j u)\rangle,$$

$$A_2(u) = \sum_{j=1}^{N} \sum_{|\alpha|=|\beta|=m} \left(\langle [\zeta_j, \partial^\alpha]u \,|\, \zeta_j a_{\alpha\beta}\partial^\beta u\rangle + \langle \partial^\alpha(\zeta_j u) \,|\, [\zeta_j, \partial^\beta]u\rangle\right)$$

Since $\zeta_j u$ is supported in $B_\delta(x_j)$, we know that

$$\operatorname{Re} A_1(u) \geq \frac{C_1}{2} \sum_{1}^{N} \|\zeta_j u\|_{m,\Omega}^2.$$

But

$$\sum_{1}^{N} \|\zeta_j u\|_{m,\Omega}^2 \geq \sum_{1}^{N} \sum_{|\alpha|=m} \langle \partial^\alpha(\zeta_j u) \,|\, \partial^\alpha(\zeta_j u)\rangle$$

$$= \sum_{1}^{N} \sum_{|\alpha|=m} \langle \zeta_j \partial^\alpha u \,|\, \zeta_j \partial^\alpha u\rangle$$

$$+ \sum_{1}^{N} \sum_{|\alpha|=m} \left(\langle [\partial^\alpha, \zeta_j]u \,|\, \partial^\alpha(\zeta_j u)\rangle + \langle \zeta_j \partial^\alpha u \,|\, [\partial^\alpha, \zeta_j]u\rangle\right).$$

The first sum on the right is just

$$\sum_{|\alpha|=m} \langle \partial^\alpha u \,|\, \partial^\alpha u\rangle = \sum_{|\alpha|=m} \|\partial^\alpha u\|_{0,\Omega}^2,$$

which, for $u \in H_m^0(\Omega)$, dominates $\|u\|_{m,\Omega}^2$ by Proposition (6.15). On the other hand, $[\partial^\alpha, \zeta_j]$ is a differential operator of order $m-1$, so the second sum is dominated in absolute value by $\|u\|_{m,\Omega}\|u\|_{m-1,\Omega}$. For the same reason, $A_2(u)$ is dominated by $\|u\|_{m,\Omega}\|u\|_{m-1,\Omega}$. In short, for some constants $C_2, C_3 \geq 0$,

$$\operatorname{Re} D_m(u) \geq C_2\|u\|_{m,\Omega}^2 - C_3\|u\|_{m,\Omega}\|u\|_{m-1,\Omega},$$

and combining this with (7.16) we obtain

$$\operatorname{Re} D(u,u) \geq C_2\|u\|_{m,\Omega}^2 - (C_0 + C_3)\|u\|_{m,\Omega}\|u\|_{m-1,\Omega}.$$

But, as in the proof of Theorem (7.13), we have

$$(C_0 + C_3)\|u\|_{m,\Omega}\|u\|_{m-1,\Omega} \leq \frac{C_2}{2}\|u\|_{m,\Omega}^2 + C_4\|u\|_{m-1,\Omega}^2,$$

where $C_4 = (C_0 + C_3)^2/2C_2$. Also, since $\|\cdot\|_{k,\Omega}$ is equivalent to $\|\cdot\|_k$ on $H_m^0(\Omega)$ for $k \leq m$, by Lemma (6.11) there is a constant $C_5 \geq 0$ such that

$$C_4\|u\|_{m-1,\Omega}^2 \leq \frac{C_2}{4}\|u\|_{m,\Omega}^2 + C_5\|u\|_{0,\Omega}^2.$$

Finally, combining these estimates yields

$$\mathrm{Re}\, D(u,u) \geq \frac{C_2}{4}\|u\|_{m,\Omega}^2 - C_5\|u\|_{0,\Omega}^2,$$

and the proof is complete. ∎

Gårding's inequality also has a converse:

(7.17) Theorem.
If the Dirichlet form D of order m is coercive over $H_m^0(\Omega)$, then D is strongly elliptic.

Proof: Given $\phi \in C_c^\infty(\Omega)$, $\xi \neq 0 \in \mathbb{R}^n$, and $\tau > 0$, let $u(x) = \phi(x)e^{i\tau\xi\cdot x}$. Then $\partial^\alpha u(x) = (i\tau\xi)^\alpha u(x)$ modulo terms of order less than $|\alpha|$ in τ, so

$$D(u,u) = \tau^{2m} \sum_{|\alpha|=|\beta|=m} \langle \xi^\alpha u \mid a_{\alpha\beta}\xi^\beta u \rangle + \cdots$$

$$= \tau^{2m} \sum_{|\alpha|=|\beta|=m} \xi^{\alpha+\beta} \int \overline{a}_{\alpha\beta}(x)|\phi(x)|^2\, dx + \cdots,$$

where the dots indicate terms of lower order in τ. Likewise,

$$\|u\|_{m,\Omega}^2 = \sum_{|\alpha|\leq m} \|\partial^\alpha u\|_{0,\Omega}^2 = \tau^{2m} \sum_{|\alpha|=m} \xi^{2\alpha} \int |\phi(x)|^2\, dx + \cdots,$$

and $\|u\|_{0,\Omega} = \|\phi\|_{0,\Omega}$. Hence, if we apply the coercive estimate (7.12), divide by τ^{2m}, and let $\tau \to \infty$, we obtain

$$\mathrm{Re} \sum_{|\alpha|=|\beta|=m} \xi^{\alpha+\beta} \int a_{\alpha\beta}(x)|\phi(x)|^2\, dx \geq C \sum_{|\alpha|=m} \xi^{2\alpha} \int |\phi(x)|^2\, dx$$

(7.18)

$$\geq C'|\xi|^{2m} \int |\phi(x)|^2\, dx.$$

Now, given $x_0 \in \Omega$, choose $\psi \in C_c^\infty$ with $\int |\psi|^2 = 1$ and take $\phi(x) = \epsilon^{-n/2} \psi(\epsilon^{-1}(x - x_0))$ where $\epsilon > 0$ is small enough so that ϕ is supported in Ω. Then as $\epsilon \to 0$, $|\phi|^2$ approaches the delta-function at x_0 (cf. the proof of Theorem (0.13)), so in the limit, (7.18) becomes

$$\mathrm{Re} \sum_{|\alpha|=|\beta|=m} \xi^{\alpha+\beta} a_{\alpha\beta}(x_0) \geq C'|\xi|^{2m}.$$

This inequality holds for all $x_0 \in \Omega$, hence for all $x_0 \in \overline{\Omega}$ by continuity; that is, D is strongly elliptic on $\overline{\Omega}$.   ∎

We note that if D is coercive over $\mathfrak{X} \supset H_m^0(\Omega)$, then a fortiori D is coercive over $H_m^0(\Omega)$. Thus strong ellipticity is always a necessary condition for coerciveness.

EXERCISES

1. Show that the Dirichlet form $D(v, u) = \langle \Delta v \, | \, \Delta u \rangle$ for Δ^2 on $\mathbb{R}^n$ is not coercive over $H_2(\Omega)$ for any $\Omega \subset \mathbb{R}^n$. (The idea is similar to the example in the text.)

2. Explain why the proof of Theorem (7.13) breaks down for forms of higher order or for forms whose coefficients b_{ij} are not all real.

E. Existence, Uniqueness, and Eigenvalues

We now confront the questions of existence and uniqueness of solutions for the (D, X) boundary value problem, where D is assumed to be coercive over X. We have set up the problem in such a way that the answers to these questions are obtained very easily. We need only one more tool.

(7.19) The Lax-Milgram Lemma.
Let $\mathcal{H}$ be a Hilbert space with inner product $\langle \cdot \, | \, \cdot \rangle$ and norm $\| \cdot \|$, and let $D : \mathcal{H} \times \mathcal{H} \to \mathbb{C}$ be a sesquilinear form on $\mathcal{H}$ (not necessarily Hermitian symmetric). Suppose there are constants $C_1, C_2 > 0$ such that

$$|D(v, u)| \leq C_1 \|v\| \, \|u\|, \qquad |D(u, u)| \geq C_2 \|u\|^2$$

for all $u, v \in \mathcal{H}$. Then there exist invertible bounded operators Φ and Ψ on $\mathcal{H}$ such that for all $v, w \in \mathcal{H}$,

$$\langle v \, | \, w \rangle = D(v, \Phi w) = \overline{D(\Psi w, v)}.$$

Proof: Given $u \in \mathcal{H}$, the map $v \to D(v, u)$ is a bounded linear functional on $\mathcal{H}$, so there is a unique $Ru \in \mathcal{H}$ such that $\langle v \mid Ru \rangle = D(v, u)$ for all v. We have $\|Ru\| \leq C_1\|u\|$, so R is a bounded linear operator on $\mathcal{H}$; moreover,

$$\|Ru\|\,\|u\| \geq |\langle u \mid Ru \rangle| = |D(u, u)| \geq C_2\|u\|^2,$$

so $\|Ru\| \geq C_2\|u\|$. Therefore R is injective and the range of R is closed, for the convergence of $\{Ru_j\}$ implies the convergence of $\{u_j\}$. On the other hand, the range of R is dense, for if $u \in \mathcal{H}$ is orthogonal to it we have $D(u, u) = \langle u \mid Ru \rangle = 0$ and hence $u = 0$. Hence R is invertible, so if we set $\Phi = R^{-1}$ we have $\langle v \mid w \rangle = D(v, \Phi w)$ for all $v, w \in \mathcal{H}$, and $\|\Phi w\| \leq C_2^{-1}\|w\|$. Similarly, if we define the operator B by $\langle v \mid Bu \rangle = \overline{D(u, v)}$, then B is invertible and we can take $\Psi = B^{-1}$. $\blacksquare$

Our first main result is the following:

(7.20) Theorem.
Let $\mathcal{X}$ be a closed subspace of $H_m(\Omega)$ that contains $H_m^0(\Omega)$, and let D be a Dirichlet form of order m that is strictly coercive over $\mathcal{X}$. There is a bounded injective operator $A : H_0(\Omega) \to \mathcal{X}$ that solves the $(D, \mathcal{X})$ problem; that is, $D(v, Af) = \langle v \mid f \rangle$ for all $v \in \mathcal{X}$ and $f \in H_0(\Omega)$.

Proof: $\mathcal{X}$ is a Hilbert space with norm $\|\cdot\|_{m,\Omega}$, and D satisfies the hypotheses of the Lax-Milgram lemma on $\mathcal{X}$; hence there is a bounded linear operator Φ on $\mathcal{X}$ such that $\langle v \mid w \rangle_{m,\Omega} = D(v, \Phi w)$ for all $v, w \in \mathcal{X}$. On the other hand, if $f \in H_0(\Omega)$, the map $v \to \langle v \mid f \rangle$ (L^2 inner product) is a bounded linear functional on $\mathcal{X}$:

$$|\langle v \mid f \rangle| \leq \|f\|_{0,\Omega}\|v\|_{0,\Omega} \leq \|f\|_{0,\Omega}\|v\|_{m,\Omega}.$$

Hence there is a unique $Rf \in \mathcal{X}$ such that $\langle v \mid Rf \rangle_{m,\Omega} = \langle v \mid f \rangle$ for all $v \in \mathcal{X}$, and $\|Rf\|_{m,\Omega} \leq \|f\|_{0,\Omega}$. R is thus a bounded linear map from $H_0(\Omega)$ to $\mathcal{X}$, and the desired solution operator is $A = \Phi \circ R$. $\blacksquare$

We can also consider the adjoint Dirichlet form

$$D^*(v, u) = \overline{D(u, v)} \qquad (u, v \in H_m(\Omega)).$$

Clearly if D is a Dirichlet form for L then D^* is a Dirichlet form for the formal adjoint L^*, and the $(D^*, \mathcal{X})$ problem is a boundary value problem for L^* which is called the **adjoint** of the $(D, \mathcal{X})$ problem. The $(D, \mathcal{X})$ problem is called **self-adjoint** if $D = D^*$. This implies that $L = L^*$ and also that the free boundary conditions for the two problems are the same.

In any event, if D is strictly coercive over X then so is D^*, so Theorem (7.20) yields linear maps A and B from $H_0(\Omega)$ to X that solve the (D, X) and (D^*, X) problems. Let us denote by T and S the compositions of A and B, respectively, with the inclusion map $X \to H_0(\Omega)$. Thus T and S are operators on $H_0(\Omega)$. There are two crucial observations to be made:
(i) T and S are compact. This follows from Rellich's theorem (6.46).
(ii) $S = T^*$. Indeed, for any $f, g \in H_0(\Omega)$ we have

$$D(Bg, Af) = \langle Bg \mid f \rangle = \langle Sg \mid f \rangle,$$

and on the other hand,

$$D(Bg, Af) = \overline{D^*(Af, Bg)} = \overline{\langle Af \mid g \rangle} = \overline{\langle Tf \mid g \rangle} = \langle g \mid Tf \rangle,$$

so that $\langle Sg \mid f \rangle = \langle g \mid Tf \rangle$.

With this, we are ready to handle the general case.

(7.21) Theorem.
Let X be a closed subspace of $H_m(\Omega)$ that contains $H_m^0(\Omega)$, and let D be a Dirichlet form of order m that is coercive over X. Define

$$V = \left\{ u \in X : D(v, u) = 0 \text{ for all } v \in X \right\},$$
$$W = \left\{ u \in X : D(u, v) = 0 \text{ for all } v \in X \right\}.$$

Then $\dim V = \dim W < \infty$. Moreover, if $f \in H_0(\Omega)$, there exists $u \in X$ such that $D(v, u) = \langle v \mid f \rangle$ for all $v \in X$ if and only if f is orthogonal to W in $H_0(\Omega)$, in which case the solution u is unique modulo V. In particular, if $V = W = \{0\}$ the solution always exists and is unique.

Proof: By assumption,

$$|D(u, u)| \geq C\|u\|_{m,\Omega}^2 - \lambda\|u\|_{0,\Omega}^2 \qquad (u \in X),$$

where we can assume $\lambda > 0$ since the case $\lambda = 0$ is covered by Theorem (7.20). Let

$$D'(v, u) = D(v, u) + \lambda\langle v \mid u \rangle.$$

Then D' is strictly coercive over X, so there is a compact operator T on $H_0(\Omega)$ whose range is in X such that $D'(v, Tf) = \langle v \mid f \rangle$ for all $v \in X$ and all $f \in H_0(\Omega)$. Now,

$$D(v, u) = \langle v \mid f \rangle \quad \Longleftrightarrow \quad D'(v, u) = \langle v \mid f \rangle + \lambda\langle v \mid u \rangle = \langle v \mid f + \lambda u \rangle.$$

For fixed f and u, the latter equation holds for all $v \in \mathcal{X}$ if and only if $T(f + \lambda u) = u$; thus,

$$D(v, u) = \langle v \mid f \rangle \text{ for all } v \in \mathcal{X} \quad \Longleftrightarrow \quad \lambda^{-1} u - Tu = \lambda^{-1} Tf.$$

In particular, taking $f = 0$ we see that $\mathcal{V} = \{u \in H_0(\Omega) : Tu = \lambda^{-1} u\}$ (the condition $Tu = \lambda^{-1} u$ automatically implies $u \in \mathcal{X}$). Likewise, in view of the remarks preceding the theorem, $\mathcal{W} = \{u \in H_0(\Omega) : T^* u = \lambda^{-1} u\}$. The first assertion therefore follows from Theorem (0.38). Moreover, if $w \in \mathcal{W}$, for any f we have $\langle Tf \mid w \rangle = \langle f \mid T^* w \rangle = \lambda^{-1} \langle f \mid w \rangle$, so $f \perp \mathcal{W}$ if and only if $Tf \perp \mathcal{W}$. The second assertion therefore follows from Corollary (0.42). ∎

In case D is self-adjoint, we can say more.

(7.22) Theorem.
Suppose D is coercive over $\mathcal{X}$ and $D = D^*$. There exists an orthonormal basis $\{u_j\}$ of $H_0(\Omega)$ consisting of eigenfunctions for the $(D, \mathcal{X})$ problem; that is, for each j we have $u_j \in \mathcal{X}$ and there is a real constant μ_j such that $D(v, u_j) = \mu_j \langle v \mid u_j \rangle$ for all $v \in \mathcal{X}$. Moreover, $\mu_j > -\lambda$ for all j where λ is the constant in (7.12), $\lim_{j \to \infty} \mu_j = +\infty$, and $u_j \in C^\infty(\Omega)$ for all j.

Proof: Let T be the solution operator for D' as in the proof of Theorem (7.21) (we no longer assume $\lambda > 0$). T is compact and self-adjoint; it is also injective, being the composition of an invertible map from $H_0(\Omega)$ to $\mathcal{X}$ with the inclusion map from $\mathcal{X}$ to $H_0(\Omega)$. Hence, by the spectral theorem (0.44) there is an orthonormal basis $\{u_j\}$ for $H_0(\Omega)$ and a sequence of nonzero real numbers $\{\alpha_j\}$ with $\lim_{j \to \infty} \alpha_j = 0$ such that $Tu_j = \alpha_j u_j$. (In particular, $u_j = \alpha_j^{-1} Tu_j \in \mathcal{X}$ for all j.) Since

$$\alpha_j \|u_j\|^2 = \langle Tu_j \mid u_j \rangle = D'(Tu_j, Tu_j) = \operatorname{Re} D'(Tu_j, Tu_j) > 0,$$

we have $\alpha_j > 0$ for all j. Thus, if we set $\mu_j = \alpha_j^{-1} - \lambda$, we have $\mu_j > -\lambda$, $\lim_{j \to \infty} \mu_j = +\infty$, and $D(v, u_j) = \mu_j \langle v \mid u_j \rangle$ for all $v \in \mathcal{X}$. Finally, if L is the elliptic operator associated to D, u_j is a distribution solution of $(L - \mu_j) u_j = 0$ on Ω, so $u_j \in C^\infty(\Omega)$ by Corollary (6.34). ∎

Now let us interpret this theorem in some specific instances. First, the Dirichlet problem.

(7.23) Theorem.
Suppose that L is a strongly elliptic operator of order $2m$ on $\overline{\Omega}$ that satisfies $L = L^*$ and (7.2). There is an orthonormal basis $\{u_j\}$ for $H_0(\Omega)$ consisting of eigenfunctions for L which are C^∞ on $\overline{\Omega}$ and satisfy the Dirichlet conditions $\partial_\nu^i u_j = 0$ on $\partial\Omega$ for $0 \le i < m$. The eigenvalues are real and accumulate only at $+\infty$. If L has a self-adjoint Dirichlet form that is strictly coercive over $H_m^0(\Omega)$, the eigenvalues are all positive.

    Proof: Let D be a Dirichlet form for L. Then D^* is a Dirichlet form for $L^* = L$. Since for the Dirichlet problem the choice of Dirichlet form is at our disposal, we can use the self-adjoint form $\frac{1}{2}(D + D^*)$, which is coercive over $H_m^0(\Omega)$ by Gårding's inequality. All the assertions then follow from Theorem (7.22) except the claim that u_j extends smoothly to the boundary. We shall prove this in §7F for the case $m = 1$; for the general case, see the references in §7G.   ■

(7.24) Corollary.
There is an orthonormal basis for $H_0(\Omega)$ consisting of eigenfunctions for the Laplacian such that $u_j \in C^\infty(\overline{\Omega})$ and $u_j = 0$ on $\partial\Omega$ for all j. The eigenvalues are all negative.

    Proof: The Dirichlet form $D(v, u) = \sum_1^n \langle \partial_j v \mid \partial_j u \rangle$ for $-\Delta$ is strictly coercive over $H_1^0(\Omega)$ by Theorem (7.14).   ■

    Next, let

$$L = -\sum_{ij} a_{ij}\partial_i\partial_j + \sum_j a_j \partial_j + a$$

be a second order elliptic operator with real coefficients, where (a_{ij}) is positive definite. A simple calculation shows that

$$L^* = -\sum_{ij} a_{ij}\partial_i\partial_j - 2\sum_{ij}(\partial_i a_{ij})\partial_j - \sum_j a_j \partial_j - \sum_{ij}(\partial_i\partial_j a_{ij}) - \sum_j (\partial_j a_j) + a.$$

Hence $L = L^*$ if and only if

$$a_j = -\sum_i \partial_i(a_{ij}) \qquad (j = 1, \ldots, n).$$

In this case we have

(7.26)
$$\begin{aligned}
L = L^* &= -\sum_{ij} a_{ij}\partial_i\partial_j - \sum_{ij}(\partial_i a_{ij})\partial_j + a \\
&= -\sum_{ij} \partial_i(a_{ij}\partial_j) + a,
\end{aligned}$$

so L has the self-adjoint Dirichlet form

$$(7.27) \qquad D(v, u) = \sum_{ij} \langle \partial_i v \,|\, a_{ij} \partial_j u \rangle + \langle v \,|\, au \rangle.$$

According to the discussion in (7.11), if we take $\mathfrak{X} = H_1(\Omega)$, the $(D, \mathfrak{X})$ problem for this choice of D is the Neumann problem for L. Hence:

(7.28) Theorem.
Let L be given by (7.26), where the coefficients are real and (a_{ij}) is symmetric positive definite. There is an orthonormal basis $\{u_j\}$ for $H_0(\Omega)$ consisting of eigenfunctions for L which are C^∞ on $\overline{\Omega}$ and satisfy the Neumann condition $\partial_\nu \cdot u_j = 0$ on $\partial\Omega$. The eigenvalues are real and bounded below and accumulate only at $+\infty$. If $a \geq 0$ (resp. $a > 0$), the eigenvalues are all nonnegative (resp. positive).

   Proof: The Dirichlet form (7.27) is coercive over $H_1(\Omega)$ by Theorem (7.13), so the existence of the eigenbasis and the first statement about the eigenvalues follow from Theorem (7.22). To prove the last statement, we re-examine the proof of Theorem (7.13). Since $\sum a_{ij}\xi_i \overline{\xi}_j \geq C|\xi|^2$ for all $\xi \in \mathbb{C}^n$ where $C > 0$, we have

$$D(u, u) \geq C \sum_1^n \|\partial_i u\|_{0,\Omega}^2 + \langle u \,|\, au \rangle \geq \langle u \,|\, au \rangle.$$

Hence if u_j has eigenvalue μ_j,

$$\mu_j = \mu_j \|u_j\|^2 = \langle u_j \,|\, Lu_j \rangle = D(u_j, u_j) \geq \langle u_j \,|\, au_j \rangle.$$

If $a \geq 0$ the last expression is nonnegative, and if $a > 0$ it is positive.   ∎

   For the Laplacian, in particular, the eigenvalues are all nonpositive. In fact, the only eigenfunctions with eigenvalue zero are the locally constant functions, for $D(u, u) = \sum \|\partial_j u\|_{0,\Omega}^2 > 0$ unless ∇u vanishes identically.

F. Regularity at the Boundary: the Second Order Case

Let L be a strongly elliptic operator of order $2m$ on $\overline{\Omega}$. If $f \in H_k(\Omega)$ and u is a solution of $Lu = f$, the local regularity theorem (6.33) guarantees that $u \in H^{\text{loc}}_{2m+k}(\Omega)$. If in addition u satisfies certain kinds of boundary conditions, it will turn out that u is actually in $H_{2m+k}(\Omega)$. In this section we shall assume that $m = 1$ and prove this assertion for u a solution of the $(D, \mathcal{X})$ problem, where $\mathcal{X}$ is either $H^0_1(\Omega)$ or $H_1(\Omega)$ and D is a Dirichlet form for L that is coercive over $\mathcal{X}$. This setup includes the Dirichlet problem for second order strongly elliptic operators and the Neumann and oblique derivative problems for second order elliptic operators with real coefficients. We shall make some comments on the higher order case, with references to the literature, in the next section.

Most of the labor of the proof will be performed in small open sets near the boundary $S = \partial\Omega$ which look like half-balls. Specifically, let V be an open set intersecting S such that $V \cap \Omega$ can be represented as

$$V \cap \Omega = \{x \in V : x_i < \phi(x_1, \ldots, x_{i-1}, x_{i+1}, \ldots, x_n)\}$$

for some i, where ϕ is a C^∞ function. Define new coordinates on V by

$$y_j = \begin{cases} x_j & \text{for } j < i, \\ x_{j+1} & \text{for } i \leq j < n, \\ x_i - \phi(x_1, \ldots) & \text{for } j = n. \end{cases}$$

Then $V \cap \Omega$ is represented in the y coordinates by the condition $y_n < 0$. By a translation we may assume that $y = 0$ lies on $V \cap S$, and we fix $r > 0$ so small that the set where $|y| < r$ lies in V. For any $\rho \leq r$, we then set

$$N(\rho) = \{y : |y| < \rho \text{ and } y_n < 0\} = \Omega \cap B_\rho(0),$$

as in (6.42).

Now let D be a Dirichlet form for the second order operator L on Ω. Since $|\det(\partial y_j / \partial x_k)| \equiv 1$, Lebesgue measure is preserved by the change of coordinates, and hence so are L^2 inner products. We assume that in the y coordinates D and L have the form

$$D(v, u) = \sum_{|\alpha|, |\beta| \leq 1} \langle \partial^\alpha u \mid a_{\alpha\beta} \partial^\beta u \rangle, \qquad L = \sum_{|\alpha| \leq 2} a_\alpha \partial^\alpha,$$

where $\partial^\alpha = (\partial/\partial y)^\alpha$.

Also, let $\mathcal{X}$ be either $H^0_1(\Omega)$ or $H_1(\Omega)$. We set

$$\mathcal{X}[r] = \{u \in \mathcal{X} : u = 0 \text{ on } \Omega \setminus N(\rho) \text{ for some } \rho < r\}.$$

The space $\mathcal{X}[r]$ has the following two crucial properties (valid for either $\mathcal{X} = H^0_1(\Omega)$ or $\mathcal{X} = H_1(\Omega)$):

i. If $\zeta \in C_c^\infty(\overline{\Omega})$ vanishes on $\Omega \setminus N(\rho)$ for some $\rho < r$, then $\zeta u \in \mathcal{X}[r]$ for any $u \in \mathcal{X}$.

ii. If $u \in \mathcal{X}[r]$ vanishes on $\Omega \setminus N(\rho)$, then for any $h \in \mathbb{R}$ with $|h| < r - \rho$ and any $j < n$, the function

$$u_{he_j}(y) = u(y_1, \ldots, y_{j-1}, y_j + h, y_{j+1}, \ldots, y_n)$$

belongs to $\mathcal{X}$ and vanishes on $\Omega \setminus N(\rho + h)$, hence belongs to $\mathcal{X}[r]$. Therefore the difference quotients $\Delta_h^\alpha u$, as defined before Theorem (6.19), belong to $\mathcal{X}[r]$ provided $|h| < r - \rho$ and $\alpha_n = 0$.

In this setup, then, we have the following local regularity theorem at the boundary.

(7.29) Theorem.
Suppose D is coercive over $\mathcal{X}$, $f \in H_k(N(r))$ for some $k \geq 0$, $u \in \mathcal{X}$, and $D(v, u) = \langle v \mid f \rangle$ for all $v \in \mathcal{X}[r]$. Then for any $\rho < r$, $u \in H_{k+2}(N(\rho))$ and there is a constant $C > 0$, depending only on ρ and k, such that

$$\|u\|_{k+2, N(\rho)} \leq C\big(\|f\|_{k, N(r)} + \|u\|_{1, N(r)}\big).$$

Proof: The proof is accomplished in three steps, of which the first is the most substantial.

Assertion 1. If $\rho < r$, $j \leq k+1$, and γ is a multi-index with $|\gamma| = j$ and $\gamma_n = 0$, then $\partial^\gamma u \in H_1(N(\rho))$ and there is a constant $C > 0$, depending only on ρ and j, such that

$$\|\partial^\gamma u\|_{1, N(\rho)} \leq C\big(\|f\|_{k, N(r)} + \|u\|_{1, N(r)}\big).$$

Proof: By induction on j, the case $j = 0$ being trivial. Suppose the assertion is true with j replaced by $0, \ldots, j - 1$. Set $t = (2\rho + r)/3$ and $s = (\rho + 2r)/3$, so $\rho < t < s < r$. Then by inductive hypothesis, with ρ replaced by s, we have $\partial^\delta u \in H_1(N(s))$ for $|\delta| \leq j - 1$ and $\delta_n = 0$, and

$$(7.30) \qquad \|\partial^\delta u\|_{1, N(s)} \leq C\big(\|f\|_{k, N(r)} + \|u\|_{1, N(r)}\big),$$

where C is the largest of the constants obtained in the proof for $0, \ldots, j-1$ with ρ replaced by $s = (\rho + 2r)/3$, so that C depends only on j and ρ. Fix $\zeta \in C_c^\infty(\overline{\Omega})$ with $\zeta = 0$ on $\Omega \setminus N(t)$ and $\zeta = 1$ on $N(\rho)$. If γ is a multi-index with $|\gamma| = j$ and $\gamma_n = 0$, we wish to consider the difference quotients $\Delta_h^\gamma(\zeta u)$ for $|h| < s - t$, which belong to $\mathcal{X}[s]$. Choosing some index $i < n$ with $\gamma_i \neq 0$, we factor one difference operator in the i direction out of $\Delta_{\mathbf{h}}^\gamma$ and write $\Delta_{\mathbf{h}}^\gamma = \Delta_h^i \Delta_{\mathbf{h}'}^{\gamma'}$.

We claim that for any $v \in \mathcal{X}[s]$,

$$(7.31) \qquad \left| D(v, \Delta_{\mathbf{h}}^\gamma \zeta u) \right| \leq C \|v\|_{1,N(s)} \left(\|f\|_{k,N(r)} + \|u\|_{1,N(r)} \right),$$

where C depends only on ρ and j (and ζ, which is fixed). (Here and in what follows, differential and difference operators act on everything to the right of them unless separated by parentheses; e.g., $\Delta_{\mathbf{h}}^\gamma \zeta u$ means $\Delta_{\mathbf{h}}^\gamma(\zeta u)$.) To prove (7.31), we shall commute various operators and move various quantities from one side of the inner product to the other, obtaining

$$
\begin{aligned}
D(v, \Delta_{\mathbf{h}}^\gamma \zeta u) &= \sum \langle \partial^\alpha v \mid a_{\alpha\beta} \partial^\beta \Delta_{\mathbf{h}}^\gamma \zeta u \rangle \\
&= \sum \langle \partial^\alpha v \mid \Delta_{\mathbf{h}}^\gamma a_{\alpha\beta} \partial^\beta \zeta u \rangle + E_1 \\
&= \sum \langle \partial^\alpha v \mid \Delta_{\mathbf{h}}^\gamma a_{\alpha\beta} \zeta \partial^\beta u \rangle + E_1 + E_2 \\
&= (-1)^j \sum \langle \zeta \Delta_{-\mathbf{h}}^\gamma \partial^\alpha v \mid a_{\alpha\beta} \partial^\beta u \rangle + E_1 + E_2 \\
&= (-1)^j \sum \langle \partial^\alpha \zeta \Delta_{-\mathbf{h}}^\gamma v \mid a_{\alpha\beta} \partial^\beta u \rangle + E_1 + E_2 + E_3 \\
&= (-1)^j D(\zeta \Delta_{-\mathbf{h}}^\gamma v, u) + E_1 + E_2 + E_3 \\
&= (-1)^j \langle \zeta \Delta_{-\mathbf{h}}^\gamma v \mid f \rangle + E_1 + E_2 + E_3 \\
&= E_1 + E_2 + E_3 + E_4,
\end{aligned}
$$

where

$$E_1 = \sum \langle \partial^\alpha v \mid [a_{\alpha\beta}, \Delta_{\mathbf{h}}^\gamma] \partial^\beta \zeta u \rangle,$$

$$E_2 = \sum_{|\beta|=1} \langle \partial^\alpha v \mid \Delta_{\mathbf{h}}^\gamma a_{\alpha\beta} (\partial^\beta \zeta) u \rangle,$$

$$E_3 = (-1)^{j+1} \sum_{|\alpha|=1} \langle (\partial^\alpha \zeta) \Delta_{-\mathbf{h}}^\gamma v \mid a_{\alpha\beta} \partial^\beta u \rangle,$$

$$E_4 = -\langle \Delta_{-\mathbf{h}}^i v \mid \Delta_{\mathbf{h}}^{\gamma'} \zeta f \rangle.$$

We have used the facts that $\Delta_{\mathbf{h}}^\gamma$ commutes with ∂^α and ∂^β, that the adjoint of $\Delta_{\mathbf{h}}^\gamma$ is $(-1)^{|\gamma|} \Delta_{-\mathbf{h}}^\gamma$, and — in the formulas for E_2 and E_3 — that $|\alpha| \leq 1$ and $|\beta| \leq 1$.

We claim that the terms $E_1, \ldots, E_4$ all satisfy the estimate (7.31), and to prove this we make use of Theorem (6.51), Proposition (6.52), and the product rule for derivatives:

$$|E_1| \leq C'\|v\|_{1,N(s)} \sum_{\delta < \gamma,\ |\beta| \leq 1} \|\partial^\delta \partial^\beta \zeta u\|_{0,N(s)}$$

$$\leq C'\|v\|_{1,N(s)} \sum_{\delta < \gamma} \|\partial^\delta \zeta u\|_{1,N(s)}$$

$$\leq C''\|v\|_{1,N(s)} \sum_{|\delta| \leq j-1,\ \delta_n=0} \|\partial^\delta u\|_{1,N(s)},$$

$$|E_2| \leq C'\|v\|_{1,N(s)} \sum_{\alpha,\beta} \|\partial^\gamma a_{\alpha\beta}(\partial^\beta \zeta)u\|_{0,N(s)}$$

$$\leq C''\|v\|_{1,N(s)} \sum_{|\delta| \leq j,\ \delta_n=0} \|\partial^\delta u\|_{0,N(s)}$$

$$\leq C''\|v\|_{1,N(s)} \sum_{|\delta| \leq j-1,\ \delta_n=0} \|\partial^\delta u\|_{1,N(s)},$$

$$|E_3| = \left| \sum_{\alpha,\beta} \langle \Delta^i_{-h} v \mid \Delta^{\gamma'}_h (\partial^\alpha \zeta) a_{\alpha\beta} \partial^\beta u \rangle \right|$$

$$\leq C'\|v\|_{1,N(s)} \sum_{\alpha,\beta} \|\partial^{\gamma'}(\partial^\alpha \zeta) a_{\alpha\beta} \partial^\beta u\|_{0,N(s)}$$

$$\leq C''\|v\|_{1,N(s)} \sum_{|\delta| \leq j-1,\ \delta_n=0} \|\partial^\delta u\|_{1,N(s)}.$$

The desired estimate for E_1, E_2, and E_3 then follows from (7.30). Finally,

$$|E_4| \leq C'\|v\|_{1,N(s)}\|\partial^{\gamma'}\zeta f\|_{1,N(s)} \leq C''\|v\|_{1,N(s)}\|f\|_{k,N(r)},$$

since $s < r$ and $|\gamma'| = j - 1 \leq k$. Thus (7.31) is established.

Now since $\Delta^\gamma_h \zeta u \in X[s]$ we can set $v = \Delta^\gamma_h \zeta u$ in (7.31) and apply the coercive estimate:

$$\|\Delta^\gamma_h \zeta u\|^2_{1,N(s)} \leq C_0\big(|D(\Delta^\gamma_h \zeta u, \Delta^\gamma_h \zeta u)| + \|\Delta^\gamma_h \zeta u\|_{0,N(s)}\big)$$

$$\leq C_1\|\Delta^\gamma_h \zeta u\|_{1,N(s)}\big(\|f\|_{k,N(r)} + \|u\|_{1,N(r)} + \|\Delta^\gamma_h \zeta u\|_{0,N(s)}\big).$$

But, writing $\Delta^\gamma_h = \Delta^i_h \Delta^{\gamma'}_{h'}$ and applying (7.30) and Theorem (6.51), we have

$$\|\Delta^\gamma_h \zeta u\|_{0,N(s)} \leq \|\partial^{\gamma'}\zeta u\|_{1,N(s)} \leq C\big(\|f\|_{k,N(r)} + \|u\|_{1,N(r)}\big).$$

Therefore,

$$\|\Delta^\gamma_h \zeta u\|_{1,N(s)} \leq C_2\big(\|f\|_{k,N(r)} + \|u\|_{1,N(r)}\big),$$

where C_2 depends only on ρ and j. Assertion 1 then follows from Theorem (6.51) together with the observation that $\|\partial^\gamma u\|_{1,N(\rho)} \leq \|\partial^\gamma \zeta u\|_{1,N(s)}$ since $\zeta = 1$ on $N(\rho)$.

Assertion 2. If $\rho < r$, $j \le k+2$, and γ is a multi-index with $|\gamma| \le k+2$ and $\gamma_n = j$, then $\partial^\gamma u \in H_0(N(\rho))$ and there is a constant $C > 0$, depending only on j and ρ, such that

$$\|\partial^\gamma u\|_{0,N(\rho)} \le C(\|f\|_{k,N(r)} + \|u\|_{1,N(r)}).$$

Proof: By induction on j. Assertion 1 shows that Assertion 2 is true for $j = 0, 1$. If $j \ge 2$, set

$$\gamma' = (\gamma_1, \ldots, \gamma_{n-1}, \gamma_n - 2),$$

so $|\gamma'| = j - 2 \le k$. Since $L = \sum a_\alpha \partial^\alpha$ is elliptic, the coefficient $\tilde{a} = a_{(0,\ldots,0,2)}$ of ∂_n^2 is never zero, so the equation $Lu = f$ means that

$$\partial_n^2 u = \tilde{a}^{-1}\left(f - \sum_{\alpha_n < 2} a_\alpha \partial^\alpha u\right).$$

Thus

$$\partial^\gamma u = \partial^{\gamma'} \partial_n^2 u = \partial^{\gamma'}\left[\tilde{a}^{-1}\left(f - \sum_{\alpha_n < 2} a_\alpha \partial^\alpha u\right)\right].$$

On the right, after performing the differentiation $\partial^{\gamma'}$ we have a sum of smooth functions times derivatives of f of order $\le k$, plus a sum of smooth functions times derivatives $\partial^\beta u$ with $|\beta| \le k+2$ and $\beta_n \le j-1$. Taking L^2 norms on $N(\rho)$, the first sum is dominated by $\|f\|_{k,N(r)}$, and the second sum is dominated by $\|f\|_{k,N(r)} + \|u\|_{1,N(r)}$ by inductive hypothesis.

Now we can complete the proof of Theorem (7.29). Let C be the largest of the constants in Assertion 2 as j ranges from 0 to $k + 2$. Then all derivatives $\partial^\gamma u$ with $|\gamma| \le k + 2$ are in $H_0(N(\rho))$ and satisfy the estimate

$$\|\partial^\gamma u\|_{0,N(\rho)} \le C(\|f\|_{k,N(r)} + \|u\|_{1,N(r)}).$$

Therefore $u \in W_{k+2}(N(\rho))$ and

$$\|u\|_{k+2,N(\rho)} \le C'(\|f\|_{k,N(r)} + \|u\|_{1,N(r)}).$$

Since $N(\rho)$ clearly has the segment property, the theorem follows from Theorem (6.39).   ∎

If we transform back to the original coordinates in which $\partial\Omega$ need not be flat, we see that we have established the following local regularity results for solutions of the $(D, \mathcal{X})$ problem. Suppose V is an open subset of Ω; suppose $u \in \mathcal{X}$ and $f \in H_0(\Omega)$ satisfy $D(v, u) = \langle v \mid f \rangle$ for all $v \in \mathcal{X}$; and suppose $f \in H_k(V)$. Then:

i. If W is an open set with $\overline{W} \subset V$, then $u \in H_{k+2}(W)$.

ii. If $x_0 \in \partial\Omega$ has a neighborhood W such that $W \cap \Omega \subset V$, then there is a neighborhood $W' \subset W$ of x_0 such that $u \in H_{k+2}(W' \cap \Omega)$.

(i) follows from Theorem (6.33), and (ii) follows from Theorem (7.29). We now combine these results to obtain the global regularity theorem for solutions of the $(D, \mathcal{X})$ problem with an optimal estimate.

(7.32) Theorem.
Suppose that $\mathcal{X}$ is either $H_1^0(\Omega)$ or $H_1(\Omega)$ and that D is a Dirichlet form of order 1 for the operator L which is coercive over $\mathcal{X}$. Suppose further that $u \in \mathcal{X}$ and $f \in H_0(\Omega)$ satisfy $D(v, u) = \langle v \mid f \rangle$ for all $v \in \mathcal{X}$. If $f \in H_k(\Omega)$ $(k = 0, 1, 2, \ldots)$, then $u \in H_{k+2}(\Omega)$ and there is a constant $C > 0$, independent of u and f, such that

$$\|u\|_{k+2,\Omega} \le C(\|f\|_{k,\Omega} + \|u\|_{0,\Omega}).$$

Proof: We begin by taking a covering of $\partial\Omega$ by open sets $V_1, \ldots, V_M$ such that each $V_m \cap \Omega$ can be mapped to a half-ball $N(r)$ as in the discussion preceding Theorem (7.29). We can then find an open covering $W_0, \ldots, W_M$ of $\overline{\Omega}$ such that $\overline{W}_0 \subset \Omega$ and $\overline{W}_j \subset V_j$ for $j \ge 1$.

We know that $u \in H_{k+2}(W_0)$ by Theorem (6.33). To obtain an estimate for $\|u\|_{k+2,W_0}$, let $U_0 = \Omega$ and $U_{k+1} = W_0$, and interpolate a sequence of open sets $U_1, \ldots, U_k$ such that $U_j \supset \overline{U}_{j+1}$ for $0 \le j \le k$. For each such j, choose $\zeta_j \in C_c^\infty(U_j)$ with $\zeta_j = 1$ on U_{j+1}. By Corollary (6.31) (with $t = 0$),

$$\begin{aligned}
\|u\|_{j+2,U_{j+1}} &\le \|\zeta_j u\|_{j+2} \le C_1(\|L\zeta_j u\|_j + \|\zeta_j u\|_0) \\
&\le C_1(\|\zeta_j Lu\|_j + \|[L, \zeta_j]u\|_j + \|\zeta_j u\|_0) \\
&\le C_2(\|Lu\|_{j,\Omega} + \|u\|_{j+1,U_j} + \|u\|_{0,\Omega}),
\end{aligned}$$

since $[L, \zeta_j]$ is a first order operator with coefficients supported in U_j. Combining these estimates for $0 \le j \le k$ with the equation $Lu = f$ yields

$$\|u\|_{k+2,W_0} \le C_3(\|f\|_{k,\Omega} + \|u\|_{1,\Omega}).$$

On the other hand, when $V_j \cap \Omega$ is mapped to a half-ball $N(r)$, $W_j \cap \Omega$ will be mapped into a half-ball $N(\rho)$ with $\rho < r$. Hence, by Theorem (7.29) we have $u \in H_{k+2}(W_j \cap \Omega)$ for $j \geq 1$, and

$$\|u\|_{k+2,W_j\cap\Omega} \leq C_4\big(\|f\|_{k,V_j\cap\Omega} + \|u\|_{1,V_j\cap\Omega}\big) \leq C_4\big(\|f\|_{k,\Omega} + \|u\|_{1,\Omega}\big).$$

Since $\Omega = W_0 \cup \bigcup_1^M (W_j \cap \Omega)$, therefore, $u \in H_{k+2}(\Omega)$ and

$$\|u\|_{k+2,\Omega} \leq C_5\big(\|f\|_{k,\Omega} + \|u\|_{1,\Omega}\big).$$

It remains to obtain the sharper estimate with $\|u\|_{1,\Omega}$ replaced by $\|u\|_{0,\Omega}$. This follows from the coercive estimate:

$$\|u\|_{1,\Omega}^2 \leq C\big(|D(u,u)| + \|u\|_{0,\Omega}^2\big) \leq C\big(|\langle u \mid f\rangle| + \|u\|_{0,\Omega}^2\big)$$
$$\leq C\|u\|_{0,\Omega}\big(\|f\|_{0,\Omega} + \|u\|_{0,\Omega}\big) \leq C\|u\|_{1,\Omega}\big(\|f\|_{0,\Omega} + \|u\|_{0,\Omega}\big),$$

so that

$$\|u\|_{1,\Omega} \leq C\big(\|f\|_{0,\Omega} + \|u\|_{0,\Omega}\big) \leq C\big(\|f\|_{k,\Omega} + \|u\|_{0,\Omega}\big).$$

The proof is complete. ∎

(7.33) Corollary.
If $f \in C^\infty(\overline{\Omega})$ then $u \in C^\infty(\overline{\Omega})$.

Proof: Use the Sobolev lemma (6.45). ∎

(7.34) Corollary.
If $u \in X$ is an eigenfunction for D, i.e., if for some constant λ we have $D(v,u) = \lambda\langle v \mid u\rangle$ for all $v \in X$, then $u \in C^\infty(\overline{\Omega})$.

Proof: Apply Corollary (7.33), with $f = 0$, to the Dirichlet form $D'(v,u) = D(v,u) - \lambda\langle v \mid u\rangle$ for $L - \lambda$. ∎

EXERCISES

1. Show that the Dirichlet form $D(v,u) = \langle \Delta v \mid \Delta u\rangle$ is coercive over $H_2(\Omega) \cap H_1^0(\Omega)$, for any Ω. (Cf. Exercise 1 in §7C and Exercise 1 in §7D.)

2. Let L be a formally self-adjoint strongly elliptic operator of order 2 on Ω, let $\{u_j\}$ be the eigenbasis for the Dirichlet problem as in Theorem (7.23), and let $\{\mu_j\}$ be the corresponding eigenvalues.

a. Suppose $f = \sum c_j u_j \in H_0(\Omega)$ and $c_j = 0$ whenever $\mu_j = 0$. Show that the solution of the Dirichlet problem $Lu = f$, $u \in H_1^0(\Omega)$, is $u = \sum \mu_j^{-1} c_j u_j$.

b. Suppose $u \in H_2(\Omega)$ and $f = Lu \in H_0(\Omega)$ have the expansions $u = \sum b_j u_j$ and $f = \sum c_j u_j$. Show that the following are equivalent: (i) $u \in H_1^0(\Omega)$, (ii) $c_j = \mu_j b_j$, (iii) $\sum |\mu_j b_j|^2 < \infty$.

c. Suppose $u = \sum b_j u_j \in H_0(\Omega)$. Show that the series $\sum b_j u_j$ converges in the norm of $H_{2k}(\Omega)$ ($k \geq 1$) if and only if $u \in H_{2k}(\Omega)$ and $L^j u \in H_1^0(\Omega)$ for $0 \leq j < k$.

G. Further Results and Techniques

The regularity theorem (7.32) is true for much more general boundary value problems for elliptic operators of any even order $2m$ if one replaces the subscript $k+2$ by $k+2m$. For example, it is true for the Dirichlet problem for an arbitrary strongly elliptic operator. Proofs of this fact, in the same spirit as our proof of Theorem (7.32), can be found in Agmon [3], Friedman [21], and Bers, John, and Schechter [7]. The first two of these deal explicitly with the more general case where the coefficients of the operator are not assumed to be C^∞, in which case one only obtains regularity up to the order of smoothness of the coefficients; however, this extension is straightforward and involves no essentially new ideas. Agmon [3] also contains some interesting material on the spectral theory of elliptic operators.

More generally, suppose $L = \sum a_\alpha \partial^\alpha$ is an elliptic operator (with C^∞ coefficients) of order $2m$ on $\overline{\Omega}$. Let J be a subset of $[0, 2m-1]$ of cardinality m, and let $\{B_j\}_{j \in J}$ be a normal J-system on $S = \partial\Omega$. Consider the boundary value problem

(7.35) $Lu = f$ on Ω, $B_j u = 0$ on S for $j \in J$.

To describe the regularity theorem for (7.35) we need to introduce three definitions:

i. L is **properly elliptic** if for every $x \in \overline{\Omega}$ and every pair ξ_1, ξ_2 of linearly independent vectors in $\mathbb{R}^n$, the polynomial $P(z) = \chi_L(x, \xi_1 + z\xi_2)$ has m roots with positive imaginary part and m roots with negative imaginary part. (Recall that $\chi_L(x,\xi) = \sum_{|\alpha|=2m} a_\alpha(x)\xi^\alpha$ is the characteristic form of L as defined in §1A.) Every strongly elliptic operator is properly elliptic, and the proof of Proposition (7.1) shows that every elliptic operator on $\overline{\Omega} \subset \mathbb{R}^n$ is properly elliptic if $n \geq 3$.

ii. Let L be properly elliptic. The operators B_j **cover** L (or satisfy the **complementing condition** or the **Lopatinski-Shapiro condition**) if the following condition holds. Let x be any point on S, ν the normal to S at x, and ξ any tangent vector to S at x. Let $z_1(x,\xi),\ldots,z_m(x,\xi)$ be the roots of the polynomial $P(z) = \chi_L(x, \xi + z\nu)$ with positive imaginary part, let $P_+(z) = \prod_1^m (z - z_j(x,\xi))$, and let $I(P_+)$ denote the ideal in the polynomial ring $\mathbb{C}[z]$ generated by P_+. Then the polynomials $Q_j(z) = \chi_{B_j}(x, \xi + z\nu)$ $(j \in J)$ are linearly independent modulo $I(P_+)$.

iii. Let us call the problem (7.35) **regular** if there are finite dimensional subspaces $\mathcal{V}$ and $\mathcal{W}$ of $C^\infty(\overline{\Omega})$ and a sequence of positive constants $C_0, C_1, \ldots$ with the following properties. Let $\mathcal{V}^\perp$ denote the orthogonal complement of $\mathcal{V}$ in $H_0(\Omega)$. Then for any $f \in \mathcal{V}^\perp \cap H_k(\Omega)$ there is a solution $u \in H_{k+2m}(\Omega)$ of (7.35); u is unique modulo $\mathcal{W}$ and satisfies

$$\|u\|_{k+2m,\Omega} \leq C_k (\|f\|_{k,\Omega} + \|u\|_{0,\Omega}).$$

We then have the following theorem, due to Agmon, Douglis, and Nirenberg:

(7.36) Theorem.
Suppose L is properly elliptic. The problem (7.35) is regular if and only if the operators $\{B_j\}_{j \in J}$ cover L.

An exposition of this result along the lines of the original arguments, including a detailed discussion of the generalization to inhomogeneous boundary conditions, can be found in Lions and Magenes [34]. See also Miranda [37] for a historical discussion. However, the best proofs of Theorem (7.36) now available use the technology of pseudodifferential operators; see Hörmander [27, vol. III], Taylor [48], or Treves [53, vol. I].

It follows from results of Agmon [2] that if the problem (7.35) arises as a $(D, \mathcal{X})$ problem as in (7.10), the coerciveness of D over $\mathcal{X}$ implies that the operators B_j cover L, so that Theorem (7.36) can be applied. Theorem (7.36) is much more general, however. It should also be mentioned that not all interesting $(D, \mathcal{X})$ problems are coercive. The most important and intensively studied non-coercive problem is the $\overline{\partial}$-Neumann problem that arises in complex analysis in several variables; see Folland and Kohn [18] for the basic theory and Greiner and Stein [23] for more refined results; also Stein [46, §XIII.6].

Although L^2 methods tend to be relatively simple and elegant, they are not the alpha and omega of elliptic regularity theory. For example, they do not yield the solution $u \in C(\overline{\Omega})$ of the Dirichlet problem

$$\Delta u = 0 \text{ on } \Omega, \qquad u = f \text{ on } S \quad (f \in C(S))$$

which we obtained in Chapter 3. For another example, suppose we wish to solve $\Delta u = f$ where f is (say) C^k with compact support. Theorem (6.33) guarantees that every solution u is in H^{loc}_{k+2}, and hence (by the Sobolev lemma) in C^j for any $j < k + 2 - \frac{1}{2}n$. But Theorem (2.28) shows that u is in $C^{k+1+\beta}$ for any $\beta < 1$ and is in $C^{k+2+\alpha}$ $(0 < \alpha < 1)$ provided that $f \in C^{k+\alpha}$. In general this is a considerable improvement over the result obtained by L^2 methods.

To get the best possible results, one has to use a variety of function spaces, the most important of which are the L^p Sobolev spaces L^p_s that were introduced at the end of §6A and the Hölder or Lipschitz spaces $C^{k+\alpha}$, $0 < \alpha < 1$. (See Stein [45], Adams [1], and Nirenberg [38] for discussions of these spaces and their relationships.) As we mentioned at the end of §6C, an analogue of the local regularity theorem (6.33) holds for these spaces, and the same is true of Theorem (7.36). That is, if the problem (7.35) is regular, then $u \in L^p_{k+2m}(\Omega)$ whenever $f \in L^p_k(\Omega)$, and $u \in C^{k+2m+\alpha}(\Omega)$ whenever $f \in C^{k+\alpha}(\Omega)$. These facts can be established by using the solution of (7.35) via pseudodifferential operators together with the L^p and Hölder estimates for these operators; see Taylor [48].

H. Epilogue: the Return of the Green's Function

At last we can tie up the loose ends in §2E. The result we need is the following:

(7.37) Proposition.
Let Ω be a bounded domain with C^∞ boundary S, and suppose $f \in C^\infty(\overline{\Omega})$. Then the solution u of the Dirichlet problem

$$\Delta u = 0 \text{ on } \Omega, \qquad u = f \text{ on } S$$

(which exists and is unique by Theorem (3.40)) is in $C^\infty(\overline{\Omega})$.

Proof: Let v be the solution of

$$\Delta v = \Delta f \text{ on } \Omega, \qquad v = 0 \text{ on } S.$$

v exists and is unique by Theorem (7.20), since the Dirichlet problem for $-\Delta$ is strictly coercive by Theorem (7.14). Moreover, Corollary (7.33) shows that $v \in C^\infty(\overline{\Omega})$. But $\Delta(f - v) = 0$ on Ω and $f - v = f$ on S, so by the uniqueness theorem (2.15), $u = f - v \in C^\infty(\overline{\Omega})$. ∎

Now we can construct the Green's function on Ω. Let N be the fundamental solution for Δ given by (2.18), and for each $x \in \Omega$ let u_x be the solution of the Dirichlet problem

$$\Delta u_x = 0 \text{ on } \Omega, \qquad u_x(y) = N(x - y) \text{ for } y \in S.$$

u_x exists, is unique, and is C^∞ on $\overline{\Omega}$ by Proposition (7.37), since $N(x - \cdot)$ agrees on S with a function $f_x \in C^\infty(\overline{\Omega})$. (For example, take $f_x(y) = [1 - \phi_x(y)]N(x - y)$ where $\phi_x \in C_c^\infty(\Omega)$ and $\phi_x = 1$ on a neighborhood of x.) Then the Green's function is

$$G(x, y) = N(x - y) - u_x(y).$$

$G(x, \cdot)$ is thus C^∞ on $\overline{\Omega} \setminus \{x\}$, so Claim (2.35) is established.
We now restate and prove Claim (2.38):

(7.38) Proposition.
Let g be a continuous function on S, and define w on Ω by

$$w(x) = \int_S g(y) \partial_{\nu_y} G(x, y) \, d\sigma(y).$$

Then w extends continuously to $\overline{\Omega}$ and solves the Dirichlet problem

$$\Delta w = 0 \text{ on } \Omega, \qquad w = g \text{ on } S.$$

Proof: The solution to this Dirichlet problem exists and is unique by Theorem (3.40); call it u. We need only show that $u = w$ on Ω. By the Weierstrass approximation theorem (4.9), we can find a sequence of polynomials $\{g_j\}$ that converges uniformly to g on S. For each j let u_j be the solution of the Dirichlet problem

$$\Delta u_j = 0 \text{ on } \Omega, \qquad u_j = g_j \text{ on } S.$$

By Proposition (7.37), $u_j \in C^\infty(\overline{\Omega})$. The remarks in §2E preceding Claim (2.38) then show that for all $x \in \Omega$,

$$u_j(x) = \int_S u_j(y) \partial_{\nu_y} G(x,y) \, d\sigma(y) = \int_S g_j(y) \partial_{\nu_y} G(x,y) \, d\sigma(y).$$

Now let $j \to \infty$. On the one hand, it is clear that for each $x \in \Omega$,

$$\int_S g_j(y) \partial_{\nu_y} G(x,y) \, d\sigma(y) \to \int_S g(y) \partial_{\nu_y} G(x,y) \, d\sigma(y) = w(x).$$

On the other hand, the maximum principle (2.13) implies that $u_j \to u$ uniformly on $\overline{\Omega}$. Thus $u = w$, and we are done.   ∎

Chapter 8
PSEUDODIFFERENTIAL OPERATORS

The theory of pseudodifferential operators was initiated around 1964 by
Kohn and Nirenberg, although it has roots in earlier work in the theory
of singular integrals and Fourier analysis, and it was subsequently refined
and extended by a number of other authors, notably Hörmander. It has
become one of the most essential tools in the modern theory of differential equations, as it offers a powerful and flexible way of applying Fourier
techniques to the study of variable-coefficient operators and singularities of
distributions. In this chapter we shall present the elements of this theory
together with a few applications. More comprehensive accounts, including many more applications, can be found in Taylor [48], Treves [53], and
Hörmander [27, vol. III]; see also Saint Raymond [42] for a detailed elementary treatment and Stein [46] for related recent developments.

A. Basic Definitions and Properties

Throughout this chapter we shall adopt the notational convention for derivatives introduced in §6D, namely

$$D = \frac{1}{2\pi i}\partial, \quad \text{i.e.,} \quad D_j = \frac{1}{2\pi i}\partial_j \text{ and } D^\alpha = \frac{1}{(2\pi i)^{|\alpha|}}\partial^\alpha.$$

The convenience in this lies in the Fourier transform formula $(D^\alpha u)\widehat{\ }(\xi) =
\xi^\alpha \hat{u}(\xi)$. Any linear differential operator with C^∞ coefficients on an open
set $\Omega \subset \mathbb{R}^n$ can then be written in the form

$$L = \sum_{|\alpha| \le k} a_\alpha(x)D^\alpha \qquad (a_\alpha \in C^\infty(\Omega)),$$

and the function
$$\sigma_L(x,\xi) = \sum_{|\alpha|\leq k} a_\alpha(x)\xi^\alpha$$

is called the **symbol** of L. It can be used to represent the operator L as a Fourier integral; namely, on applying the Fourier inversion theorem to the formula $(D^\alpha u)\widehat{}(\xi) = \xi^\alpha \widehat{u}(\xi)$, for $u \in C_c^\infty(\Omega)$ we obtain

$$Lu(x) = \sum a_\alpha(x) \int e^{2\pi i x\cdot\xi}\xi^\alpha\widehat{u}(\xi)\,d\xi = \int e^{2\pi i x\cdot\xi}\sigma_L(x,\xi)\widehat{u}(\xi)\,d\xi.$$

To rephrase this: Differential operators with C^∞ coefficients on Ω are the operators of the form

$$(8.1) \qquad Lu(x) = \int e^{2\pi i x\cdot\xi}p(x,\xi)\widehat{u}(\xi)\,d\xi \qquad (u\in C_c^\infty(\Omega)),$$

where $p(x,\xi)$ is a polynomial in ξ with coefficients that are C^∞ functions of $x\in\Omega$.

The idea of pseudodifferential operators, then, is to consider operators of the form (8.1) where p is a more general sort of function. Actually, if one allows general functions or distributions p in (8.1), one obtains an enormous family of operators that is much too diverse to support an interesting theory. (For example, any continuous linear map from $S(\mathbb{R}^n)$ to $S'(\mathbb{R}^n)$ can be represented in the form (8.1) where p is a tempered distribution on $\mathbb{R}^n\times\mathbb{R}^n$ and the integral is interpreted in the sense of distributions. See Exercise 1 in §8B.) We shall restrict attention to functions $p(x,\xi)$ that behave qualitatively like polynomials or homogeneous functions of ξ as $\xi\to\infty$ — that is, they grow or decay like powers of $|\xi|$, and differentiation in ξ lowers the order of growth.

To be precise, suppose Ω is an open set in $\mathbb{R}^n$ and m is a real number. (*Note:* In this chapter, in contrast to the preceding one, we do not assume that Ω is bounded.) The set of **symbols of order** m on Ω, denoted by $S^m(\Omega)$, is the space of functions $p\in C^\infty(\Omega\times\mathbb{R}^n)$ such that for all multi-indices α and β and every compact set $K\subset\Omega$ there is a constant $C_{\alpha,\beta,K}$ such that

$$(8.2) \qquad \sup_{x\in K}|D_x^\beta D_\xi^\alpha p(x,\xi)| \leq C_{\alpha,\beta,K}(1+|\xi|)^{m-|\alpha|}.$$

A **pseudodifferential operator** (or ΨDO for short) **of order** m on Ω is a linear map from $C_c^\infty(\Omega)$ to $C^\infty(\Omega)$ of the form (8.1) where p is a symbol of order m on Ω. We shall generally denote the map in (8.1) by $p(x,D)$,

$$(8.3) \qquad p(x,D)u(x) = \int e^{2\pi i x\cdot\xi}p(x,\xi)\widehat{u}(\xi)\,d\xi,$$

and we denote the set of pseudodifferential operators of order m on Ω by $\Psi^m(\Omega)$:

$$\Psi^m(\Omega) = \{p(x, D) : p \in S^m(\Omega)\}.$$

Warning: Our placement of the factor of 2π in (8.3) flies in the face of convention. It is much more common in the literature on pseudodifferential operators to define the Fourier transform without the 2π in the exponent, and accordingly to modify (8.3) by omitting the 2π from the exponent and inserting a factor of $(2\pi)^{-n}$ in front of the integral. In particular, this entails defining D as $(1/i)\partial$ rather than $(1/2\pi i)\partial$.

Example 1.
If $p(x, \xi) = \sum_{|\alpha| \le k} a_\alpha(x)\xi^\alpha$, then $p \in S^k(\Omega)$. In other words, every differential operator with C^∞ coefficients on Ω is a ΨDO on Ω.

Example 2.
Let us say that a function $p \in C^\infty(\Omega \times \mathbb{R}^n)$ is **homogeneous of degree m for large ξ** $(m \in \mathbb{R})$ if there exists $c \ge 0$ such that $p(x, t\xi) = t^m p(x, \xi)$ whenever $t \ge 1$ and $|\xi| \ge c$, or in other words, if p agrees with a function that is positively homogeneous of degree m in ξ on the set $|\xi| \ge c$. (The utility of this notion stems from the fact that homogeneous functions are never C^∞ at the origin unless they are actually polynomials.) If p is homogeneous of degree m for large ξ, then $D_x^\beta D_\xi^\alpha p$ is homogeneous of degree $m - |\alpha|$ for large ξ, and it follows that $p \in S^m(\Omega)$. More generally, if $p = \sum_1^J p_j$ where p_j is homogeneous of degree m_j for large ξ, then $p \in S^m(\Omega)$ where $m = \max\{m_j\}$.

Example 3.
The function $(x, \xi) \to (1 + |\xi|^2)^{s/2}$ belongs to $S^s(\mathbb{R}^n)$ (Exercise 1), and hence the operator Λ^s defined by (6.4), which plays an essential role in the theory of Sobolev spaces, belongs to $\Psi^s(\mathbb{R}^n)$.

The preceding examples of symbols are functions that are homogeneous in ξ or asymptotically homogeneous, in a suitable sense, as $\xi \to \infty$. The symbol classes $S^m(\Omega)$ also contain other sorts of functions, however; see Exercises 2 and 3.

If $p \in S^{m_1}(\Omega)$ and $q \in S^{m_2}(\Omega)$, it is clear that $p + q \in S^{\max(m_1, m_2)}(\Omega)$, and a simple calculation with the product rule shows that $pq \in S^{m_1 + m_2}(\Omega)$. Moreover, if $m_1 < m_2$, we clearly have $S^{m_1}(\Omega) \subset S^{m_2}(\Omega)$, so it is natural to set

$$S^\infty(\Omega) = \bigcup_{m \in \mathbb{R}} S^m(\Omega), \qquad S^{-\infty}(\Omega) = \bigcap_{m \in \mathbb{R}} S^m(\Omega).$$

$S^\infty(\Omega)$, the set of symbols of arbitrary order on Ω, is thus a filtered algebra. Likewise, we set

$$\Psi^\infty(\Omega) = \bigcup_{m \in \mathbb{R}} \Psi^m(\Omega), \qquad \Psi^{-\infty}(\Omega) = \bigcap_{m \in \mathbb{R}} \Psi^m(\Omega).$$

$\Psi^\infty(\Omega)$, the set of pseudodifferential operators on Ω, is also a filtered algebra, but the fact that if $P \in \Psi^{m_1}(\Omega)$ and $Q \in \Psi^{m_2}(\Omega)$ then $PQ \in \Psi^{m_1+m_2}(\Omega)$ requires some proof that we shall not give until §8D.

Let us discuss the domain and range of pseudodifferential operators a little more fully. If $p \in S^m(\Omega)$, the integral (8.3) converges for $x \in \Omega$ whenever $u \in \mathcal{S}$. However, it is more suitable to think of the domain of $p(x, D)$ as consisting of functions living on Ω, so to begin with we take the domain to be $C_c^\infty(\Omega)$. If $u \in C_c^\infty(\Omega)$, the differentiated integrals

$$D^\alpha[p(x, D)u(x)] = \int D_x^\alpha[e^{2\pi i x \cdot \xi} p(x, \xi)]\widehat{u}(\xi)\, d\xi$$

converge absolutely and uniformly on compact subsets of Ω since $\widehat{u} \in \mathcal{S}$ and the derivatives of $p(x, \xi)$ and $e^{2\pi i x \cdot \xi}$ have polynomial growth in ξ. It follows easily that $p(x, D)u \in C^\infty(\Omega)$, and moreover that $p(x, D)$ is a continuous linear map from $C_c^\infty(\Omega)$ to $C^\infty(\Omega)$. That is, if $u_1, u_2, \ldots$ are elements of $C_c^\infty(\Omega)$ that are supported in a common compact subset of Ω and $D^\alpha u_j \to D^\alpha u$ uniformly for all α, then $D^\alpha[p(x, D)u_j] \to D^\alpha[p(x, D)u]$ uniformly on compact subsets of Ω for all α.

We would like to relax the requirements that u be smooth and compactly supported. To deal fully with the issue of compact support requires a restriction on the symbol p that we shall discuss in §8B, but the smoothness condition can easily be dispensed with, as there is a natural way of defining $p(x, D)u$ as a distribution on Ω whenever u is a distribution with compact support on Ω. Indeed, if $u \in C_c^\infty(\Omega)$, for any $\phi \in C_c^\infty(\Omega)$ we have

$$\langle p(x, D)u, \phi \rangle = \iint e^{2\pi i x \cdot \xi} p(x, \xi)\widehat{u}(\xi)\phi(x)\, d\xi\, dx = \int g_\phi(\xi)\widehat{u}(\xi)\, d\xi,$$

where

(8.4) $$g_\phi(\xi) = \int e^{2\pi i x \cdot \xi} p(x, \xi)\phi(x)\, dx.$$

(8.5) Lemma.
If $p \in S^m(\Omega)$ and $\phi \in C_c^\infty(\Omega)$, the function g_ϕ defined by (8.4) decays at infinity more rapidly than any power of ξ.

Proof: For any $\xi, \eta \in \mathbb{R}^n$ we have

$$\eta^\alpha \int e^{2\pi i x \cdot \eta} p(x, \xi) \phi(x)\, dx = \int D_x^\alpha (e^{2\pi i x \cdot \eta}) p(x, \xi) \phi(x)\, dx$$

$$= (-1)^{|\alpha|} \int e^{2\pi i x \cdot \eta} D_x^\alpha [p(x, \xi) \phi(x)]\, dx.$$

Since the derivatives of $p(x, \xi)$ are all dominated by $(1+|\xi|)^m$ for $x \in \operatorname{supp} \phi$,

$$\left| \eta^\alpha \int e^{2\pi i x \cdot \eta} p(x, \xi) \phi(x)\, dx \right| \le C_\alpha (1 + |\xi|)^m$$

for all α, so

$$(1 + |\eta|)^N \left| \int e^{2\pi i x \cdot \eta} p(x, \xi) \phi(x)\, dx \right| \le C_N (1 + |\xi|)^m$$

for all N. Setting $\eta = \xi$, we obtain the desired result:

$$|g_\phi(\xi)| \le C_N (1 + |\xi|)^{m-N}. \qquad \blacksquare$$

Now, if $u \in \mathcal{E}'(\Omega)$, or more generally if $u \in H_s$ for some $s \in \mathbb{R}$ (cf. Corollary (6.8)), then $\hat{u}$ is a function that is square-integrable with respect to $(1 + |\xi|^2)^{-s}\, d\xi$ for some $s \in \mathbb{R}$; hence Lemma (8.5) shows that the integral $\int g_\phi \hat{u}$ is convergent. Moreover, the proof of Lemma (8.5) shows that if $\phi_j \to \phi$ in $C_c^\infty(\Omega)$ then $(1+|\xi|)^N g_{\phi_j}(\xi) \to (1+|\xi|)^N g_\phi(\xi)$ uniformly on $\mathbb{R}^n$ for all ξ, so that $\int g_{\phi_j} \hat{u} \to \int g_\phi \hat{u}$; that is, $\phi \to \int g_\phi \hat{u}$ is a continuous linear functional on $C_c^\infty(\Omega)$. In short, we can define $p(x, D)u \in \mathcal{D}'(\Omega)$ for any $u \in \mathcal{E}'(\Omega)$ by

$$\langle p(x, D)u, \phi \rangle = \langle g_\phi, \hat{u} \rangle, \qquad g_\phi(\xi) = \int e^{2\pi i x \cdot \xi} p(x, \xi) \phi(x)\, dx.$$

It follows easily from this construction that if $u_j \to u$ weakly in $\mathcal{E}'(\Omega)$ then $p(x, D)u_j \to p(x, D)u$ weakly in $\mathcal{D}'(\Omega)$.

We conclude this section by remarking that the theory of pseudodifferential operators has been extended to include classes of symbols and operators that are wider than $S^\infty(\Omega)$ and $\Psi^\infty(\Omega)$. The operators in $\Psi^\infty(\Omega)$ have earned the name of **classical** pseudodifferential operators; they are the operators that have come to be a standard tool for studying all sorts of differential equations, and they are the only ones we shall consider in this book. However, more general symbol classes have been found useful for various problems. The ones the reader is most likely to encounter are

Hörmander's classes $S_{\rho,\delta}^m(\Omega)$ ($m \in \mathbb{R}$, $0 \le \delta \le \rho \le 1$), which are defined just like $S^m(\Omega)$ except that condition (8.2) is replaced by

$$\sup_{x \in K} |D_x^\beta D_\xi^\alpha p(x,\xi)| \le C_{\alpha,\beta,K}(1 + |\xi|)^{m - \rho|\alpha| + \delta|\beta|}.$$

(Thus, $S^m(\Omega) = S_{1,0}^m(\Omega)$.) Most of the theory that we shall develop for $S^m(\Omega)$ generalizes to $S_{\rho,\delta}^m(\Omega)$ in a fairly straightforward way provided that $\delta < \rho$. The theory for $\rho = \delta < 1$ is more subtle but also richer in interesting applications. (The cases $\delta > \rho$ and $\delta = \rho = 1$ are pathological.) The reader may find an account of the theory of ΨDO with symbols in $S_{\rho,\delta}^m(\Omega)$ in Taylor [48] or Treves [53, vol. I], and extensions of the theory to much more general symbol classes in Beals [5] and Hörmander [27, vol. III].

EXERCISES

1. Let $p(x,\xi) = (1 + |\xi|^2)^{s/2}$ ($s \in \mathbb{R}$). Show that $p \in S^s(\mathbb{R}^n)$.

2. Pick $\phi \in C_c^\infty(\mathbb{R}^n)$ with $\phi = 1$ on a neighborhood of the origin, and let $p(x,\xi) = [1 - \phi(\xi)]\sin\log|\xi|$. Show that $p \in S^0(\mathbb{R}^n)$.

3. Suppose $p \in S^0(\Omega)$ and ψ is a C^∞ function on a neighborhood of the closure of the range of p (a subset of $\mathbb{C}$). Show that $\psi \circ p \in S^0(\Omega)$.

4. Show that if $p \in S^{-\infty}(\Omega)$ then $p(x,D)$ maps $\mathcal{E}'(\Omega)$ into $C^\infty(\Omega)$.

B. Kernels of Pseudodifferential Operators

Suppose T is an integral operator on some space of functions on Ω that includes $C_c^\infty(\Omega)$:

$$Tu(x) = \int_\Omega K(x,y)u(y)\, dy \qquad (u \in C_c^\infty(\Omega)).$$

If $v \in C_c^\infty(\Omega)$, we have

$$\int Tu(x)v(x)\, dx = \iint_{\Omega \times \Omega} K(x,y)v(x)u(y)\, dy\, dx,$$

or in the language of distributions,

$$\langle Tu, v \rangle = \langle K, v \otimes u \rangle,$$

where $v \otimes u \in C_c^\infty(\Omega \times \Omega)$ is defined by

$$(v \otimes u)(x, y) = v(x)u(y).$$

There is now an obvious generalization: Suppose T is a continuous linear map from $C_c^\infty(\Omega)$ to $\mathcal{D}'(\Omega)$. If there is a distribution K on $\Omega \times \Omega$ such that

$$\langle Tu, v \rangle = \langle K, v \otimes u \rangle \qquad (u, v \in C_c^\infty(\Omega)),$$

then K is called the **distribution kernel** of T. K is uniquely determined by T because the linear span of the functions of the form $v \otimes u$ is dense in $C_c^\infty(\Omega)$. For example, the distribution kernel of the identity operator is the delta-function $\delta(x - y)$.

In fact, every continuous linear $T : C_c^\infty(\Omega) \to \mathcal{D}'(\Omega)$ has a distribution kernel. This is one version of the Schwartz kernel theorem, for the proof of which we refer to Treves [49] or Hörmander [27, vol. I]. Another one is that if T is a continuous linear map from $\mathcal{S}(\mathbb{R}^n)$ to $\mathcal{S}'(\mathbb{R}^n)$, there is a $K \in \mathcal{S}'(\mathbb{R}^n \times \mathbb{R}^n)$ such that $\langle Tu, v \rangle = \langle K, v \otimes u \rangle$ for all $u, v \in \mathcal{S}(\mathbb{R}^n)$. These facts provide a helpful background for the discussion of distribution kernels, but we shall have no specific need for them.

It is easy to compute the distribution kernel of a pseudodifferential operator. Indeed, if $p \in S^m(\Omega)$,

$$\langle p(x, D)u, v \rangle = \iint e^{2\pi i x \cdot \xi} p(x, \xi) \widehat{u}(\xi) v(x) \, d\xi \, dx$$

$$= \iiint e^{2\pi i (x-y) \cdot \xi} p(x, \xi) v(x) u(y) \, dy \, d\xi \, dx,$$

from which it follows that the kernel K of $p(x, D)$ is

(8.6) $$K(x, y) = p_2^\vee(x, \, x - y) \qquad (x, y \in \Omega),$$

where $p_2^\vee$ denotes the inverse Fourier transform of p in its second variable. Of course, (8.6) is to be interpreted in the sense of distributions. That is, $p(x, \cdot)$ is a tempered distribution depending smoothly on x, so the same is true of $p_2^\vee(x, \cdot)$. In particular, $p_2^\vee$ defines a distribution on $\Omega \times \mathbb{R}^n$; composing it with the self-inverse linear transformation $(x, y) \to (x, \, x - y)$ yields another such distibution, and K is the restriction of the latter to $\Omega \times \Omega$.

The distribution K turns out to be nicer than one might initially expect: It is a C^∞ function off the diagonal

(8.7) $$\Delta_\Omega = \{(x, y) \in \Omega \times \Omega : x = y\},$$

and its singularities along the diagonal can be killed by multiplying it by suitable powers of $x - y$. More precisely:

(8.8) Theorem.
Suppose $p \in S^m(\Omega)$ and K is the distribution on $\Omega \times \Omega$ defined by (8.6).

a. *If $|\alpha| > m + n + j$, the function $f_\alpha(x, z) = z^\alpha p_2^\vee(x, z)$ is of class C^j on $\Omega \times \mathbb{R}^n$. Moreover, f_α and its derivatives of order $\leq j$ are bounded on $A \times \mathbb{R}^n$ for any compact $A \subset \Omega$.*

b. *If $|\alpha| > m + n + j$, $(x - y)^\alpha K(x, y)$ and its derivatives of order $\leq j$ are continuous functions on $\Omega \times \Omega$. In particular, K is a C^∞ function on $(\Omega \times \Omega) \setminus \Delta_\Omega$.*

Proof: $z^\alpha p_2^\vee(x, z)$ is the inverse Fourier transform (in the second variable) of $(-1)^{|\alpha|} D_\xi^\alpha p(x, \xi)$. For x in any compact set A, the latter function is dominated by $(1 + |\xi|)^{m - |\alpha|}$; in particular, if $|\alpha| > m + n$, it is integrable as a function of ξ, so the Fourier transform can be interpreted in the classical sense:

$$f_\alpha(x, z) = (-1)^{|\alpha|} \int e^{2\pi i z \cdot \xi} D_\xi^\alpha p(x, \xi) \, d\xi \qquad (|\alpha| > m + n).$$

Thus f_α is a bounded continuous function on $A \times \mathbb{R}^n$. Moreover, if the integrand is differentiated no more than j times with respect to x or z, it is still dominated by $(1 + |\xi|)^{|\alpha| - m + j}$ for $x \in A$, so if $|\alpha| > m + n + j$ one can differentiate under the integral and conclude that the derivatives of f_α of order $\leq j$ are bounded and continuous on $A \times \mathbb{R}^n$. This proves (a), and (b) then follows by virtue of (8.6). ∎

From this result we can deduce an important regularity property of pseudodifferential operators. Some terminology: The **singular support** of a distribution $u \in \mathcal{D}'(\Omega)$ is the complement (in Ω) of the largest open set on which u is a C^∞ function; it is denoted by sing supp u. A linear map $T : \mathcal{E}'(\Omega) \to \mathcal{D}'(\Omega)$ is called **pseudolocal** if sing supp $Tu \subset$ sing supp u for all $u \in \mathcal{E}'(\Omega)$. (The motivation for this name is as follows: A linear operator T on functions is called **local** if $Tu_1 = Tu_2$ on any open set where $u_1 = u_2$; this is equivalent to the requirement that supp $Tu \subset$ supp u for all u. Pseudolocality is the analogue with support replaced by singular support.) For example, all differential operators with C^∞ coefficients are pseudolocal (and also local); hypoellipticity of a differential operator L means precisely that the reverse inclusion sing supp $u \subset$ sing supp Tu also holds.

(8.9) Theorem.
Every pseudodifferential operator is pseudolocal.

Proof: Suppose $P \in \Psi^m(\Omega)$ and $u \in \mathcal{E}'(\Omega)$. Given a neighborhood V of sing supp u in Ω, choose $\phi \in C_c^\infty(V)$ such that $\phi = 1$ on sing supp u, and set $u_1 = \phi u$ and $u_2 = (1 - \phi)u$. Then $u = u_1 + u_2$, supp $u_1 \subset V$, and $u_2 \in C_c^\infty(\Omega)$. If K is the distribution kernel of P, by Theorem (8.8) $K(x, y)$ is a C^∞ function for $x \notin V$ and $y \in V$. Hence for $x \notin V$ we can write

$$Pu_1(x) = \langle K(x, \cdot), u_1 \rangle = \int_{\text{supp } u_1} K(x, y)u_1(y)\, dy,$$

from which it is clear that $D^\alpha Pu_1(x) = \langle D_x^\alpha K(x, \cdot), u_1 \rangle$ and hence that Pu_1 is C^∞ outside of V. On the other hand, $Pu_2 \in C^\infty(\Omega)$ since $u_2 \in C_c^\infty(\Omega)$. Thus Pu is C^∞ outside V, and since V is an arbitrary neighborhood of sing supp u, Pu is C^∞ outside sing supp u. ∎

We shall call a linear map $T : \mathcal{E}'(\Omega) \to \mathcal{D}'(\Omega)$ a **smoothing operator** if the range of T is actually in $C^\infty(\Omega)$, that is, if sing supp $Tu = \varnothing$ for all $u \in \mathcal{E}'(\Omega)$. As another easy consequence of Theorem (8.8), we have the following.

(8.10) Proposition.
Every $P \in \Psi^{-\infty}(\Omega)$ is a smoothing operator.

Proof: By Theorem (8.8), the distribution kernel K of P is C^∞ on $\Omega \times \Omega$, so as in the proof of Theorem (8.9) we see that $Pu(x) = \langle K(x, \cdot), u \rangle$ is a C^∞ function for every $u \in \mathcal{E}'(\Omega)$. ∎

Not every smoothing operator is pseudodifferential, however. For example, if $p(x, \xi) = \phi(x)\psi(\xi)$ where $\phi \in C^\infty(\mathbb{R}^n)$ and $\psi \in \mathcal{E}'(\mathbb{R}^n)$ then the operator $p(x, D)$ defined by (8.3) is smoothing (in fact, $p(x, D)u = \phi(u * \psi^\vee)$, which is C^∞ since $\psi^\vee$ is C^∞), but p does not belong to our symbol classes unless ψ is C^∞.

We next address a point glossed over earlier, namely, the injectivity of the symbol-operator correspondence. Suppose $p \in S^m(\Omega)$. If we regard the domain of the operator $P = p(x, D)$ as $\mathcal{S}(\mathbb{R}^n)$ then p is completely determined by P, for in the integrals

$$p(x, D)u(x) = \int e^{2\pi i x \cdot \xi} p(x, \xi)\hat{u}(\xi)\, d\xi$$

we can choose u so that $\hat{u}$ approximates the point mass at an arbitrary point of $\mathbb{R}^n$. However, if we restrict P to $C_c^\infty(\Omega)$, as we have agreed to do, P does not completely determine p unless Ω is very large. The situation is explained in the following proposition.

(8.11) Proposition.
Suppose $p \in S^m(\Omega)$ and $p(x, D) = 0$ as a map from $C_c^\infty(\Omega)$ to $C^\infty(\Omega)$.
  a. *If the set $\Omega - \Omega = \{x - y : x, y \in \Omega\}$ is dense in $\mathbb{R}^n$ then p is necessarily zero, but otherwise p may be nonzero.*
  b. *In any case, $p \in S^{-\infty}(\Omega)$.*

Proof: p is related to the distribution kernel K of $p(x, D)$ by (8.6):
$K(x, y) = p_2^\vee(x, x - y)$ for $x, y \in \Omega$, so if $p(x, D) = 0$ then $p_2^\vee(x, z) = 0$
for $x \in \Omega$ and $z \in \Omega - \Omega$. Since $\Omega - \Omega$ is an open set containing the
origin, Theorem (8.8a) shows that $p_2^\vee(x, \cdot)$ is a Schwartz class function
depending smoothly on x, so the same is true of $p(x, \cdot)$. This means that
$p \in S^{-\infty}(\Omega)$, so (b) is proved. Moreover, if $\Omega - \Omega$ is dense in $\mathbb{R}^n$, the
restriction of $p_2^\vee$ to $\Omega \times (\Omega - \Omega)$ determines $p_2^\vee$, and hence p, completely.
On the other hand, if $\Omega - \Omega$ is not dense, we can take $p(x, \xi) = \psi(x)\widehat{\phi}(\xi)$,
where $\psi \in C_c^\infty(\Omega)$, $\phi \in C_c^\infty(\mathbb{R}^n)$, and supp ϕ is disjoint from $\Omega - \Omega$; then
$K(x, y) = \psi(x)\phi(x - y) \equiv 0$ on $\Omega \times \Omega$ and hence $p(x, D) = 0$. $\blacksquare$

One drawback to pseudodifferential operators is that their domains consist of compactly supported functions or distributions while their ranges do not, so in general they cannot be composed with one another. It is therefore often convenient to impose an additional condition on them to overcome this difficulty. The full solution to this problem will be achieved in §8D; the discussion that follows is meant to lay the groundwork for it.

Suppose Ω is an open subset of $\mathbb{R}^n$, and denote by π_x and π_y the projection maps from $\Omega \times \Omega$ onto the first and second factors, respectively. A subset W of $\Omega \times \Omega$ is called **proper** if $\pi_x^{-1}(A) \cap W$ and $\pi_y^{-1}(A) \cap W$ are compact whenever A is a compact subset of Ω. For example, the diagonal Δ_Ω (see (8.7)) is proper; most of the proper sets we shall consider will be neighborhoods of Δ_Ω. (See Exercise 2.)

Next, suppose T is a linear map from $C_c^\infty(\Omega)$ to $C^\infty(\Omega)$ with distribution kernel K. T will be called **properly supported** if supp K is a proper subset of $\Omega \times \Omega$.

(8.12) Proposition.
$T : C_c^\infty(\Omega) \to C^\infty(\Omega)$ is properly supported if and only if the following two conditions hold:
  i. *For every compact $A \subset \Omega$ there exists a compact $B \subset \Omega$ such that supp $Tu \subset B$ whenever supp $u \subset A$.*
  ii. *For every compact $A \subset \Omega$ there exists a compact $C \subset \Omega$ such that $Tu = 0$ on A whenever $u = 0$ on C.*

Proof: If K is the distribution kernel of T and supp K is proper, (i) and (ii) hold if we take

$$B = \pi_x\big[\pi_y^{-1}(A) \cap \operatorname{supp} K\big], \qquad C = \pi_y\big[\pi_x^{-1}(A) \cap \operatorname{supp} K\big].$$

This follows easily from the formula $Tu(x) = \int K(x,y)u(y)\,dy$; see Figure 8.1. For the same reason, if (i) and (ii) hold and $A \subset \Omega$ is compact, we have $\pi_y^{-1}(A) \cap \operatorname{supp} K \subset B \times A$ and $\pi_x^{-1}(A) \cap \operatorname{supp} K \subset A \times C$. We leave the details to the reader as Exercise 3.   ∎

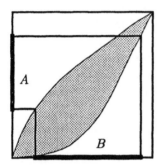 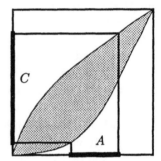

Figure 8.1. Schematic representation of Properties (i) and (ii) in Proposition (8.12). The square represents $\Omega \times \Omega$ and the shaded region represents supp K.

Property (i) clearly implies that T maps $C_c^\infty(\Omega)$ into itself, and property (ii) implies that T has a continuous extension to a linear map from $C^\infty(\Omega)$ to itself. Indeed, if $u \in C^\infty(\Omega)$, we define Tu on an arbitrary compact subset A of Ω as follows. Let C be as in (ii), pick $\phi \in C_c^\infty(\Omega)$ with $\phi = 1$ on C, and set $Tu(x) = T[\phi u](x)$ for $x \in A$. This is independent of the choice of ϕ, for if ϕ' is another such function we have $\phi u - \phi' u = 0$ on C and hence $T[\phi u] - T[\phi' u] = 0$ on A. It follows that the definitions of Tu on two different A's agree on their intersection (take a ϕ that equals 1 on both C's). This procedure defines Tu as a C^∞ function on Ω, and the continuity of T on $C^\infty(\Omega)$ follows from its continuity on $C_c^\infty(\Omega)$.

(8.13) Corollary.
If T and S are properly supported, one can form their composition ST as an operator on either $C_c^\infty(\Omega)$ or $C^\infty(\Omega)$. Moreover, ST is also properly supported.

Proof: It is easily verified that properties (i) and (ii) of Proposition (8.12) are preserved under composition.   ∎

Every differential operator is properly supported. In fact, the distribution kernel of $\sum a_\alpha(x) D^\alpha$ is $\sum a_\alpha(x) D^\alpha \delta(x-y)$, whose support is a subset of the diagonal Δ_Ω. In §8D we shall show that every pseudodifferential operator can be modified in a rather harmless way so as to become properly supported, and that the properly supported pseudodifferential operators form an algebra under composition.

For the moment, we observe that if T is a properly supported ΨDO, then T maps $\mathcal{E}'(\Omega)$ into itself and extends to a continuous map of $\mathcal{D}'(\Omega)$ into itself. The first assertion holds because any $u \in \mathcal{E}'(\Omega)$ is the limit of a sequence of functions $u_n \in C_c^\infty(\Omega)$ supported in a common compact neighborhood of supp u (convolve u with a suitable approximate identity); by Property (i), the Tu_n are supported in a common compact set B and hence supp $Tu \subset B$. A similar approximation argument that we leave to the reader (Exercise 4) shows that the procedure we used to define T on $C^\infty(\Omega)$ can be extended to define T on $\mathcal{D}'(\Omega)$.

We conclude with a couple of technical results on cutoff functions related to proper sets which will be needed in later sections.

(8.14) Proposition.
If W is a proper subset of $\Omega \times \Omega$, there exists $\phi \in C^\infty(\Omega \times \Omega)$ such that supp ϕ is proper and $\phi = 1$ on a neighborhood of W.

Proof: We begin with the following fact whose proof we leave to the reader (Exercise 5): There exist sequences $\{U_j\}_1^\infty$ and $\{V_j\}_1^\infty$ of open sets such that $\bigcup_1^\infty U_j = \Omega$, $\overline{U}_j \subset V_j$, $\overline{V}_j$ is compact, and each point in Ω has a neighborhood that intersects only finitely many $\overline{V}_j$. Let

$$\mathfrak{J} = \big\{ (j,k) : W \cap (U_j \times U_k) \neq \varnothing \big\}.$$

We claim that for each j there are only finitely many k such that $(j,k) \in \mathfrak{J}$. If not, there exists a j and a point $(x_k, y_k) \in W \cap (U_j \times U_k)$ for all k in some infinite set K. Since $\pi_x^{-1}(\overline{U}_j) \cap W$ is compact, the sequence $\{(x_k, y_k)\}_{k \in K}$ must have a cluster point. But this is impossible, for every $y \in \Omega$ has a neighborhood that meets only finitely many U_k and hence contains at most finitely many y_k.

Next, let

$$X = \bigcup_{(j,k) \in \mathfrak{J}} \overline{V}_j \times \overline{V}_k.$$

We claim that X is proper. It suffices to show that $\pi_x^{-1}(\overline{U}_j) \cap X$ is compact for all j, since every compact set in Ω is contained in a finite union of $\overline{U}_j$'s. (The same argument will also work with π_x replaced by π_y.) But

$$\pi_x^{-1}(\overline{U}_{j_0}) \cap X = \pi_x^{-1}(\overline{U}_{j_0}) \cap \bigcup \{\overline{V}_j \times \overline{V}_k : \overline{V}_j \cap \overline{U}_{j_0} \neq \varnothing, \ (j,k) \in \mathfrak{I}\}.$$

The union on the right is finite, for there are only finitely many $\overline{V}_j$ that meet the compact set $\overline{U}_{j_0}$, and for each such j there are only finitely many k with $(j,k) \in \mathfrak{I}$. Since the sets $\overline{V}_j \times \overline{V}_k$ are compact and $\pi_x^{-1}(\overline{U}_{j_0})$ is closed, we are done.

Finally we can construct the desired function ϕ. For each j, pick a nonnegative $\psi_j \in C_c^\infty(\Omega)$ such that $\psi_j = 1$ on U_j and $\operatorname{supp} \psi_j \subset \overline{V}_j$, and let $\psi(x,y) = \sum_{(j,k)\in\mathfrak{I}} \psi_j(x)\psi_k(y)$. Each (x,y) has a neighborhood on which only finitely many terms of this sum are nonzero, so $\psi \in C^\infty(\Omega \times \Omega)$. Also, $\operatorname{supp} \psi$ is proper since it is contained in X, and $\psi \geq 1$ on $\bigcup_{(j,k)\in\mathfrak{I}}(U_j \times U_k)$, which is a neighborhood of W. Hence we may take $\phi = \zeta \circ \psi$ where $\zeta : \mathbb{R} \to [0,1]$ is a C^∞ function such that $\zeta(t) = 0$ for $t \leq 0$ and $\zeta(t) = 1$ for $t \geq 1$.   ∎

(8.15) Proposition.
If N is a neighborhood of the diagonal Δ_Ω, there exists $\phi \in C^\infty(\Omega \times \Omega)$ such that:
a. $\operatorname{supp} \phi$ is proper and contained in N,
b. $\phi = 1$ on a neighborhood of Δ_Ω.

Proof: Let U_j and V_j be as in the preceding proof. Each $x \in \Omega$ has neighborhoods O_x^1 and O_x^2 such that $\overline{O_x^1} \subset O_x^2$ and $\overline{O_x^2} \times \overline{O_x^2} \subset N$. Each $\overline{V}_j$ can be covered by finitely many O_x^1's, say $\overline{V}_j = \bigcup \overline{V}_j \cap O_{x_k}^1$. Let $U_{jk} = U_j \cap O_{x_k}^1$ and $V_{jk} = V_j \cap O_{x_k}^2$; then $\bigcup_{j,k} U_{jk} \times U_{jk}$ is a neighborhood of Δ_Ω and $\bigcup_{j,k} \overline{V}_{jk} \times \overline{V}_{jk} \subset N$. We therefore obtain ϕ as in the preceding proof by modifying $\psi(x,y) = \sum_{j,k} \psi_{jk}(x)\psi_{jk}(y)$, where now $\psi_{jk} = 1$ on U_{jk} and $\operatorname{supp} \psi_{jk} \subset \overline{V}_{jk}$.   ∎

EXERCISES

1. Use the Schwartz kernel theorem to show that every continuous linear map T from $\mathcal{S}(\mathbb{R}^n)$ to $\mathcal{S}'(\mathbb{R}^n)$ can be represented in the form (8.3) where $p \in \mathcal{S}'(\mathbb{R}^n \times \mathbb{R}^n)$ and the integral is interpreted in the distributional sense. (Hint: Use (8.6) to define p.)

2. Show that $\{(x, y) : x^2 \leq y \leq x^{1/2}\}$ is a proper subset of $(0, 1) \times (0, 1)$.

3. Complete the proof of Proposition (8.12).

4. Show that every properly supported ΨDO on Ω extends to a continuous linear operator on $\mathcal{D}'(\Omega)$.

5. Prove the topological lemma that begins the proof of Proposition (8.14). (Hint: First construct a sequence $\{S_j\}$ of open subsets of Ω such that $\overline{S}_j$ is compact, $\overline{S}_j \subset S_{j+1}$, and $\Omega = \bigcup_1^\infty S_j$. Set $U_j = S_j \setminus \overline{S}_{j-1}$ and $V_j = S_{j+1} \setminus \overline{S}_{j-2}$.)

C. Asymptotic Expansions of Symbols

Much of the calculus of pseudodifferential operators consists of performing calculations with "highest order terms" and keeping careful track of lower order "error terms." For this purpose, the following notion of asymptotic expansion is extremely useful.

Suppose $\{m_j\}_0^\infty$ is a strictly decreasing sequence of real numbers such that $\lim m_j = -\infty$, and suppose $p_j \in S^{m_j}(\Omega)$ for each j. We say that the formal series $\sum_0^\infty p_j$ is an **asymptotic expansion** of the symbol $p \in S^{m_0}(\Omega)$ if

$$p - \sum_{j<k} p_j \in S^{m_k}(\Omega) \text{ for all } k > 0,$$

in which case we write

$$p \sim \sum_0^\infty p_j.$$

We emphasize that the series $\sum_0^\infty p_j$ need not be convergent, and if it is, its sum need not be p. The "sum" $\sum_0^\infty p_j$ is simply a convenient way of keeping track of the *sequence* $\{p_j\}_0^\infty$.

For most purposes in the theory of pseudodifferential operators, it is only asymptotic behavior that really counts. If two symbols p and q have the same asymptotic expansion $\sum p_j$, then

$$p - q = \left(p - \sum_{j<k} p_j\right) - \left(q - \sum_{j<k} p_j\right) \in S^{m_k}(\Omega)$$

for all k, and hence $p - q \in S^{-\infty}(\Omega)$; on the other hand, if $p - q \in S^{-\infty}(\Omega)$ then the series $q + 0 + 0 + \cdots$ is an asymptotic expansion of p. Accordingly, symbols of order $-\infty$ are usually regarded as negligible. Likewise, on the level of operators, ΨDO of order $-\infty$ — or more generally, smoothing operators of any sort — are usually regarded as negligible. In this sense, the symbol-operator correspondence $p \to p(x, D)$ is essentially a bijection for any Ω; see Proposition (8.11).

It is an important fact that asymptotic series can be used to build symbols:

(8.16) Theorem.
Suppose $p_j \in S^{m_j}(\Omega)$ for each $j \geq 0$, where m_j strictly decreases to $-\infty$. Then there exists $p \in S^{m_0}(\Omega)$ (unique modulo $S^{-\infty}(\Omega)$) such that $p \sim \sum_0^\infty p_j$.

Proof: Let $\{\Omega_j\}_1^\infty$ be an increasing sequence of open subsets of Ω with compact closure whose union is Ω, and choose $\phi \in C^\infty(\mathbb{R}^n)$ such that $\phi(\xi) = 1$ for $|\xi| \geq 1$ and $\phi(\xi) = 0$ for $|\xi| \leq \frac{1}{2}$.

Claim: There is a sequence $\{t_j\}_0^\infty$ of positive numbers converging to 0 such that for each $j > 0$,

(8.17)
$$\left| D_x^\beta D_\xi^\alpha [\phi(t_j\xi)p_j(x, \xi)] \right| \leq 2^{-j}(1 + |\xi|)^{m_{j-1} - |\alpha|}$$
for $x \in \Omega_j$ and $|\alpha| + |\beta| \leq j$.

Taking this for granted for the moment, we define

$$p(x, \xi) = \sum_{j=0}^\infty \phi(t_j\xi)p_j(x, \xi).$$

Since $t_j \to 0$, for ξ in any bounded set there are only finitely many nonzero terms in this sum, and it follows that $p \in C^\infty(\Omega \times \mathbb{R}^n)$. To see that $p \in S^{m_0}(\Omega)$, suppose B is a compact subset of Ω; we wish to estimate $D_x^\beta D_\xi^\alpha p(x, \xi)$ for $x \in B$. Choose k large enough so that $B \subset \Omega_k$ and $|\alpha| + |\beta| \leq k$, and write

$$p(x, \xi) = \sum_{j \leq k} \phi(t_j\xi)p_j(x, \xi) + \sum_{j > k} \phi(t_j\xi)p_j(x, \xi).$$

Since $p_j \in S^{m_j}$ and $\phi(t_j\xi) = 1$ for $|\xi|$ large, we clearly have

$$\left| \sum_{j \leq k} D_x^\beta D_\xi^\alpha \phi(t_j\xi)p_j(x, \xi) \right| \leq \sum_{j \leq k} C_j(1 + |\xi|)^{m_j - |\alpha|} \leq C(1 + |\xi|)^{m_0 - |\alpha|}.$$

On the other hand, by (8.17) we have

$$\left| \sum_{j > k} D_x^\beta D_\xi^\alpha \phi(t_j\xi)p_j(x, \xi) \right| \leq \sum_{j > k} 2^{-j}(1 + |\xi|)^{m_{j-1} - |\alpha|} \leq (1 + |\xi|)^{m_0 - |\alpha|}.$$

Thus $p \in S^{m_0}(\Omega)$. The same argument shows that $\sum_k^\infty \phi(t_j \xi) p_j(x, \xi) \in S^{m_k}(\Omega)$ for all k; since $\phi(t_j \xi) p_j(x, \xi) = p_j(x, \xi)$ for ξ large, it follows that $p \sim \sum_0^\infty p_j$.

It remains to prove the claim. First, let us observe that $D_\xi^\gamma [\phi(t\xi)] = t^{|\gamma|} [D_\xi^\gamma \phi](t\xi)$ and that, for $\gamma \neq 0$, $[D_\xi^\gamma \phi](t\xi) = 0$ except when $|\xi|$ is comparable to t^{-1}. If $t \leq 1$, this implies that $1 + |\xi|$ is also comparable to t^{-1}, and it follows that

$$|D_\xi^\gamma [\phi(t\xi)]| \leq C_\gamma (1 + |\xi|)^{-|\gamma|} \text{ for all } t \leq 1.$$

The product rule and the fact that $p_j \in S^{m_j}(\Omega)$ now easily imply that for some $C_j > 0$,

$$|D_x^\beta D_\xi^\alpha [\phi(t\xi) p_j(x, \xi)]| \leq C_j (1 + |\xi|)^{m_j - |\alpha|} \text{ for } x \in \Omega_j, \ |\alpha| + |\beta| \leq j, \ t \leq 1.$$

Of course the expression on the left is actually zero if $|\xi| \leq (2t)^{-1}$. It therefore suffices to pick t_j small enough so that (a) $t_j \to 0$ and (b) if $|\xi| \geq (2t_j)^{-1}$ then $C_j (1 + |\xi|)^{m_j - m_{j-1}} \leq 2^{-j}$. ∎

In the next section we shall need a technical result which shows that a series $\sum p_j$ is an asymptotic expansion of a symbol p if certain apparently weaker conditions hold. Before stating and proving it, we need a lemma from calculus:

(8.18) Lemma.
If $f : \mathbb{R} \to \mathbb{R}$ is twice differentiable and f, f', f'' are bounded,

$$\left(\sup_t |f'(t)| \right)^2 \leq 4 \left(\sup_t |f(t)| \right) \left(\sup_t |f''(t)| \right).$$

Proof: By considering $g(t) = a^{-1} f(\sqrt{a/b}\, t)$ where $a = \sup |f(t)|$ and $b = \sup |f''(t)|$, we are reduced to proving that if $\sup |g(t)| \leq 1$ and $\sup |g''(t)| \leq 1$ then $\sup |g'(t)| \leq 2$. Suppose instead that $|g'(t_0)| > 2$. By the mean value theorem, $|g'(t) - g'(t_0)| \leq 1$ for $|t - t_0| \leq 1$ and hence $|g'(t)| > 1$ for $|t - t_0| \leq 1$. By the mean value theorem again, $|g(t_0 + 1) - g(t_0 - 1)| > 2$, which contradicts $\sup |g(t)| \leq 1$. ∎

Remark: The optimal constant in Lemma (8.18) is 2, not 4; see Schoenberg [43].

(8.19) Lemma.
Let V be a compact subset of $\mathbb{R}^n$ and W a neighborhood of V. There is a constant $C > 0$ such that for all $f \in C^2(W)$,

$$\left(\sup_{x \in V} |\partial_j f(x)| \right)^2 \le C \left(\sup_{x \in W} |f(x)| \right) \left(\sum_{k=0}^{2} \sup_{x \in W} |\partial_j^k f(x)| \right).$$

Proof: Pick $\phi \in C_c^\infty(W)$ such that $\phi = 1$ on a neighborhood of V. The result follows by applying Lemma (8.18) to the real and imaginary parts of ϕf, considered as functions of x_j alone (Exercise 1). ∎

(8.20) Theorem.
Suppose $p_j \in S^{m_j}(\Omega)$ for $j \ge 0$, where m_j decreases to $-\infty$, and $p \in C^\infty(\Omega \times \mathbb{R}^n)$. Then $p \in S^{m_0}(\Omega)$ and $p \sim \sum_0^\infty p_j$ provided that the following conditions hold:

i. *There exists a sequence of real numbers μ_k with $\lim \mu_k = -\infty$ such that for every compact $B \subset \Omega$,*

$$\sup_{x \in B} \left| p(x,\xi) - \sum_{j<k} p_j(x,\xi) \right| \le C_{B,k}(1 + |\xi|)^{\mu_k}.$$

ii. *If α and β are multi-indices, there is a real number $\mu(\alpha, \beta)$ such that for every compact $B \subset \Omega$,*

$$\sup_{x \in B} |D_x^\beta D_\xi^\alpha p(x,\xi)| \le C_{B,\alpha,\beta}(1 + |\xi|)^{\mu(\alpha,\beta)}.$$

Proof: By Theorem (8.16), there exists $q \in S^{m_0}(\Omega)$ such that $q \sim \sum_0^\infty p_j$, and it will suffice to show that $p - q \in S^{-\infty}(\Omega)$.

First, by condition (i), for any compact $B \subset \Omega$ and $k > 0$, if $x \in B$ we have

$$|p(x,\xi) - q(x,\xi)| \le \left| p(x,\xi) - \sum_{j<k} p_j(x,\xi) \right| + \left| q(x,\xi) - \sum_{j<k} p_j(x,\xi) \right|$$

$$\le C_{B,k}\left[(1 + |\xi|)^{\mu_k} + (1 + |\xi|)^{m_k} \right].$$

Since $\mu_k \to -\infty$ and $m_k \to -\infty$,

$$(8.21) \qquad \sup_{x \in B} |(p - q)(x,\xi)| \le C_{B,N}(1 + |\xi|)^{-N}$$

for all N.

We need to prove that estimates of this sort hold also for $D_x^\beta D_\xi^\alpha (p-q)$. For the case $|\alpha| + |\beta| = 1$, for each ξ we apply Lemma (8.19) to the function $f(x, \eta) = (p-q)(x, \xi + \eta)$, taking $V = B \times \{0\}$ and $W = B' \times \{\eta : |\eta| < 1\}$ where B' is a compact neighborhood of B in Ω. By condition (ii) and the fact that $q \in S^{m_0}(\Omega)$, for some $\mu \in \mathbb{R}$ we have

$$\sup_{x \in B', \, |\eta| \leq 1} |D_x^\beta D_\xi^\alpha (p-q)(x, \xi + \eta)| \leq C_{B'}(1 + |\xi|)^\mu \qquad (|\alpha| + |\beta| \leq 2).$$

Lemma (8.19) together with (8.21) therefore gives

$$\sup_{x \in B} |D_x^\beta D_\xi^\alpha (p-q)(x, \xi)| \leq C'_{B,N}(1 + |\xi|)^{\mu - N} \qquad (|\alpha| + |\beta| = 1)$$

for all N, as desired. The proof is now completed by induction on $|\alpha| + |\beta|$: assuming that estimates of the type (8.21) for $D_x^\beta D_\xi^\alpha (p-q)$ with $|\alpha| + |\beta| = k-1$, Lemma (8.19) together with condition (ii) yields estimates of the type (8.21) for $D_x^\beta D_\xi^\alpha (p-q)$ with $|\alpha| + |\beta| = k$. ∎

EXERCISES

1. Complete the proof of Lemma (8.19).

2. Adapt the proof of Theorem (8.16) to show that every formal power series $\sum_{|\alpha| \geq 0} c_\alpha x^\alpha$ ($c_\alpha \in \mathbb{C}$) is the Taylor series of some C^∞ function.

D. Amplitudes, Adjoints, and Products

Our definition of pseudodifferential operators is based on the usual practice of writing partial differential operators as differentiations followed by multiplication by coefficients:

$$Pu(x) = \sum a_\alpha(x) D^\alpha u(x) = \int \sum e^{2\pi i x \cdot \xi} a_\alpha(x) \xi^\alpha \hat{u}(\xi) \, d\xi.$$

However, in some situations such as computing adjoints, it is preferable to reverse the order of these operations:

$$Qu(x) = \sum D^\alpha (a_\alpha u)(x) = \int \sum e^{2\pi i x \cdot \xi} \xi^\alpha (a^\alpha u)^\gamma(\xi) \, d\xi.$$

Setting $p(x,\xi) = \sum a_\alpha(x)\xi^\alpha$ and writing out the Fourier transform explicitly, these become

$$Pu(x) = \iint e^{2\pi i(x-y)\cdot\xi} p(x,\xi)u(y)\,dy\,d\xi,$$

$$Qu(x) = \iint e^{2\pi i(x-y)\cdot\xi} p(y,\xi)u(y)\,dy\,d\xi.$$

This suggests that we could also define pseudodifferential operators by using more general symbols p in the last formula for Q.

Actually, for maximum flexibility it is preferable to allow coefficients on both sides of the differentiations. We are therefore led to consider operators of the form

$$(8.22) \qquad P_a u(x) = \iint e^{2\pi i(x-y)\cdot\xi} a(x,\xi,y)u(y)\,dy\,d\xi,$$

where a belongs to the class

$$A^m(\Omega) = \big\{a \in C^\infty(\Omega \times \mathbb{R}^n \times \Omega) : \text{ for every compact } B \subset \Omega,$$
$$\sup_{x,y\in B} |D_x^\beta D_y^\gamma D_\xi^\alpha a(x,\xi,y)| \le C_{\alpha,\beta,\gamma,B}(1+|\xi|)^{m-|\alpha|}\big\}.$$

The elements of the class $A^m(\Omega)$ are called **amplitudes** of order m on Ω. A few remarks on these definitions are in order:

i. The integral in (8.22) must always be interpreted as an iterated integral in the indicated order; it is usually not absolutely convergent as a double integral.

ii. The operator P_a defined by (8.22) is to be interpreted, as in the case of ΨDO, as a linear map from $C_c^\infty(\Omega)$ to $C^\infty(\Omega)$.

iii. The class $A^m(\Omega)$ includes $S^m(\Omega)$ if we regard symbols as amplitudes that are independent of the third variable, and hence the corresponding class of operators $\{P_a : a \in A^m(\Omega)\}$ includes $\Psi^m(\Omega)$. We shall shortly see that these two classes actually coincide modulo smoothing operators and that the properly supported operators in the two classes are the same.

iv. In contrast to the symbol-operator correspondence $p \to p(x,D)$, the amplitude-operator correspondence $a \to P_a$ is highly non-injective, a fact which is useful in some ways and awkward in others. For example, if $\phi_1, \phi_2 \in C^\infty(\Omega)$ then the function $a(x,\xi,y) = \phi_1(x)\phi_2(y)$ belongs to $A^0(\Omega)$, and $P_a u = \phi_1\phi_2 u$. In particular, if $\operatorname{supp}\phi_1 \cap \operatorname{supp}\phi_2 = \varnothing$ then

$P_a = 0$, and if $\psi \in C^\infty(\Omega)$, there are at least as many amplitudes for the operator $u \to \psi u$ as there are ways of factoring ψ as $\phi_1 \phi_2$.

v. There is another way of defining pseudodifferential operators that deals with differentiation and multiplication by coefficients in a more symmetric way, the so-called Weyl calculus; see Folland [15].

By an argument like the one in §8B, if $a \in A^m(\Omega)$ the distribution kernel K of P_a is given by

$$\langle K, w \rangle = \iiint e^{2\pi i (x-y) \cdot \xi} a(x, \xi, y) w(x, y) \, dy \, d\xi \, dx \qquad (w \in C_c^\infty(\Omega \times \Omega)).$$

In other words, K is given by the inverse Fourier transform of a in its second variable:

$$(8.23) \qquad\qquad K(x, y) = a_2^\vee(x,\, x - y,\, y),$$

interpreted in the appropriate distributional sense. From this it is clear that

$$\operatorname{supp} K \subset \Sigma_a,$$

where

$$(8.24) \qquad \Sigma_a = \big\{ (x, y) \in \Omega \times \Omega : (x, \xi, y) \in \operatorname{supp} a \text{ for some } \xi \in \mathbb{R}^n \big\}.$$

(8.25) Proposition.
If $a \in A^m(\Omega)$, there exists $b \in A^m(\Omega)$ such that P_b is properly supported and $P_a - P_b$ is a smoothing operator.

Proof: By Proposition (8.14) we can find a properly supported $\phi \in C^\infty(\Omega)$ such that $\phi = 1$ on a neighborhood of the diagonal Δ_Ω. Let $b(x, \xi, y) = \phi(x, y) a(x, \xi, y)$. It is easily verified that $b \in A^m(\Omega)$. If K is the distribution kernel of P_a, it follows from (8.23) that the distribution kernel of P_b is ϕK, which is properly supported since ϕ is. Moreover, the same argument that proves Theorem (8.8) shows that K is C^∞ away from Δ_Ω and hence that $(1 - \phi)K$, the distribution kernel of $P_a - P_b$, is C^∞ everywhere. Hence $P_a - P_b$ is a smoothing operator.   ∎

(8.26) Proposition.
If $a \in A^m(\Omega)$ and P_a is properly supported, there exists $b \in A^m(\Omega)$ such that:

a. *$b(x, \xi, y) = a(x, \xi, y)$ for all $\xi \in \mathbb{R}^n$ and x, y in some neighborhood of the diagonal Δ_Ω;*

b. *Σ_b (defined by (8.24)) is a proper subset of $\Omega \times \Omega$;*

c. *$P_b = P_a$.*

Proof: Let K be the distribution kernel of P_a. By Proposition (8.14) we can find a properly supported $\phi \in C^\infty(\Omega \times \Omega)$ such that $\phi = 1$ on a neighborhood of $(\text{supp } K) \cup \Delta_\Omega$. Set $b(x, \xi, y) = \phi(x, y)a(x, \xi, y)$. As in the preceding proof, $b \in A^m(\Omega)$ and the distribution kernel of P_b is $\phi K = K$, so $P_b = P_a$. Finally, Σ_b is proper since it is contained in $\text{supp } \phi$.   ∎

Suppose that $a \in A^m(\Omega)$ and P_a is properly supported. By Proposition (8.12) and the discussion following it, P_a extends to a continuous operator on $C^\infty(\Omega)$. In particular, we can apply P_a to the function $E_\xi(x) = e^{2\pi i x \cdot \xi}$ for any $\xi \in \mathbb{R}^n$. In what follows, by a slight abuse of notation we shall denote the function $P_a(E_\xi)(x)$ by $P_a(e^{2\pi i x \cdot \xi})$.

(8.27) Theorem.
Suppose $a \in A^m(\Omega)$ and P_a is properly supported. Let

$$(8.28) \qquad p(x, \xi) = e^{-2\pi i x \cdot \xi} P_a(e^{2\pi i x \cdot \xi}).$$

Then $p \in S^m(\Omega)$ and $P_a = p(x, D)$. Moreover,

$$(8.29) \qquad p(x, \xi) \sim \sum_{|\alpha| \geq 0} \frac{1}{\alpha!} \partial_\xi^\alpha D_y^\alpha a(x, \xi, y)\big|_{y=x}.$$

Remark: The last statement means, more precisely, that

$$p \sim \sum_0^\infty p_j \quad \text{where} \quad p_j(x, \xi) = \sum_{|\alpha|=j} \frac{1}{\alpha!} \partial_\xi^\alpha D_y^\alpha a(x, \xi, y)\big|_{y=x}.$$

p_j clearly belongs to $S^{m-j}(\Omega)$. (Similar remarks will apply to the asymptotic expansions in Theorems (8.36) and (8.37) below.)

Proof: If $u \in C_c^\infty(\Omega)$, we write $u(x) = \int e^{2\pi i x \cdot \xi} \widehat{u}(\xi) \, d\xi$. As the reader may verify (Exercise 1), this Fourier integral is the limit of Riemann sums $\sum e^{2\pi i x \cdot \xi_j} \widehat{u}(\xi_j) \Delta \xi_j$ that converge in the topology of $C^\infty(\Omega)$ (i.e., the sums and their derivatives converge uniformly on compact sets to u and its derivatives). Since P_a is linear and continuous on $C^\infty(\Omega)$, we have

$$P_a u(x) = \int P_a(e^{2\pi i x \cdot \xi}) \widehat{u}(\xi) \, d\xi = \int e^{2\pi i x \cdot \xi} p(x, \xi) \widehat{u}(\xi) \, d\xi,$$

so $P_a = p(x, D)$. We shall show that $p \in S^m(\Omega)$ and that (8.29) holds by using Theorem (8.20).

By Proposition (8.26), we can modify a so that Σ_a is proper without affecting P_a or the behavior of a along Δ_Ω (which is all that matters in (8.29)), and we henceforth assume that this has been done. We then have

$$p(x, \eta) = e^{-2\pi i x \cdot \eta} \iint e^{2\pi i (x-y) \cdot \xi} a(x, \xi, y) e^{2\pi i y \cdot \eta} \, dy \, d\xi,$$

where the inner integral converges nicely since the function $y \to a(x, \xi, y)$ belongs to $C_c^\infty(\Omega)$ for each x and ξ. If we set

$$b(x, \xi, y) = a(x, \xi, x + y)$$

this becomes

$$p(x, \eta) = \iint e^{2\pi i (x-y) \cdot \xi} b(x, \xi, y - x) e^{2\pi i (y-x) \cdot \eta} \, d\eta \, d\xi$$

$$= \iint e^{-2\pi i z \cdot (\xi - \eta)} b(x, \xi, z) \, dz \, d\xi$$

$$= \int \widehat{b}_3(x, \xi, \xi - \eta) \, d\xi,$$

or

(8.30) $$p(x, \eta) = \int \widehat{b}_3(x, \xi + \eta, \xi) \, d\xi,$$

where $\widehat{b}_3$ denotes the Fourier transform of b in its third variable.

Since $a \in A^m(\Omega)$ and $b(x, \eta, \cdot) \in C_c^\infty$ for each x and η, for x in any fixed compact subset of Ω we have

(8.31) $$|D_x^\beta D_\eta^\alpha \widehat{b}_3(x, \eta, \xi)| \le C_{\alpha, \beta, N}(1 + |\eta|)^{m-|\alpha|}(1 + |\xi|)^{-N}.$$

In particular,

$$|D_x^\beta D_\eta^\alpha \widehat{b}_3(x, \xi + \eta, \xi)| \le C_{\alpha, \beta, N}(1 + |\xi + \eta|)^{m-|\alpha|}(1 + |\xi|)^{-N}.$$

The triangle inequality implies that

$$1 + |\xi + \eta| \le (1 + |\xi|)(1 + |\eta|),$$

and raising both sides to the (positive!) power $|m| + |\alpha|$ shows that

$$|D_x^\beta D_\eta^\alpha \widehat{b}_3(x, \xi + \eta, \xi)| \le C_{\alpha, \beta, N}(1 + |\eta|)^{|m|+|\alpha|}(1 + |\xi|)^{|m|+|\alpha|-N}.$$

If we apply this estimate to (8.30) with $N = |m| + |\alpha| + n + 1$, we see that

$$|D_x^\beta D_\eta^\alpha p(x, \eta)| \le C_{\alpha,\beta}(1 + |\eta|)^{|m|+|\alpha|},$$

which is condition (ii) in Theorem (8.20).

To establish condition (i), we apply Taylor's theorem to the function $\zeta \to \widehat{b}_3(x, \eta + \zeta, \xi)$, set $\zeta = \xi$, and apply (8.31):

$$\left| \widehat{b}_3(x, \xi + \eta, \xi) - \sum_{|\alpha|<k} \partial_\eta^\alpha \widehat{b}_3(x, \eta, \xi) \frac{\xi^\alpha}{\alpha!} \right|$$

$$\le c_k \sup_{|\alpha|=k, \ 0 \le t \le 1} |\xi|^k |\partial_\eta^\alpha \widehat{b}_3(x, \eta + t\xi, \xi)|$$

$$\le c_{k,N} \sup_{0 \le t \le 1} |\xi|^k (1 + |\eta + t\xi|)^{m-k} (1 + |\xi|)^{-N}.$$

(These estimates, like the preceding ones, hold for x in a fixed compact subset of Ω.) If $|\xi| < \frac{1}{2}|\eta|$, we take $N = k$ and obtain

$$\left| \widehat{b}_3(x, \xi + \eta, \xi) - \sum_{|\alpha|<k} \partial_\eta^\alpha \widehat{b}_3(x, \eta, \xi) \frac{\xi^\alpha}{\alpha!} \right| \le c_k'(1 + |\eta|)^{m-k}.$$

If $|\xi| \ge \frac{1}{2}|\eta|$, we take $N = 2k + n + \max(m - k, 0)$ and obtain

$$\left| \widehat{b}_3(x, \xi + \eta, \xi) - \sum_{|\alpha|<k} \partial_\eta^\alpha \widehat{b}_3(x, \eta, \xi) \frac{\xi^\alpha}{\alpha!} \right| \le c_k''(1 + |\xi|)^{-n-k}.$$

Substituting these results into (8.30) and integrating in polar coordinates, we see that

$$\left| p(x, \eta) - \sum_{|\alpha| \le k} \int \partial_\eta^\alpha \widehat{b}_3(x, \eta, \xi) \frac{\xi^\alpha}{\alpha!} \, d\xi \right|$$

$$\le c_k' \int_{|\xi|<|\eta|/2} (1 + |\eta|)^{m-k} \, d\xi + c_k'' \int_{|\xi| \ge |\eta|/2} (1 + |\xi|)^{-n-k} \, d\xi$$

$$\le C_k \left[(1 + |\eta|)^{m+n-k} + (1 + |\eta|)^{-k} \right].$$

But

$$\int \partial_\eta^\alpha \widehat{b}_3(x, \eta, \xi) \xi^\alpha \, d\xi = \left[D_y^\alpha \int e^{2\pi i y \cdot \xi} \partial_\eta^\alpha \widehat{b}_3(x, \eta, \xi) \, d\xi \right]_{y=0}$$

$$= D_y^\alpha \partial_\eta^\alpha b(x, \eta, y)\big|_{y=0}$$

$$= D_y^\alpha \partial_\eta^\alpha a(x, \eta, y)\big|_{y=x},$$

so

$$\left| p(x, \eta) - \sum_{|\alpha|<k} \frac{1}{\alpha!} D_y^\alpha \partial_\eta^\alpha a(x, \eta, y)\big|_{y=x} \right| \le 2C_k(1 + |\eta|)^{\mu_k},$$

where $\mu_k = \max(m + n, 0) - k$. Condition (i) of Theorem (8.20) is thus fulfilled, and we are done.   ∎

(8.32) Corollary.
If $a \in A^m(\Omega)$, there is a properly supported $Q \in \Psi^m(\Omega)$ such that $P_a - Q$ is a smoothing operator. If $a \in S^m(\Omega)$ (so that $P_a = a(x, D)$), there is a properly supported $Q \in \Psi^m(\Omega)$ such that $a(x, D) - Q \in \Psi^{-\infty}(\Omega)$.

Proof: By Proposition (8.25) there exists $b \in A^m(\Omega)$ such that P_b is properly supported and $P_a - P_b$ is smoothing, and by Theorem (8.27), $P_b \in \Psi^m(\Omega)$; this proves the first assertion. Moreover, the b in Proposition (8.25) is constructed so that the distribution kernel of $P_a - P_b$ vanishes on a neighborhood of the diagonal. Thus if $a \in S^m(\Omega)$ and q is the symbol of P_b given by (8.28), by (8.6) $(a - q)_2^\vee(x, z)$ vanishes near $z = 0$ for each x. By Theorem (8.8a), $(a - q)_2^\vee(x, \cdot)$ is a Schwartz class function for each x, with estimates depending uniformly on x for x in a compact set, and hence the same is true of $(a - q)(x, \cdot)$. Thus $a - q \in \Psi^{-\infty}(\Omega)$; in other words, $P_a - P_b \in \Psi^{-\infty}(\Omega)$. ∎

As we observed in §8B, whether $P \in \Psi^m(\Omega)$ has a unique symbol or not depends on the nature of Ω. However, if P is properly supported, (8.28) gives a canonical choice of symbol for P, and we shall denote it by σ_P:

$$(8.33) \qquad \sigma_P(x, \xi) = e^{-2\pi i x \cdot \xi} P(e^{2\pi i x \cdot \xi}).$$

We can now obtain a perfect correspondence between symbols and operators if we follow the philosophy introduced in §8C of regarding symbols and operators of order $-\infty$ as negligible, i.e., of considering equivalence classes of symbols and operators modulo symbols and operators of order $-\infty$.

(8.34) Corollary.
The map $p \to p(x, D)$ induces a bijection from $S^\infty(\Omega)/S^{-\infty}(\Omega)$ to $\Psi^\infty(\Omega)/\Psi^{-\infty}(\Omega)$, and the map $P \to \sigma_P$ induces a bijection from $\Psi^\infty(\Omega)/\Psi^{-\infty}(\Omega)$ to $S^\infty(\Omega)/S^{-\infty}(\Omega)$. These bijections are mutually inverse.

Proof: Each equivalence class in $\Psi^\infty(\Omega)/\Psi^{-\infty}(\Omega)$ contains a properly supported operator, by Corollary (8.32). The assertions therefore follow from Proposition (8.11). ∎

We are now in a good position to compute transposes, adjoints and products of pseudodifferential operators. The terminology we employ is as follows: If T and S are linear maps from $C_c^\infty(\Omega)$ to $C^\infty(\Omega)$, we say that S is the **transpose** of T and write $S = T'$ if $\langle Tu, v \rangle = \langle u, Sv \rangle$ for all $u, v \in C_c^\infty(\Omega)$; we say that S is the **adjoint** of T and write $S = T^*$ if $\langle Tu, \overline{v} \rangle = \langle u, \overline{Sv} \rangle$ for all $u, v \in C_c^\infty(\Omega)$. If T has the distribution kernel $K(x, y)$, then T' and T^* have the distribution kernels $K'(x, y) = K(y, x)$ and $K^*(x, y) = \overline{K(y, x)}$ respectively. It follows that if T is properly supported, then so are T' and T^*.

(8.35) Lemma.
Suppose $p \in S^m(\Omega)$. Then $p(x, D)' = P_a$ where $a(x, \xi, y) = p(y, -\xi)$, and $p(x, D)^* = P_b$ where $b(x, \xi, y) = \overline{p(y, \xi)}$.

Proof: We consider only $p(x, D)'$; the argument for $p(x, D)^*$ is essentially identical. Suppose $u, v \in C_c^\infty(\Omega)$. Since

$$\langle p(x, D)u, \, v \rangle = \iiint e^{2\pi i (x-y) \cdot \xi} p(x, \xi) u(x) v(y) \, dy \, d\xi \, dx,$$

the proof is just a matter of reversing the order of integration. This is not a routine application of Fubini's theorem, however, since the triple integral is (usually) not absolutely convergent. Instead, we apply Fubini's theorem to the absolutely convergent double integral

$$\langle p(x, D)u, \, v \rangle = \iint e^{2\pi i x \cdot \xi} p(x, \xi) \widehat{u}(\xi) v(x) \, d\xi \, dx$$

to obtain

$$\langle p(x, D)u, \, v \rangle = \int g(\xi) \widehat{u}(\xi) \, d\xi,$$

where

$$g(\xi) = \int e^{2\pi i x \cdot \xi} p(x, \xi) v(x) \, dx.$$

g is a rapidly decreasing function by Lemma (8.5), so we have

$$\langle p(x, D)u, \, v \rangle = \langle \widehat{u}, g \rangle = \langle u, \widehat{g} \rangle.$$

Therefore

$$p(x, D)'v(y) = \widehat{g}(y) = \iint e^{2\pi i (x-y) \cdot \xi} p(x, \xi) v(x) \, dx \, d\xi$$

$$= \iint e^{2\pi i (y-x) \cdot \xi} p(x, -\xi) v(x) \, dx \, d\xi,$$

so that $p(x, D)' = P_a$ with $a(x, \xi, y) = p(y, -\xi)$ as claimed. ∎

(8.36) Theorem.
If $P \in \Psi^m(\Omega)$ is properly supported, then $P', P^* \in \Psi^m(\Omega)$, and

$$\sigma_{P'}(x,\xi) \sim \sum_{|\alpha| \geq 0} \frac{(-1)^{|\alpha|}}{\alpha!} \partial_\xi^\alpha D_x^\alpha \sigma_P(x, -\xi),$$

$$\sigma_{P^*}(x,\xi) \sim \sum_{|\alpha| \geq 0} \frac{1}{\alpha!} \partial_\xi^\alpha D_x^\alpha \overline{\sigma_P(x,\xi)}.$$

Proof: This is an immediate consequence of Theorem (8.27), Corollary (8.34), and Lemma (8.35). ∎

(8.37) Theorem.
If $P \in \Psi^m(\Omega)$ and $Q \in \Psi^{m'}(\Omega)$ are properly supported, then $QP \in \Psi^{m+m'}(\Omega)$ and

$$\sigma_{QP}(x,\xi) \sim \sum_{|\alpha| \geq 0} \frac{1}{\alpha!} \partial_\xi^\alpha \sigma_Q(x,\xi) \cdot D_x^\alpha \sigma_P(x,\xi).$$

Proof: Since $P = (P')'$, by Lemma (8.35) we have

$$Pu(x) = \iint e^{2\pi i(x-y)\cdot\xi} \sigma_{P'}(y, -\xi)u(y)\, dy\, d\xi,$$

that is,

$$\widehat{Pu}(\xi) = \int e^{-2\pi iy\cdot\xi} \sigma_{P'}(y, -\xi)u(y)\, dy.$$

Therefore,

$$QPu(x) = \iint e^{2\pi i(x-y)\cdot\xi} \sigma_Q(x,\xi)\sigma_{P'}(y, -\xi)u(y)\, dy\, d\xi,$$

which means that

$$QP = P_a \quad \text{where} \quad a(x,\xi,y) = \sigma_Q(x,\xi)\sigma_{P'}(y, -\xi).$$

Clearly $a \in A^{m+m'}(\Omega)$, and QP is properly supported by Corollary (8.13). Hence, by Theorem (8.27), $QP \in \Psi^{m+m'}(\Omega)$ and

$$\sigma_{QP}(x,\xi) \sim \sum_{|\alpha| \geq 0} \frac{1}{\alpha!} \partial_\xi^\alpha D_y^\alpha [\sigma_Q(x,\xi)\sigma_{P'}(y, -\xi)]\Big|_{y=x}.$$

Now, by the product rule,

$$\partial_\xi^\alpha D_y^\alpha [\sigma_Q(x,\xi)\sigma_{P'}(y,-\xi)]\big|_{y=x} = \sum_{\beta+\gamma=\alpha} \frac{\alpha!}{\beta!\gamma!} [\partial_\xi^\beta \sigma_Q(x,\xi)]\cdot\partial_\xi^\gamma D_x^\alpha[\sigma_{P'}(x,-\xi)],$$

so

$$\sigma_{QP}(x,\xi) \sim \sum_{\beta,\gamma} \frac{1}{\beta!\gamma!}\partial_\xi^\beta \sigma_Q(x,\xi)\cdot\partial_\xi^\gamma D_x^{\beta+\gamma}[\sigma_{P'}(x,-\xi)],$$

with the understanding that, for each $j \geq 0$, all the terms with $|\beta|+|\gamma| = j$ (of which there are a finite number) are to be grouped together to form a single term $p_j(x,\xi) \in S^{m+m'-j}$ in the asymptotic sum. With the same understanding, Theorem (8.36) then implies that

$$\sigma_{QP}(x,\xi) \sim \sum_{\beta,\gamma,\delta} \frac{(-1)^{|\delta|}}{\beta!\gamma!\delta!}\partial_\xi^\beta \sigma_Q(x,\xi)\cdot\partial_\xi^{\gamma+\delta} D_x^{\beta+\gamma+\delta}\sigma_P(x,\xi)$$

$$= \sum_{\beta,\lambda}\left[\sum_{\gamma+\delta=\lambda}\frac{(-1)^{|\delta|}}{\gamma!\delta!}\right]\frac{1}{\beta!}\partial_\xi^\beta \sigma_Q(x,\xi)\cdot\partial_\xi^\lambda D_x^{\beta+\lambda}\sigma_P(x,\xi).$$

But by the binomial theorem, if $\mathbf{x} = (1,1,\cdots,1) \in \mathbb{R}^n$,

$$\sum_{\gamma+\delta=\lambda}\frac{(-1)^{|\delta|}}{\gamma!\delta!} = (\mathbf{x}-\mathbf{x})^\lambda = \begin{cases} 1 & \text{if } \lambda = 0, \\ 0 & \text{if } \lambda \neq 0. \end{cases}$$

Therefore,

$$\sigma_{QP}(x,\xi) \sim \sum_\beta \frac{1}{\beta!}\partial_\xi^\beta \sigma_Q(x,\xi)\cdot D_x^\beta \sigma_P(x,\xi),$$

and we are done.   ∎

$S^\infty(\Omega)$ is a $*$-algebra (i.e., algebra with involution) under pointwise multiplication and the involution $p \to \overline{p}$. $S^{-\infty}(\Omega)$ is a $*$-ideal, so the quotient $S^\infty(\Omega)/S^{-\infty}(\Omega)$ inherits the structure of a $*$-algebra. On the other hand, by Theorems (8.36) and (8.37) the set of properly supported operators in $\Psi^\infty(\Omega)$ is a $*$-algebra under operator composition and the involution $P \to P^*$. Again, the operators of order $-\infty$ form a $*$-ideal, so in view of Corollary (8.32), $\Psi^\infty(\Omega)/\Psi^{-\infty}(\Omega)$ inherits the structure of a $*$-algebra. The correspondence $p \to p(x,D)$ and its inverse $P \to \sigma_P$ (as in Corollary (8.34)) are not homomorphisms — after all, the symbol algebra is commutative whereas the operator algebra is not — but they become homomorphisms if we neglect all but the highest order terms in the asymptotic expansions of Theorems (8.36) and (8.37). Indeed, a glance at these expansions immediately yields the following result.

(8.38) Corollary.
*The bijections of Corollary (8.34) are *-algebra homomorphisms modulo lower order terms. More precisely:*
a. *If $P \in \Psi^m(\Omega)$ and $Q \in \Psi^{m'}(\Omega)$ are properly supported then then*

$$\sigma_{PQ} = \sigma_P \sigma_Q \pmod{S^{m+m'-1}(\Omega)},$$
$$\sigma_{P^*} = \overline{\sigma_P} \pmod{S^{m-1}(\Omega)}.$$

b. *If $p \in S^m(\Omega)$ and $q \in S^{m'}(\Omega)$ and $p(x, D)$ and $q(x, D)$ are properly supported then*

$$p(x, D)q(x, D) = (pq)(x, D) \pmod{\Psi^{m+m'-1}(\Omega)},$$
$$p(x, D)^* = \overline{p}(x, D) \pmod{\Psi^{m-1}(\Omega)}.$$

There is yet another algebra structure that is preserved by the symbol-operator correspondence up to lower order terms. Namely, for C^∞ functions on $\Omega \times \mathbb{R}^n$ one has the **Poisson bracket**

$$\{p, q\} = \sum \left(\frac{\partial p}{\partial \xi_j} \frac{\partial q}{\partial x_j} - \frac{\partial q}{\partial \xi_j} \frac{\partial p}{\partial x_j} \right),$$

which makes $S^\infty(\Omega)$ and $S^\infty(\Omega)/S^{-\infty}(\Omega)$ into Lie algebras. (More precisely, if $p \in S^m(\Omega)$ and $q \in S^{m'}(\Omega)$ then $\{p, q\} \in S^{m+m'-1}(\Omega)$.) On the other hand, by Theorem (8.37) and Corollary (8.32), $\Psi^\infty(\Omega)/\Psi^{-\infty}(\Omega)$ inherits a Lie algebra structure from the commutator of operators, $[P, Q] = PQ - QP$. If we subtract the asymptotic expansion for the symbol of QP from that of PQ, we see that the highest order terms cancel and that the first remaining terms give the Poisson bracket, except for a factor of $2\pi i$ becasue one has D_x instead of ∂_x. In other words:

(8.39) Corollary.
If $P \in \Psi^m(\Omega)$ and $Q \in \Psi^{m'}(\Omega)$ are properly supported then $[P, Q] \in \Psi^{m+m'-1}(\Omega)$ and

$$\sigma_{[P,Q]} = \frac{1}{2\pi i} \{\sigma_P, \sigma_Q\} \pmod{S^{m+m'-2}(\Omega)}.$$

If $p \in S^m(\Omega)$ and $q \in S^{m'}(\Omega)$ and $p(x, D)$ and $q(x, D)$ are properly supported then

$$[p(x, D), q(x, D)] = 2\pi i \{p, q\}(x, D) \pmod{\Psi^{m+m'-2}(\Omega)}.$$

The symbol-operator correspondence is related to the problem of defining a correspondence between observables in classical mechanics and observables in quantum mechanics. For the latter purpose one should insert factors of $\hbar$ (Planck's constant) in various places in our formulas, and the asymptotic expansions of the theorems in this section then become expansions in powers of $\hbar$. The correspondence between Poisson brackets and commutators is of particular importance in this setting. See Folland [15].

EXERCISES

1. Show that if $u \in \mathcal{S}$, there is a sequence of finite sums of the form $S_N(x) = \sum_{j=1}^{J(N)} c_j^N e^{2\pi i x \cdot \xi_j^N} \widehat{u}(\xi_j^N)$ such that $\partial^\alpha S_N \to \partial^\alpha u$ uniformly on compact sets for every α.

2. Our hypotheses about proper support are sometimes more stringent than necessary. For example, the product of two ΨDO is well-defined if at least one of them is properly supported. Formulate and prove versions of Theorem (8.37) and Corollaries (8.38) and (8.39) (as corollaries of these results themselves) that apply in this more general situation.

3. Since $\partial = 2\pi i D$, the asymptotic formula for $\sigma_{P\bullet}$ in Theorem (8.36) can be written formally as

$$\sigma_{P\bullet}(x, \xi) \sim e^{2\pi i D_x \cdot D_\xi} \sigma_P(x, \xi).$$

Another interpretation of $e^{2\pi i D_x, D_\xi}$ is available, as an operator on $\mathcal{S}(\mathbb{R}^{2n})$ defined by the Fourier transform:

$$[e^{2\pi i D_x \cdot D_\xi} f]\widehat{}(\eta, y) = e^{2\pi i \eta \cdot y} \widehat{f}(\eta, y).$$

Show that if $P = p(x, D)$ where $p \in \mathcal{S}(\mathbb{R}^{2n})$ (so $P \in \Psi^{-\infty}(\mathbb{R}^n)$), then $\sigma_{P\bullet} = e^{2\pi i D_x \cdot D_\xi} p$ (exact equality!). (Hint: Use (8.6).)

4. (A product rule for ΨDO) Suppose $p \in S^m(\Omega)$ and $v \in C^\infty(\Omega)$. Show that for every positive integer k there exists $R_k \in \Psi^{m-k-1}(\Omega)$ such that

$$p(x, D)(vu) = \sum_{|\alpha| \le k} \frac{1}{\alpha!} D^\alpha v \cdot (\partial_\xi^\alpha p)(x, D)u + R_k u.$$

In certain cases one can obtain an exact formula for the product of two ΨDO, with no error term. The following exercises examine three such cases; in all of them, one should work directly with the definition (8.3) rather than trying to apply the results of this section.

5. Suppose $p, q, r \in S^\infty(\Omega)$, p is independent of ξ, and r is independent of x. What are $p(x, D)$ and $r(x, D)$ in these cases? Show that $(pq)(x, D) = p(x, D)q(x, D)$ and that $(qr)(x, D) = q(x, D)r(x, D)$.

6. Show that the result of Exercise 4 holds with $R_k = 0$ if v is a polynomial of degree $\leq k$.

7. Show that if $p \in S^m(\Omega)$,

$$D^\alpha[p(x, D)u] = \sum_{\beta + \gamma = \alpha} \frac{\alpha!}{\beta!\gamma!}(D_x^\beta p)(x, D)[D^\gamma u].$$

E. Sobolev Estimates

We now state and prove a continuity theorem for pseudodifferential operators with respect to Sobolev norms. Our argument is an elaboration of the one we used to prove Proposition (6.12).

(8.40) Theorem.
Suppose $P \in \Psi^m(\Omega)$.
a. *P maps $H_s^0(\Omega)$ continuously into $H_{s-m}^{\mathrm{loc}}(\Omega)$ for all $s \in \mathbb{R}$; that is, if $\phi \in C_c^\infty(\Omega)$ then $\|\phi Pu\|_{s-m} \leq C_{s,\phi}\|u\|_s$ for all $u \in H_s^0(\Omega)$.*
b. *If P is also properly supported, P maps $H_s^{\mathrm{loc}}(\Omega)$ continuously into $H_{s-m}^{\mathrm{loc}}(\Omega)$ for all $s \in \mathbb{R}$; that is, for every $\phi \in C_c^\infty(\Omega)$ there is a $\psi \in C_c^\infty(\Omega)$ such that $\|\phi Pu\|_{s-m} \leq C_{\phi,s}\|\psi u\|_s$.*

Proof: Let $P = p(x, D)$. Then the map $u \to \phi Pu$ is $q(x, D)$ where $q(x, \xi) = \phi(x)p(x, \xi)$. To prove (a), it therefore suffices to show that if $q(x, \xi) \in S^m(\Omega)$ and $q(x, \xi) = 0$ for x outside some fixed compact set B, then $Q = q(x, D)$ is bounded from $H_s^0(\Omega)$ to H_{s-m}.

Suppose then that $u \in H_s^0(\Omega)$. Qu is defined by the recipe for the action of Q on distributions in §8A, and Qu has compact support — in fact, supp $Qu \subset B$. It follows that

$$\widehat{Qu}(\eta) = \langle Qu, e^{-2\pi i(\cdot)\cdot\eta}\rangle = \iint e^{2\pi i x\cdot(\xi-\eta)}q(x, \xi)\widehat{u}(\xi)\, dx\, d\xi$$

$$= \int \widehat{q}_1(\eta - \xi, \xi)\widehat{u}(\xi)\, d\xi,$$

where $\hat{q}_1$ denotes the Fourier transform of q in its first variable. Hence, if $v \in \mathcal{S}$,

$$\langle Qu \mid v \rangle = \langle \widehat{Qu} \mid \hat{v} \rangle = \iint K(\eta, \xi) f(\xi) \overline{g(\eta)} \, d\xi \, d\eta$$

where

$$f(\xi) = (1 + |\xi|^2)^{s/2} \hat{u}(\xi), \qquad g(\eta) = (1 + |\eta|^2)^{(m-s)/2} \hat{v}(\eta),$$
$$K(\eta, \xi) = \hat{q}_1(\eta - \xi, \xi)(1 + |\xi|^2)^{-s/2}(1 + |\eta|^2)^{(s-m)/2}.$$

Since $q(x, \xi)$ has compact support in x, for any multi-index α we have

$$|\zeta^\alpha \hat{q}_1(\zeta, \xi)| = \left| \int [D_x^\alpha e^{-2\pi i x \cdot \zeta}] q(x, \xi) \, dx \right|$$
$$= \left| \int e^{-2\pi i x \cdot \zeta} D_x^\alpha q(x, \xi) \, dx \right| \le C_\alpha (1 + |\xi|^2)^{m/2},$$

so that for any positive integer N,

$$|\hat{q}_1(\zeta, \xi)| \le C_N (1 + |\xi|^2)^{m/2}(1 + |\zeta|^2)^{-N}.$$

Hence, by Lemma (6.10),

$$|K(\eta, \xi)| \le C_N (1 + |\xi|^2)^{(m-s)/2}(1 + |\eta|^2)^{(s-m)/2}(1 + |\eta - \xi|^2)^{-N}$$
$$\le C'_N (1 + |\eta - \xi|^2)^{-N + |s - m|/2}.$$

If we choose $N > \frac{1}{2}n + |s - m|$, we see that

$$\int |K(\eta, \xi)| \, d\eta \le C_s, \qquad \int |K(\eta, \xi)| \, d\xi \le C_s.$$

Therefore, by (0.10) and the Schwarz inequality,

$$|\langle Qu \mid v \rangle| \le C_s \|f\|_{L^2} \|g\|_{L^2} = C_s \|u\|_s \|v\|_{m-s},$$

so the duality of H_{s-m} and H_{m-s} implies that $\|Qu\|_{s-m} \le C_s \|u\|_s$, as desired. This proves (a).

Now suppose P is properly supported, $u \in H_s^{\text{loc}}(\Omega)$, and $\phi \in C_c^\infty(\Omega)$. By Proposition (8.12) there is a compact $B \subset \Omega$ such that the values of Pu on $\text{supp}\,\phi$ depend only on the values of u on B. Thus if we pick $\psi \in C_c^\infty(\Omega)$ with $\psi = 1$ on B, we have $\phi Pu = \phi P(\psi u)$. By part (a), then, $\|\phi Pu\|_{s-m} \le C_s \|\psi u\|_s$, which proves (b). ∎

EXERCISES

1. Assume $\Omega = \mathbb{R}^n$. Find sufficient conditions on $p \in S^m(\mathbb{R}^n)$ for $p(x, D)$ to be bounded from H_s to H_{s-m} (globally, with no cutoff functions).

2. The boundedness of the function p is not necessary for the boundedness of $p(x, D)$ on $H_0 = L^2$. Show, for example, that if $p \in L^2(\Omega \times \mathbb{R}^n)$ then $p(x, D)$ (defined by (8.3), even though p may not belong to $S^\infty(\Omega)$) is a compact operator on $L^2(\Omega)$. (Hint: Theorem (0.45).)

F. Elliptic Operators

A symbol $p \in S^m(\Omega)$, or its corresponding operator $p(x, D) \in \Psi^m(\Omega)$, is said to be **elliptic of order** m if for every compact $A \subset \Omega$ there are positive constants c_A, C_A such that

$$|p(x, \xi)| \geq c_A |\xi|^m \text{ for } x \in A \text{ and } |\xi| \geq C_A.$$

This agrees with our previous definition when $p(x, D)$ is a differential operator; see the remarks at the beginning of §6C. As we did with differential operators, when we say that $P \in \Psi^m(\Omega)$ is elliptic, we shall always mean that it is elliptic of order m.

In this section we shall show that elliptic pseudodifferential operators are invertible in the algebra $\Psi^\infty(\Omega)/\Psi^{-\infty}(\Omega)$; this will yield an easy proof of the elliptic regularity theorem for pseudodifferential operators and a proof that elliptic differential operators are locally solvable. We shall also derive a version of Gårding's inequality for pseudodifferential operators. To begin with, we state the following technical lemma, whose proof we leave to the reader (Exercise 1).

(8.41) Lemma.
If $p \in S^m(\Omega)$ is elliptic, there exists $\zeta \in C^\infty(\Omega \times \mathbb{R}^n)$ with the following property: For any compact $A \subset \Omega$ there are positive constants c, C such that for $x \in A$ we have
a. $\zeta(x, \xi) = 1$ when $|\xi| \geq C$;
b. $|p(x, \xi)| \geq c|\xi|^m$ when $\zeta(x, \xi) \neq 0$.

If L is a differential operator on Ω, a left (resp. **right**) **parametrix** for L is usually defined to be an operator T (not necessarily a ΨDO), defined on some suitable space of functions or distributions on Ω, such that $TL - I$ (resp. $LT - I$) is a smoothing operator. In our context, we shall define a **parametrix** for a ΨDO $P \in \Psi^\infty(\Omega)$ to be a properly supported ΨDO $Q \in \Psi^\infty(\Omega)$ such that $PQ - I \in \Psi^{-\infty}(\Omega)$ and $QP - I \in \Psi^{-\infty}(\Omega)$. (Here and in the following discussion, we may modify any ΨDO by adding an element of $\Psi^{-\infty}(\Omega)$ to make it properly supported, as the need arises. This has no effect on our calculations, which are all performed modulo $\Psi^{-\infty}(\Omega)$.)

(8.42) Theorem.
If $P \in \Psi^m(\Omega)$ is elliptic, P has a parametrix $Q \in \Psi^{-m}(\Omega)$.

   Proof: Let $P = p(x, D)$, and let ζ be as in Lemma (8.41). Let $q_0 = \zeta/p$, with the understanding that $q_0 = 0$ wherever $\zeta = 0$. Since $q_0(x, \xi) = 1/p(x, \xi)$ for large ξ and p is elliptic, it is easily verified (Exercise 2) that $q_0 \in S^{-m}(\Omega)$; moreover, $q_0 p - 1$ has compact support in ξ and hence belongs to $S^{-\infty}(\Omega)$. Let $Q_0 = q_0(x, D)$; then by Corollary (8.38),

$$\sigma_{Q_0 P} = q_0 p \quad (\mathrm{mod}\ S^{-1}(\Omega))$$
$$= 1 - r_1 \text{ where } r_1 \in S^{-1}(\Omega).$$

Let $q_1 = \zeta r_1/p = r_1 q_0 \in S^{-m-1}(\Omega)$ and $Q_1 = q_1(x, D)$. By Corollary (8.38) again,

$$\sigma_{Q_1 P} = q_1 p \quad (\mathrm{mod}\ S^{-2}(\Omega))$$
$$= r_1 - r_2 \text{ where } r_2 \in S^{-2}(\Omega),$$

and hence

$$\sigma_{(Q_0 + Q_1)P} = \sigma_{Q_0 P} + \sigma_{Q_1 P} = 1 - r_2, \qquad r_2 \in S^{-2}(\Omega).$$

We now construct q_j inductively for $j \geq 2$. Having constructed $q_j \in S^{-m-j}(\Omega)$ for $j < k$ so that (with $Q_j = q_j(x, D)$)

$$\sigma_{(Q_0 + \cdots + Q_j)P} = 1 - r_{j+1}, \qquad r_{j+1} \in S^{-j-1}(\Omega),$$

we set $q_k = \zeta r_k/p = r_k q_0 \in S^{-m-k}(\Omega)$ and $Q_k = q_k(x, D)$. Then

$$\sigma_{Q_k P} = q_k p \quad (\mathrm{mod}\ S^{-k-1}(\Omega))$$
$$= r_k - r_{k+1} \text{ where } r_{k+1} \in S^{-k-1}(\Omega),$$

and hence

$$\sigma_{(Q_0 + \cdots + Q_k)P} = (1 - r_k) + (r_k - r_{k+1}) = 1 - r_{k+1}, \qquad r_{k+1} \in S^{-k-1}(\Omega).$$

Now, by Theorem (8.16) and Corollary (8.32), there is a properly supported $Q = q(x, D)$ such that $q \sim \sum_0^\infty q_j$, and for any k we have

$$\sigma_{QP} - 1 = \sigma_{(Q_0 + \cdots + Q_k)P} - 1 \quad (\mathrm{mod}\ S^{-k-1}(\Omega))$$
$$= 0 \quad (\mathrm{mod}\ S^{-k-1}(\Omega)),$$

so $QP - I \in \Psi^{-\infty}(\Omega)$. In exactly the same way, we can construct $Q' \in \Psi^{-m}(\Omega)$ such that $PQ' - I \in \Psi^{-\infty}(\Omega)$. But then

$$PQ - I = (PQ' - I) + P(QP - I)Q' - PQ(PQ' - I) \in \Psi^{-\infty}(\Omega),$$

so Q is a parametrix for P. ∎

As an immediate corollary, we obtain a generalization of the elliptic regularity theorem (6.33) to pseudodifferential operators. We shall derive a further refinement of this result in §8G.

(8.43) Theorem.
Suppose $P \in \Psi^m(\Omega)$ is elliptic and properly supported, and $u \in \mathcal{D}'(\Omega)$. If $Pu \in H_s^{loc}(\Omega)$, then $u \in H_{s+m}^{loc}(\Omega)$. In particular, if $Pu \in C^\infty(\Omega)$ then $u \in C^\infty(\Omega)$.

Remark: The hypothesis of proper support can be dropped if one assumes $u \in \mathcal{E}'(\Omega)$.

Proof: Let $Q \in \Psi^{-m}(\Omega)$ be a parametrix for P. We have $QPu \in H_{s+m}^{loc}(\Omega)$ by Theorem (8.40), and $(I - QP)u \in C^\infty(\Omega)$ since $I - QP \in \Psi^{-\infty}(\Omega)$. Hence $u = QPu + (I - QP)u \in H_{s+m}^{loc}(\Omega)$. The second assertion follows from Corollary (6.7). ∎

We now derive the local solvability theorem for elliptic operators. First, a technical lemma.

(8.44) Lemma.
Suppose $\mathcal{X}$ is an M-dimensional subspace of $C_c^\infty(\mathbb{R}^n)$ $(0 < M < \infty)$, and $x_0 \in \mathbb{R}^n$. For $\epsilon > 0$, let $W_\epsilon = \mathbb{R}^n \setminus B_\epsilon(x_0)$.
a. There exists $\epsilon > 0$ such that the restriction map $h \to h|W_\epsilon$ is injective.
b. Let ϵ be as in (a). For any $f \in \mathcal{E}'(\mathbb{R}^n)$ there exists $g \in C_c^\infty(W_\epsilon)$ such that $\langle f - g, h \rangle = 0$ for all $h \in \mathcal{X}$.

Proof: Let $h_1, \ldots, h_M$ be a basis for $\mathfrak{X}$. If (a) is false, for each $k \geq 1$ there are constants $b_1^k, \ldots, b_M^k$ such that $\max_m |b_m^k| = 1$ and $\sum b_m^k h_m$ is supported in $B_{1/k}(x_0)$. By passing to a subsequence we may assume that $\lim_{k \to \infty} b_m^k = b_m$ exists for each m. But then $\max_m |b_m| = 1$ and $\sum b_m h_m = 0$ (since $\text{supp}(\sum b_n h_m) \subset \{x_0\}$), which is impossible since the h_m's are linearly independent.

To prove (b), let $\mathcal{Y} = \{h|W_\epsilon : h \in \mathfrak{X}\}$ with ϵ as in (a). Then $\mathcal{Y}$ is a Hilbert subspace of $L^2(W_\epsilon)$ since $\mathfrak{X} \subset C_c^\infty$ and $\dim \mathcal{Y} < \infty$, and $\{h_m|W_\epsilon\}_{m=1}^M$ is a basis for $\mathcal{Y}$. The elements of the dual basis for $\mathcal{Y}$ can be approximated in $L^2(W_\epsilon)$ by functions in $C_c^\infty(W_\epsilon)$; hence, there exist $g_1, \ldots, g_M \in C_c^\infty(W_\epsilon)$ such that the matrix $(\langle g_l, h_m \rangle)$ is nonsingular. Given $f \in \mathcal{E}'$, then, there are unique constants $c_1, \ldots, c_M$ such that $\sum_l c_l \langle g_l, h_m \rangle = \langle f, h_m \rangle$ for $m = 1, \ldots, M$, so we can take $g = \sum c_m g_m$. ∎

(8.45) Theorem.
Suppose P is an elliptic differential operator with C^∞ coefficients on Ω. Every $x_0 \in \Omega$ has a neighborhood $U \subset \Omega$ such that the equation $Pu = f$ has a solution on U for every $f \in \mathcal{D}'(\Omega)$.

Proof: Let W be an open set such that $x_0 \in W \subset \overline{W} \subset \Omega$ and $\overline{W}$ is compact, and pick $\phi \in C_c^\infty(\Omega)$ with $\phi = 1$ on W. We first observe that it suffices to consider $f \in \mathcal{E}'(\Omega)$, for if $f \in \mathcal{D}'(\Omega)$, a solution to $Pu = \phi f$ on a neighborhood U of x_0 with $U \subset W$ also satisfies $Pu = f$ on U.

Let Q be a parametrix for P, and let $S = PQ - I$. Since P and Q are properly supported, so is S; hence, by Proposition (8.12) there is a compact $A \subset \Omega$ such that the values of Su on $\overline{W}$ depend only on the values of u on A. Pick $\psi \in C_c^\infty(\Omega)$ such that $\psi = 1$ on A, and define the operator T by $Tv = \phi S(\psi v)$; then $T \in \Psi^{-\infty}(\mathbb{R}^n)$ (it is given by the amplitude $a(x, \xi, y) = \phi(x)\sigma_S(x, \xi)\psi(y)$) and $Tv = Sv$ on W for any v. If for any $f \in \mathcal{E}'(\mathbb{R}^n)$ we can find $v \in \mathcal{D}'(\mathbb{R}^n)$ such that $(T + I)v = f$ on a neighborhood $U \subset W$ of x_0, we are done. Indeed, let $\tilde{v} = v|\Omega$; then on U we have $PQ\tilde{v} = (S + I)\tilde{v} = (T + I)v = f$, so $u = Q\tilde{v}$ does the job.

Suppose then that $f \in \mathcal{E}'(\mathbb{R}^n)$. By Corollary (6.8), $f \in H_s$ for some $s \in \mathbb{R}$. The range of T consists of C^∞ functions supported in $\text{supp } \phi$, so by Theorem (8.40) and Rellich's theorem (6.14), T is compact as an operator on H_s. By Corollary (0.42), the equation $(T+I)v = f$ has a solution $v \in H_s$ provided that f is orthogonal in H_s to the space $\mathcal{N} = \{g : (T_s^* + I)g = 0\}$, where T_s^* denotes the adjoint of T with respect to the inner product on H_s.

Now, for $g \in \mathcal{N}$ we have $g = -T_s^* g$, so

$$-\langle f \mid g \rangle_s = \langle f \mid T_s^* g \rangle_s = \langle Tf \mid g \rangle_s = \langle \Lambda^s Tf \mid \Lambda^s g \rangle_0 = \langle f, \overline{T^* \Lambda^{2s} g} \rangle,$$

where Λ^s is defined by (6.4) and T^* is the adjoint of T as defined in §8D. Thus $f \perp \mathcal{N}$ if and only if $\langle f, h \rangle = 0$ for all h in the space $X = \{\overline{T^* \Lambda^{2s} g} : g \in \mathcal{N}\}$. Note that $X \subset C_c^\infty$ since $T^* v = \overline{\psi} S^*(\overline{\phi} v)$ and S^* is smoothing.

In short, there exists $v \in H_s$ such that $(T + I)v = f$ if and only if $\langle f, h \rangle = 0$ for all h in X, a finite-dimensional subspace of C^∞. This may not be the case, but by Lemma (8.44), for some $\epsilon > 0$ there is always a $g \in C_c^\infty$ with support disjoint from $B_\epsilon(x_0)$ such that $\langle f - g, h \rangle = 0$ for all $h \in X$, and then there exists $v \in H_s$ such that $(T + I)v = f - g$. But then $(T + I)v = f$ on $B_\epsilon(x_0)$, so we can take $U = B_\epsilon(x_0) \cap W$. ∎

We now turn to Gårding's inequality for pseudodifferential operators. We shall say that a symbol $p \in S^m(\Omega)$, or its corresponding operator $p(x, D)$, is **strongly elliptic** if for every compact $A \subset \Omega$ there are positive constants c, C such that

$$(8.46) \qquad \operatorname{Re} p(x, \xi) \geq c(1 + |\xi|^2)^{m/2} \text{ for } x \in A \text{ and } |\xi| \geq C.$$

(This definition differs slightly from the one in §7A when $p(x, D)$ is a differential operator; the set we called $\overline{\Omega}$ there corresponds to the A here. The possible inclusion of lower order terms in p and the replacement of $|\xi|^m$ by $(1 + |\xi|^2)^{m/2}$ are of no consequence since (8.46) is to hold only for large $|\xi|$.)

(8.47) Theorem.
Suppose $P \in \Psi^m(\Omega)$ is strongly elliptic, and let $P_0 = \frac{1}{2}(P + P^*)$. There exists a properly supported $Q \in \Psi^{m/2}(\Omega)$ such that $P_0 - Q^* Q \in \Psi^{-\infty}(\Omega)$.

Proof: The argument is very similar to the proof of Theorem (8.42). Let $P = p(x, D)$ where p satisfies (8.46), let ζ be as in Lemma (8.41) with p replaced by $\operatorname{Re} p$ (possible since $\operatorname{Re} p$ is elliptic), and let

$$q_0(x, \xi) = \zeta(x, \xi)[\operatorname{Re} p(x, \xi)]^{1/2}, \qquad q^0(x, \xi) = \zeta(x, \xi)[\operatorname{Re} p(x, \xi)]^{-1/2}.$$

It is easy to verify (Exercise 4) that $q_0 \in S^{m/2}(\Omega)$ and $q^0 \in S^{-m/2}(\Omega)$, and $q_0 q^0 - 1$ is in $S^{-\infty}(\Omega)$ since it has compact support in ξ. Moreover, by Corollary (8.38),

$$\sigma_{P_0 - Q_0^* Q_0} = \operatorname{Re} p - q_0^2 \pmod{S^{m-1}(\Omega)}$$
$$= r_1 \in S^{m-1}(\Omega).$$

Proceeding inductively, for $j \geq 1$ we can find real $q_j \in S^{(m/2)-j}(\Omega)$ such that if $Q_j = q_j(x, D)$,

$$P_0 - (Q_0^* + \cdots + Q_{k-1}^*)(Q_0 + \cdots + Q_{k-1}) = r_k(x, D), \qquad r_k \in S^{m-k}(\Omega).$$

Indeed, if we have found $q_1, \ldots, q_{k-1}$, we observe that $r_k(x, D)^* = r_k(x, D)$ and hence, by Corollary (8.38), $r_k - \bar{r}_k \in S^{m-k-1}(\Omega)$. Thus, if $q_k \in S^{(m/2)-k}(\Omega)$ is real and $Q_k = q_k(x, D)$, Corollary (8.38) gives

$$\sigma_{P_0 - (Q_0^* + \cdots + Q_k^*)(Q_0 + \cdots + Q_k)}$$
$$= r_k - 2q_k(q_0 + \cdots + q_{k-1}) - q_k^2 \pmod{S^{m-k-1}(\Omega)}$$
$$= \operatorname{Re} r_k - 2q_0 q_k \pmod{S^{m-k-1}(\Omega)}.$$

Hence we can take $q_k = \frac{1}{2}q^0(\operatorname{Re} r_k)$.

Finally, by Theorem (8.16) and Corollary (8.32), there is a properly supported $Q = q(x, D) \in \Psi^{m/2}(\Omega)$ such that $q \sim \sum_0^\infty q_j$. Then $P_0 - Q^* Q \in \Psi^{-\infty}(\Omega)$, and we are done. ∎

(8.48) Gårding's Inequality.
Suppose $p \in S^m(\Omega)$ satisfies (8.46) and W is an open set with compact closure in Ω; let c be the constant $c_{\overline{W}}$ in (8.46). For any $\epsilon > 0$, any $s < \frac{1}{2}m$, and any open set V with $\overline{V} \subset W$ there is a constant $C \geq 0$ such that

$$\operatorname{Re}\langle p(x, D)u \mid u \rangle \geq (c - \epsilon)\|u\|_{m/2}^2 - C\|u\|_s^2, \qquad u \in C_c^\infty(V).$$

Proof: Replacing Ω by W, we may assume that $\operatorname{Re} p(x, \xi) \geq c|\xi|^m$ for all $x \in \Omega$ and $|\xi|$ sufficiently large. Let $\tilde{p}(x, \xi) = p(x, \xi) - (c - \epsilon)(1 + |\xi|^2)^{m/2}$. Then $\tilde{p} \in S^m(\Omega)$ and $\tilde{p}$ satisfies (8.46) with $c_A = \epsilon$ for all A, so we can apply Theorem (8.47) to $\tilde{P} = \tilde{p}(x, D)$ to obtain $Q \in \Psi^{m/2}(\Omega)$ with $R = \tilde{P}_0 - Q^* Q \in \Psi^{-\infty}(\Omega)$. On the other hand, $\tilde{P} = p(x, D) - (c - \epsilon)\Lambda^m$ where Λ^m is defined by (6.4), so if $u \in C_c^\infty(V)$,

$$\operatorname{Re}\langle p(x, D)u, \mid u \rangle = \operatorname{Re}\langle \tilde{P}u \mid u \rangle + (c - \epsilon)\langle \Lambda^m u \mid u \rangle$$
$$= \langle \tilde{P}_0 u \mid u \rangle + (c - \epsilon)\langle \Lambda^{m/2}u \mid \Lambda^{m/2}u \rangle$$
$$= \langle Q^* Q u \mid u \rangle + \langle Ru \mid u \rangle + (c - \epsilon)\|u\|_{m/2}^2.$$

But $\langle Ru \mid u \rangle = \langle \phi Ru \mid u \rangle$ whenever $\phi \in C_c^\infty(\Omega)$ and $\phi = 1$ on V, so by Theorem (8.40),

$$|\langle Ru \mid u \rangle| \leq \|\phi Ru\|_{-s}\|u\|_s \leq C\|u\|_s^2.$$

Also, $\langle Q^* Q u \mid u \rangle = \langle Qu \mid Qu \rangle \geq 0$, so

$$\operatorname{Re}\langle p(x, D)u \mid u \rangle \geq -|\langle Ru \mid u \rangle| + (c - \epsilon)\|u\|_{m/2}^2 \geq (c - \epsilon)\|u\|_{m/2}^2 - C\|u\|_s^2. \ ∎$$

The original Gårding inequality (7.15) follows immediately by taking the m and Ω in (7.15) to be $2m$ and V in (8.48), for the estimate in (8.48) for $u \in C_c^\infty(V)$ clearly extends to $u \in H_{m/2}^0(V)$.

Gårding's inequality implies that if $P = p(x, D)$ where p satisfies the strong positivity condition $p(x, \xi) \geq c|\xi|^m$, the symmetrized operator $P_0 = \frac{1}{2}(P + P^*)$ is a positive operator modulo a lower order error term. One may ask if a similar conclusion holds under the weaker hypothesis that $p(x, \xi) \geq 0$. That is, is it true that

$$(8.49) \qquad p \geq 0 \quad \Longrightarrow \quad \mathrm{Re}\langle p(x, D)u \,|\, u\rangle \geq -C\|u\|_s^2 \qquad (u \in C_c^\infty)$$

for some $s < \frac{1}{2}m$? If this were true, we could take $\epsilon = 0$ in (8.48) — just apply (8.49) to $\tilde{p}(x, \xi) = \mathrm{Re}\, p(x, \xi) - c(1 + |\xi|^2)^{m/2}$. For this reason, an inequality of the type (8.49) with $s < \frac{1}{2}m$ is called a "sharp Gårding inequality".

The arguments above are insufficient to yield such a result. (The proof of Theorem (8.47) breaks down if we merely assume that $p \geq 0$ because the square root function is not differentiable at the origin.) Moreover, simple examples (see Exercise 5) show that one should not hope for (8.49) to hold for $s < \frac{1}{2}m - 1$. In fact, (8.49) with $s = \frac{1}{2}m - \frac{1}{2}$ was proved by Hörmander and extended by Lax and Nirenberg to ΨDO acting on vector-valued functions; this is what is usually called the **sharp Gårding inequality**. For four different proofs of this result and a number of important applications, see Folland [15], Hörmander [27, vol. III], Taylor [48], and Treves [53, vol. I]. Moreover, Fefferman and Phong have proved the much deeper result that in the scalar case, (8.49) holds with $s = \frac{1}{2}m - 1$; see Hörmander [27, vol. III].

EXERCISES

1. Prove Lemma (8.41). (Hint: Use the ideas in Exercise 5, §8B.)

2. Show that if $p \in S^m(\Omega)$ is elliptic and ζ is as in Lemma (8.41), then $\zeta/p \in S^{-m}(\Omega)$.

3. Show that if $P \in \Psi^m(\Omega)$ has a parametrix $Q \in \Psi^{-m}(\Omega)$, then P and Q are elliptic.

4. Show that if $p \in S^m(\Omega)$ is real and strongly elliptic and ζ is as in Lemma (8.41), then $\zeta p^{1/2} \in S^{m/2}(\Omega)$ and $\zeta p^{-1/2} \in S^{-m/2}(\Omega)$.

5. Let $p \in S^2(\mathbb{R})$ be given by $p(x, \xi) = 4\pi^2 x^2 \xi^2$, so that $p(x, D)u(x) = -x^2 u''(x)$.

a. Show that for $u \in C_c^\infty(\mathbb{R})$,

$$\mathrm{Re}\langle p(x, D)u \mid u \rangle = \int |xu'(x) + \tfrac{1}{2}u(x)|^2 \, dx - \frac{3}{4} \int |u(x)|^2 \, dx.$$

b. Show that for any $\epsilon > 0$ there exists $u \in C_c^\infty(\mathbb{R})$ such that

$$\int |xu'(x) + \tfrac{1}{2}u(x)|^2 \, dx < \epsilon \int |u(x)|^2 \, dx.$$

Hint: Pick $\phi \in C_c^\infty(\mathbb{R})$ with $\phi(x) = 1$ for $x \in [\tfrac{1}{2}, 2]$ and $\mathrm{supp}\,\phi \subset [\tfrac{1}{4}, 4]$, and take $u(x) = x^{-1/2}\phi(x^\delta)$ where δ is a small constant times ϵ. (The motivation is that $f(x) = x^{-1/2}$ satisfies $xf' + \tfrac{1}{2}f = 0$, and u is a truncated version of f.)

c. Conclude that the estimate $\mathrm{Re}\langle p(x, D)u \mid u \rangle \geq -\tfrac{3}{4}\|u\|_0^2$ is sharp.

G. Introduction to Microlocal Analysis

One of the principal applications of pseudodifferential operators is in making a detailed analysis of singularities of functions and distributions, in which one considers not only the places where a distribution is singular but the directions in which it is singular. (For example, $f(x, y) = |y|$ is perfectly smooth in the x direction but is not C^1 in the y direction along the line $y = 0$.) This idea of examining things locally both in position space and in direction space goes under the name of **microlocal analysis**. In this section we give an introduction to some of the basic ideas and techniques of microlocal analysis — in particular, the notion of the wave front set of a distribution.

Roughly speaking, the wave front set of a distribution u is the set of all (x, ξ) such that u fails to be smooth at x in the direction ξ. To make this more precise, suppose for the moment that u has compact support, so $\hat{u}$ is a smooth function. We take the statement "u is smooth in the direction ξ_0," where $\xi_0 \neq 0 \in \mathbb{R}^n$, to mean that $\hat{u}(t\xi)$ is rapidly decreasing as $t \to +\infty$ for all ξ in some neighborhood of ξ_0. For a general distribution u, then, we take the statement "u is smooth at x_0 in the direction ξ_0" to mean that for some $\phi \in C_c^\infty$ with $\phi = 1$ near x_0, ϕu is smooth in the direction ξ_0. That these notions are reasonable may be seen by considering some simple examples; see Exercise 3.

For reasons of efficiency, we shall proceed somewhat differently. We begin by introducing some terminology and lemmas concerning the microlocal behavior of pseudodifferential operators. We then define wave front sets in terms of pseudodifferential operators and prove a number of results concerning them, and and finally we derive the characterization sketched above as a theorem. *In this section, the hypothesis that a pseudodifferential operator is properly supported will be assumed wherever it is convenient, without explicit comment.*

If Ω is an open subset of $\mathbb{R}^n$, we shall set

$$T^0\Omega = \Omega \times \{\xi \in \mathbb{R}^n : \xi \neq 0\}, \qquad T^1\Omega = \Omega \times \{\xi \in \mathbb{R}^n : |\xi| = 1\}.$$

(In coordinate-invariant terms, $T^0\Omega$ is the cotangent bundle of Ω with the zero section removed, and $T^1\Omega$ is the co-unit sphere bundle. See the discussion at the end of §8H.) A subset U of $\mathbb{R}^n \setminus \{0\}$ is called **conic** if $t\xi \in U$ for all $t > 0$ whenever $\xi \in U$, and a subset V of $T^0\Omega$ is called **conic** if $(x, t\xi) \in V$ for all $t > 0$ whenever $(x, \xi) \in V$. Note that a conic set in $T^0\Omega$ is completely determined by its intersection with $T^1\Omega$.

A pseudodifferential operator $P = p(x, D) \in \Psi^m(\Omega)$ is called **elliptic** (of order m) at $(x_0, \xi_0) \in T^0\Omega$ if there are positive constants c, C and a conic neighborhood V of (x_0, ξ_0) such that

$$|p(x, \xi)| \geq c|\xi|^m \text{ for all } (x, \xi) \in V \text{ with } |\xi| \geq C.$$

The **characteristic variety** of P, denoted by char P, is the complement of the set on which P is elliptic:

$$\operatorname{char} P = \{(x, \xi) \in T^0\Omega : P \text{ is not elliptic at } (x, \xi)\}.$$

At the other end of the scale, $P = p(x, D)$ is called **smoothing** at (x_0, ξ_0) if $p(x, \xi)$ is rapidly decreasing on a conic neighborhood of (x_0, ξ_0), that is, if there is a conic neighborhood V of (x_0, ξ_0) such that

$$|D_x^\beta D_\xi^\alpha p(x, \xi)| \leq C_{M,\alpha,\beta}(1 + |\xi|)^{-M} \text{ for all } M > 0, \ (x, \xi) \in V.$$

The **microsupport** of P, denoted by $\mu\operatorname{supp} P$, is the complement of the set on which P is smoothing:

$$\mu\operatorname{supp} P = \{(x, \xi) \in T^0\Omega : P \text{ is not smoothing at}(x, \xi)\}.$$

char P and $\mu\operatorname{supp} P$ are both closed conic sets. Their union is always $T^0\Omega$ since P cannot be smoothing and elliptic at the same point, but in general they have a large overlap. It is easy to see that this definition of char P agrees with our previous one when P is a differential operator; on the other hand, $\mu\operatorname{supp} P$ is of no interest when P is a differential operator. See Exercise 1. Incidentally, we follow Treves [53] in using the term "microsupport." $\mu\operatorname{supp} P$ is called the "essential support" of P in Taylor [48] and the "wave front set" of P in Hörmander [27].

(8.50) Proposition.
For any $P, Q \in \Psi^\infty(\Omega)$,

$$\mu\text{supp}(PQ) \subset (\mu\text{supp}\, P) \cap (\mu\text{supp}\, Q).$$

Proof: It follows easily from Theorem (8.37) that σ_{PQ} is rapidly decreasing on any conic set where either σ_P or σ_Q is. ∎

(8.51) Proposition.
$\mu\text{supp}\, P = \varnothing$ *if and only if* $P \in \Psi^{-\infty}(\Omega)$.

Proof: Suppose $\mu\text{supp}\, P = \varnothing$. If $A \subset \Omega$ is compact, each $(x, \xi) \in A \times S_1(0)$ has a conic neighborhood on which the symbol of P satisfies estimates of the form $|D_x^\beta D_\xi^\alpha \sigma_P(x, \xi)| \leq C(1 + |\xi|)^{-M}$. $A \times S_1(0)$ can be covered by finitely many such neighborhoods, and it follows that σ_P satisfies such estimates globally on $T^0 A$; thus $P \in \Psi^{-\infty}(\Omega)$. The converse is obvious. ∎

(8.52) Proposition.
Suppose V is a conic neighborhood of $(x_0, \xi_0) \in T^0\Omega$. There exists $p \in S^0(\Omega)$ such that $p \geq 0$, $\text{supp}\, p \subset V$ and $p(x, \xi) = 1$ for all (x, ξ) in some conic neighborhood of (x_0, ξ_0) with $|\xi| \geq 1$. In particular, $p(x, D)$ is elliptic at (x_0, ξ_0) and $\mu\text{supp}\, p(x, D) \subset V$.

Proof: Since V is conic, we may as well assume that $|\xi_0| = 1$. Let A and B be compact neighborhoods of x_0 in Ω and ξ_0 in $S_1(0) = \{\xi : |\xi| = 1\}$ such that $A \times B \subset V \cap T^1\Omega$. Pick $\phi \geq 0$ in $C_c^\infty(\Omega)$ with $\text{supp}\, \phi \subset A$ and $\phi = 1$ on a neighborhood of x_0, and pick $\psi \geq 0$ in $C^\infty(S_1(0))$ with $\text{supp}\, \psi \subset B$ and $\psi = 1$ on a neighborhood of ξ_0. Further, pick $\zeta \geq 0$ in $C^\infty(\mathbb{R})$ such that $\zeta(t) = 0$ for $t \leq \frac{1}{2}$ and $\zeta(t) = 1$ for $t \geq 1$. Then set

$$p(x, \xi) = \phi(x)\psi(|\xi|^{-1}\xi)\zeta(|\xi|).$$

$p \in S^0(\Omega)$ because it is homogeneous of degree 0 for large ξ (as defined in §8A) and C^∞ for small ξ (by virtue of the factor of ζ). ∎

We are now ready to define the **wave front set** of a distribution $u \in \mathcal{D}'(\Omega)$:

$$WF(u) = \bigcap \{\text{char}\, P : P \in \Psi^0(\Omega) \text{ and } Pu \in C^\infty\}.$$

$WF(u)$ is a closed conic subset of $T^0\Omega$. The intuition behind it is a microlocal version of the elliptic regularity theorem which we shall prove below: If

P is elliptic at (x_0, ξ_0) and Pu is smooth then u must be smooth at (x_0, ξ_0); hence $WF(u)$ consists of the points x_0 and directions ξ_0 where u fails to be C^∞.

Note. In the definition of $WF(u)$, char P is the characteristic variety of P as an operator of order zero; if $P \in \Psi^m(\Omega)$ for some $m < 0$ then char $P = T^0\Omega$, so only operators of true order zero are significant. Moreover, the condition $P \in \Psi^0(\Omega)$ in the definition of $WF(u)$ is merely a convenient normalization; we could equally well use ΨDO of other orders. Indeed, if $P \in \Psi^m(\Omega)$, choose an elliptic $Q \in \Psi^{-m}(\Omega)$ (for example, $Q = \Lambda^{-m}$ as defined by (6.4) or a properly supported variant thereof); then $QP \in \Psi^0(\Omega)$, and P is elliptic at (x_0, ξ_0) (as an operator of order m) if and only if QP is elliptic at (x_0, ξ_0) (as an operator of order 0), by Theorem (8.37).

If our definition of $WF(u)$ is to do the job it was designed for, the projection of $WF(u)$ onto Ω should be the set of points at which u is not C^∞, namely, the singular support of u. We now show that this is the case.

(8.53) Theorem.
Let $\pi : T^0\Omega \to \Omega$ be defined by $\pi(x, \xi) = x$. If $u \in \mathcal{D}'(\Omega)$, then $\pi[WF(u)] =$ sing supp u.

Proof: If $x_0 \notin$ sing supp u, there exists $\phi \in C_c^\infty(\Omega)$ such that $\phi(x_0) = 1$ and $\phi u \in C^\infty$. The operator $Pv = \phi v$ is $p(x, D)$ where $p(x, \xi) = \phi(x)$; hence $P \in \Psi^0(\Omega)$ and P is elliptic near x_0. In other words, $Pu \in C^\infty$ and $(x_0, \xi) \notin$ char P for all $\xi \neq 0$, so $x_0 \notin \pi[WF(u)]$.

Conversely, suppose $x_0 \notin \pi[WF(u)]$. For each unit vector ξ there exists $P \in \Psi^0(\Omega)$ that is elliptic on a conic neighborhood of (x_0, ξ) such that $Pu \in C^\infty$. By compactness of the unit sphere, we can find $P_1, \ldots P_M \in \Psi^0(\Omega)$ such that (i) $P_j u \in C^\infty$ for all j and (ii) for each $\xi \neq 0$ there is a j such that P_j is elliptic on a conic neighborhood of (x_0, ξ). Let $P = \sum_1^M P_j^* P_j$. Then P is elliptic on a neighborhood V of x_0 in Ω because $\sigma_P = \sum_1^M |\sigma_{P_j}|^2$ (mod $S^{-1}(\Omega)$), and $Pu \in C^\infty$. By Theorem (8.43) (with Ω replaced by V), $u \in C^\infty(V)$, so $x_0 \notin$ sing supp u. ∎

If P is a ΨDO, we showed in Theorem (8.9) that sing supp $Pu \subset$ sing supp u for any distribution u. We now derive the microlocal refinement of this result:

(8.54) Theorem.
If $P \in \Psi^\infty(\Omega)$ and $u \in \mathcal{D}'(\Omega)$ then

$$WF(Pu) \subset WF(u) \cap \mu\text{supp } P.$$

Proof: If $(x_0, \xi_0) \notin \mu\text{supp } P$, by Proposition (8.52) we can find $Q \in \Psi^0(\Omega)$ such that Q is elliptic at (x_0, ξ_0) and $\mu\text{supp } P \cap \mu\text{supp } Q = \varnothing$. Then $QP \in \Psi^{-\infty}(\Omega)$ by Propositions (8.50) and (8.51), so $QPu \in C^\infty$. But this means that $(x_0, \xi_0) \notin WF(Pu)$.

On the other hand, suppose $(x_0, \xi_0) \notin WF(u)$, so there exists $Q \in \Psi^0(\Omega)$ such that $Qu \in C^\infty$ and Q is elliptic on a conic neighborhood V of ξ_0. We may assume that $\text{Re}\,\sigma_Q \geq C > 0$ for large ξ in some conic neighborhood of (x_0, ξ_0). (Replace Q by Q^*Q if necessary; this works since $\sigma_{Q^*Q} = |\sigma_Q|^2 \pmod{S^{-1}(\Omega)}$ by Corollary (8.38).)

Claim: There exist $R, S \in \Psi^0(\Omega)$ such that $(x_0, \xi_0) \notin \text{char } R$ and $RP - SQ \in \Psi^{-\infty}(\Omega)$.

Granting the claim, we see that $SQu \in C^\infty$ since $Qu \in C^\infty$ and $(RP - SQ)u \in C^\infty$ since $RP - SQ$ is smoothing, so $RPu \in C^\infty$; hence $(x_0, \xi_0) \notin WF(Pu)$.

To prove the claim, pick an elliptic operator $\tilde{Q} \in \Psi^0(\Omega)$ such that $\sigma_{\tilde{Q}} = \sigma_Q$ on a conic neighborhood U of (x_0, ξ_0). (The assumption that $\text{Re}\,\sigma_Q \geq C$ near (x_0, ξ_0) makes this easy: take $\sigma_{\tilde{Q}} = \sigma_Q + 1 - p$ where p is as in Proposition (8.52).) Then, by Proposition (8.52) again, pick $R \in \Psi^0(\Omega)$ with $(x_0, \xi_0) \notin \text{char } R$ and $\mu\text{supp } R \subset U$. Finally, let T be a parametrix for $\tilde{Q}$ and set $S = RPT$. Since $\mu\text{supp } R \cap \mu\text{supp}(\tilde{Q} - Q) = \varnothing$, we have $RPT(\tilde{Q} - Q) \in \Psi^{-\infty}(\Omega)$ by Propositions (8.50) and (8.51), so

$$RP - SQ = RP(I - T\tilde{Q}) + RPT(\tilde{Q} - Q) \in \Psi^{-\infty}(\Omega),$$

and the claim is proved. ∎

As an immediate corollary, we obtain the microlocal refinement of the C^∞ regularity theorem for elliptic operators. (For the extension of this result to Sobolev spaces, see Exercise 2.)

(8.55) Corollary.
If $P \in \Psi^m(\Omega)$ is elliptic, then $WF(Pu) = WF(u)$ for any $u \in \mathcal{D}'(\Omega)$.

Proof: Let Q be a parametrix for P; then $QPu - u \in C^\infty$, so $WF(u) = WF(QPu)$. By Theorem (8.54), then, $WF(Pu) \subset WF(u)$ and $WF(u) = WF(QPu) \subset WF(Pu)$. ∎

Finally, we derive the more intuitive characterization of wave front sets alluded to at the beginning of this section.

(8.56) Theorem.

Suppose $u \in \mathcal{D}'(\Omega)$. Then $(x_0, \xi_0) \notin WF(u)$ if and only if there exist $\phi \in C_c^\infty(\Omega)$ such that $\phi(x_0) = 1$ and a conic neighborhood V of ξ_0 in $\mathbb{R}^n \setminus \{0\}$ such that

$$\sup_{\xi \in V}(1 + |\xi|)^M |(\phi u)\widehat{}(\xi)| < \infty \text{ for all } M > 0.$$

Proof: If such a ϕ and V exist, choose $p \in C^\infty(\mathbb{R}^n)$ such that $\operatorname{supp} p \subset V$, $p = 1$ on a conic neighborhood of ξ_0, and $p \in S^0(\Omega)$ as a function of (x, ξ) that is independent of x. (Cf. the proof of Proposition (8.52).) Then $p(\phi u)\widehat{}$ is rapidly decaying at infinity, so its inverse Fourier transform $p(D)[\phi u]$ is C^∞. But the operator $Pv = p(D)[\phi v]$ is a ΨDO of order zero whose symbol equals $\phi(x)p(\xi) \pmod{S^{-1}(\Omega)}$, so $(x_0, \xi_0) \notin \operatorname{char} P$ and $Pu \in C^\infty$. Thus $(x_0, \xi_0) \notin WF(u)$.

Conversely, suppose $(x_0, \xi_0) \notin WF(u)$. There is a neighborhood N of x_0 in Ω such that $(x, \xi_0) \notin WF(u)$ for all $x \in N$. Pick $\phi \in C_c^\infty(N)$ with $\phi(x_0) = 1$, and let

$$\Sigma = \big\{ \xi : (x, \xi) \in WF(\phi u) \text{ for some } x \big\}.$$

Since $WF(\phi u) \subset WF(u) \cap \operatorname{supp} \phi$ (this is a simple special case of Theorem (8.54)), $\xi_0 \notin \Sigma$. Moreover, Σ is a closed conic set, for if $\xi_j \in \Sigma$ for $j \geq 1$ and $\xi_j \to \xi$, there exist x_j, necessarily in the compact set $\operatorname{supp} \phi$, such that $(x_j, \xi_j) \in WF(\phi u)$; by passing to a subsequence we can assume that $x_j \to x$, so $(x, \xi) \in WF(\phi u)$ and hence $\xi \in \Sigma$. Hence, as in the preceding paragraph, we can find $p(\xi) \in S^0(\Omega)$ such that $p = 1$ on a conic neighborhood of ξ_0 and $p = 0$ on a conic neighborhood of Σ. But then $\mu\operatorname{supp} p(D) \cap WF(\phi u) = \varnothing$, so $p(D)[\phi u] \in C^\infty$ by Theorems (8.54) and (8.53).

Moreover, by Theorem (8.8a), the inverse Fourier transform $p^\vee$ of p agrees with a Schwartz class function outside a neighborhood of the origin, and the estimate (0.32) then implies that $p(D)[\phi u] = (\phi u) * p^\vee$ agrees with a Schwartz class function away from $\operatorname{supp} \phi$. Combining this with the fact that $p(D)[\phi u] \in C^\infty$, we see that $p(D)[\phi u] \in \mathcal{S}$. But then $p(\phi u)\widehat{} = (p(D)[\phi u])\widehat{} \in \mathcal{S}$, so $(\phi u)\widehat{}$ is rapidly decreasing on the set where $p = 1$. Since this is a conic neighborhood of ξ_0, we are done. ∎

EXERCISES

1. Let $P = p(x, D)$ be a differential operator on Ω.

a. Show that the definition of char P in this section agrees with the one in §1A.

b. Show that $\mu\text{supp}\, P = T^0\Omega$ unless there is a nonempty open $U \subset \Omega$ such that $p(x,\xi) \equiv 0$ for $x \in U$.

2. Suppose $u \in \mathcal{D}'(\Omega)$ and $s \in \mathbb{R}$. If $(x_0, \xi_0) \in T^0\Omega$, we say that $u \in H_s(x_0, \xi_0)$ if there exist $\phi \in C_c^\infty(\Omega)$ with $\phi(x_0) = 1$ and a conic neighborhood V of ξ_0 in $\mathbb{R}^n \setminus \{0\}$ such that $(1 + |\xi|^2)^{s/2}\widehat{\phi u}(\xi)$ is square-integrable on V.

a. Show that $u \in H_s(x_0, \xi_0)$ if and only if $u = v + w$ where $v \in H_s^0(\Omega)$ and $(x_0, \xi_0) \notin WF(w)$.

b. Suppose $P \in \Psi^m(\Omega)$ is elliptic of order m. Show that $u \in H_s(x_0, \xi_0)$ if and only if $Pu \in H_{s-m}(x_0, \xi_0)$.

3. Let $n = k + m$, and regard $\mathbb{R}^n$ as $\mathbb{R}^k \times \mathbb{R}^m$ with coordinates (x, y) and dual coordinates (ξ, η). Let μ_k be Lebesgue measure on $\mathbb{R}^k$, regarded as a distribution on $\mathbb{R}^n$ ($\langle \mu_k, \phi \rangle = \int \phi(x, 0)\, dx$).

a. Show that if $\phi \in C_c^\infty$, $(\phi\mu_k)\widehat{}(\xi, \eta) = \hat{\phi}(\xi)$.

b. Conclude from Theorem (8.56) that if $f \in C^\infty$,

$$WF(f\mu_k) = \{((x,0),\,(0,\eta)) : x \in \text{supp}\, f,\ \eta \neq 0\}.$$

c. More generally, suppose $u \in \mathcal{D}'(\mathbb{R}^k)$, and let $\iota(u)$ be the injection of u into $\mathbb{R}^n$ ($\langle i(u), \phi \rangle = \langle u, \phi | \mathbb{R}^k \rangle$). Show that $WF(\iota(u))$ is the set of all $((x,0),\,(\xi,\eta))$ with $x \in \text{supp}\, u$, $(x, \xi) \in WF(u)$, and $\eta \neq 0$.

4. Define $u \in \mathcal{D}'(\mathbb{R})$ by $u = \pi i \delta - P.V.(1/x)$, that is,

$$\langle u, \phi \rangle = \pi i \phi(0) - \lim_{\epsilon \to 0} \int_{|x| > \epsilon} \frac{\phi(x)}{x}\, dx.$$

Show that $WF(u) = \{(0, \xi) : \xi > 0\}$. Hint: Use Exercise 1, §5F.

H. Change of Coordinates

In this section we shall show how pseudodifferential operators behave under smooth changes of coordinates. Here is the setup: We suppose that Ω and Ω' are open sets in $\mathbb{R}^n$ and that $F : \Omega' \to \Omega$ is a diffeomorphism, that is, a C^∞ bijection with a C^∞ inverse. We denote the Jacobian matrix of F by J_F (i.e., $J_F = (\partial y_j / \partial x_k)$ where $y = F(x)$) and its adjoint (transpose) by J_F^*. If P is a properly supported ΨDO on Ω, we transfer P to an operator on functions on Ω' by setting

$$P^F u = [P(u \circ F^{-1})] \circ F,$$

and our aim is to show that P^F is a ΨDO on Ω'.

(8.57) Lemma.
With notation as above, there exist a neighborhood N of the diagonal $\Delta_{\Omega'}$ and a C^∞ map μ from N to the set of invertible $n \times n$ matrices such that $\mu(x, x) = J_F(x)$ for all x and

$$F(x) - F(y) = \mu(x, y)(x - y), \qquad (x, y) \in N.$$

Proof: The set N_0 of all $(x, y) \in \Omega' \times \Omega'$ such that the line segment from x to y is contained in Ω' is a neighborhood of $\Delta_{\Omega'}$. If $(x, y) \in N_0$, by the chain rule we have

$$F(x) - F(y) = \int_0^1 \frac{d}{dt} F(tx + (1-t)y)) \, dt = \mu(x, y)(x - y)$$

where

$$\mu(x, y) = \int_0^1 J_F(tx + (1-t)y) \, dt.$$

Since $J_F(tx + (1-t)y)$ is close to $J_F(x)$ when y is close to x, and $J_F(x)$ is invertible for all x, there is a neighborhood $N \subset N_0$ of $\Delta_{\Omega'}$ on which $\mu(x, y)$ is invertible, so we are done. ∎

(8.58) Theorem.
Suppose $F : \Omega' \to \Omega$ is a diffeomorphism, $P \in \Psi^m(\Omega)$ is properly supported, and $P^F : C_c^\infty(\Omega') \to C^\infty(\Omega')$ is defined by $P^F u = [P(u \circ F^{-1})] \circ F$. Then $P^F \in \Psi^m(\Omega')$, and

$$\sigma_{P^F}(x, \xi) = \sigma_P(F(x), [J_F^*(x)]^{-1}\xi) \pmod{S^{m-1}(\Omega')}.$$

Proof: First suppose that $m < -n$. Let N and μ be as in Lemma (8.57). By Proposition (8.15), we can find $\phi \in C^\infty(\Omega' \times \Omega')$ such that $\phi = 1$ on a neighborhood of $\Delta_{\Omega'}$ and supp ϕ is proper and contained in N. We then have

$$P^F u(x) = \iint e^{2\pi i(F(x)-z)\cdot\xi} p(F(x), \xi) u(F^{-1}(z)) \, dz \, d\xi$$

$$= \iint e^{2\pi i(F(x)-F(y))\cdot\xi} p(F(x), \xi) u(y) |\det J_F(y)| \, dy \, d\xi$$

$$= Qu(x) + Ru(x),$$

where

$$Qu(x) = \iint e^{2\pi i \mu(x,y)(x-y)\cdot\xi} p(F(x), \xi) u(y) |\det J_F(y)| \phi(x, y) \, dy \, d\xi,$$

$$Ru(x) = \iint e^{2\pi i(F(x)-F(y))\cdot\xi} p(F(x), \xi) u(y) |\det J_F(y)| [1 - \phi(x, y)] \, dy \, d\xi.$$

We first dispose of the error term R. With notation as in (8.6), its distribution kernel is

$$K(x, y) = p_2^{\vee}(F(x), F(x) - F(y))|\det J_F(y)|[1 - \phi(x, y)].$$

Now, $p_2^{\vee}(F(x), F(x) - F(y))$ is smooth away from $x = y$ by Theorem (8.8), and it has compact support as a function of y for each x since P is properly supported. Hence $K \in C^{\infty}$ and $K(x, \cdot) \in C_c^{\infty}$ for each x, and it follows that the function

$$r(x, \xi) = \int e^{-2\pi i z \cdot \xi} K(x, x - z) \, dz$$

belongs to $S^{-\infty}(\Omega')$. But $R = r(x, D)$ by (8.6), so $R \in \Psi^{-\infty}(\Omega')$.

Returning to the main term Q, let $\nu(x, y)$ be the inverse transpose of $\mu(x, y)$. Since we assumed $m < -n$, the double integral defining $Qu(x)$ is absolutely convergent, so we may reverse the order of integration, make the substitution $\xi = \nu(x, y)\eta$, and restore the original order to obtain

$Qu(x)$
$$= \iint e^{2\pi i (x-y) \cdot \eta} p(F(x), \nu(x, y)\eta)|\det J_F(y)| \, |\det \nu(x, y)|\phi(x, y)u(y) \, dy \, d\eta$$
$$= \iint e^{2\pi i (x-y) \cdot \eta} a(x, \eta, y)u(y) \, dy \, d\eta,$$

where

$$a(x, \eta, y) = p(F(x), \nu(x, y)\eta)|\det J_F(y)| \, |\det \nu(x, y)|\phi(x, y).$$

A simple calculation involving the chain rule shows that $a \in A^m(\Omega')$. (The point to be noted is that applying, for example, D_{y_j} to $p(F(x), \nu(x, y)\eta)$ gives $\nabla_{\eta} p(F(x), \nu(x, y)\eta) \cdot [D_{y_j}\nu(x, y)\eta]$; the first factor is dominated by $(1 + |\eta|)^{m-1}$, and the η in the second factor brings it back to $(1 + |\eta|)^m$.) Q us properly supported because of the factor of ϕ (cf. (8.23) and (8.24)), so it follows from Theorem (8.27) that $Q \in \Psi^m(\Omega')$ and that

$$\sigma_Q(x, \eta) = p(F(x), \nu(x, x)\eta)|\det J_F(x)| \, |\det \nu(x, x)| \quad (\text{mod } S^{m-1}(\Omega')).$$

But $\nu(x, x) = [J_F^*(x)]^{-1}$, so the expression on the right reduces to $p(F(x), [J_F^*(x)]^{-1}\eta)$, and the proof is complete.

This argument is still valid if $m > -n$, but the manipulation of the integral defining Q requires more justification then. Instead, we shall finesse the problem as follows. Pick an integer $M > \frac{1}{2}(n+m)$ and let $S \in \Psi^{-2M}(\Omega)$ be a parametrix for Δ^M (Δ = Laplacian) on Ω. Then $P = PS\Delta^M + T$ where $T \in \Psi^{-\infty}(\Omega)$. We can apply the preceding argument to the operators PS and T to see that $(PS)^F \in \Psi^{m-2M}(\Omega')$ and $T^F \in \Psi^{-\infty}(\Omega')$ and that

$$\sigma_{(PS)^F}(x,\xi) = p(F(x), [J_F^*(x)]^{-1}\xi)(-4\pi^2|[J_F^*(x)]^{-1}\xi|^2)^{-M}$$
$$(\text{mod } S^{m-2M-1}(\Omega')).$$

But Δ^M is a differential operator, so the elementary calculations of §1A show that $(\Delta^M)^F$ is a differential operator on Ω' and that its symbol is $(-4\pi^2|[J_F^*(x)]^{-1}\xi|^2)^M$ modulo lower order terms. Since $(PS\Delta^M)^F = (PS)^F(\Delta^M)^F$, the desired result follows from Corollary (8.38).   ∎

Remark: This proof can obviously be pushed further to obtain a complete asymptotic expansion for the symbol of P^F, but putting this expansion in a reasonably neat form requires more effort. The definitive result can be found in Hörmander [27, vol. III, Theorem 18.1.17].

Theorem (8.58) paves the way for defining pseudodifferential operators on manifolds. Namely, if M is a C^∞ manifold, a linear map $P : \mathcal{E}'(M) \to \mathcal{D}'(M)$ is called a **pseudodifferential operator** of order m if for any coordinate chart $V \subset M$ and any $\phi, \psi \in C_c^\infty(V)$, the operator $P_{\phi,\psi}u = \psi P(\phi u)$ is a ΨDO of order m with respect to some, and hence any, coordinate system on V. (More precisely, this means that if $G : V \to \mathbb{R}^n$ is a coordinate map then the transferred operator $P_{\phi,\psi}^{G^{-1}}u = [P_{\phi,\psi}(u \circ G)] \circ G^{-1}$ belongs to $\Psi^m(G(V))$.) If $M = \Omega$ is an open subset of $\mathbb{R}^n$, this class of operators is larger than $\Psi^m(\Omega)$ in that it includes all smoothing operators. (If P is a smoothing operator and $\phi, \psi \in C_c^\infty(\Omega)$ then the distribution kernel of $P_{\phi,\psi}$ belongs to C_c^∞, so it follows easily from (8.6) that $P \in \Psi^{-\infty}(\Omega)$. See also Exercise 3.)

On a manifold M, one can define the symbol class $S^m(M)$ to be the set of functions on the cotangent bundle T^*M that satisfy estimates of the form (8.2) in any local coordinate system. The precise symbol-operator correspondence only works in local coordinates, as the lower-order terms transform in complicated ways under coordinate changes, but by Theorem (8.58), the local symbols of a ΨDO on M determine a well-defined equivalence class in $S^m(M)/S^{m-1}(M)$. This is enough to show that the notion of ellipticity at a point $(x,\xi) \in T^*M$ is independent of the local coordinates. One can therefore define the characteristic variety of a ΨDO and the wave front set of a distribution on M, just as before, as closed conic subsets of T^0M (the cotangent bundle of M with the zero section removed).

A slightly better global notion of symbol is available for operators with a principal symbol. If $\Omega \subset \mathbb{R}^n$ and $P \in \Psi^m(\Omega)$, P is said to have a **principal symbol** if there is a $p^m \in C^\infty(T^0\Omega)$ such that $p^m(x, t\xi) = t^m p^m(x, \xi)$ for $t > 0$ and $\sigma_P - p^m$ agrees for large ξ with an element of $S^{m-1}(\Omega)$. (The restriction to large ξ is necessary since p^m is usually not C^∞ at $\xi = 0$.) For example, the principal symbol of a differential operator is, up to factors of $2\pi i$, what we called the characteristic form in §1A. Also, if P is elliptic with principal symbol p^m, any parametrix for P has principal symbol $1/p^m$. It is an easy consequence of Theorem (8.58) that if P is a ΨDO on a manifold M that has a principal symbol in any local coordinate system, these symbols patch together globally to make a well-defined function on T^*M.

A more comprehensive account of these matters can be found in Treves [53, vol. I].

EXERCISES

1. Suppose M is a smooth k-dimensional submanifold of $\mathbb{R}^n$, σ is surface measure on M, and $f \in C^\infty(\mathbb{R}^n)$. Compute the wave front set of the distribution $u = f \, d\sigma$ (i.e, $\langle u, \phi \rangle = \int \phi f \, d\sigma$). (Hint: Use Exercise 3 in §8G and Theorem (8.58).)

The next exercise uses the following extension of Theorem (0.19), which may be proved using the ideas in the proof of Proposition (8.14) (see also Rudin [41, Theorem 6.20] and Folland [14, Exercise 4.56]): Suppose $\{\Omega_\alpha\}$ is a collection of open sets in $\mathbb{R}^n$ and $\Omega = \bigcup \Omega_\alpha$. There exist sequences $\{\phi_j\}$ and $\{\psi_j\}$ in $C_c^\infty(\Omega)$ such that (i) $\psi_j = 1$ on supp ϕ_j; (ii) each ψ_j is supported in some Ω_α; each $x \in \Omega$ has a neighborhood that intersects only finitely many supp ψ_j; (iv) $\sum \phi_j \equiv 1$ on Ω.

2. Suppose $\{\Omega_\alpha\}$ is a collection of open sets in $\mathbb{R}^n$ and $\Omega = \bigcup \Omega_\alpha$. Suppose $P : \mathcal{E}'(\Omega) \to \mathcal{D}'(\Omega)$ is a linear map with the following properties: (i) For each α, the map $P_\alpha : \mathcal{E}'(\Omega_\alpha) \to \mathcal{D}'(\Omega_\alpha)$ defined by $P_\alpha u = Pu|\Omega_\alpha$ belongs to $\Psi^m(\Omega_\alpha)$. (ii) sing supp $Pu \subset$ supp u for all $u \in \mathcal{E}'(\Omega)$. Show that $P = Q + R$ where $Q \in \Psi^m(\Omega)$ and R is smoothing. (Hint: Let ϕ_j and ψ_j be as in the remarks above, and let $Qu = \sum \psi_j P(\phi_j u)$.)

3. Suppose Ω is an open set in $\mathbb{R}^n$ and $P : \mathcal{E}'(\Omega) \to \mathcal{D}'(\Omega)$ is a linear map such that for any $\phi, \psi \in C_c^\infty(\Omega)$, the map $u \to \psi P(\phi u)$ belongs to $\Psi^m(\Omega)$. Show that P satisfies the hypotheses of Exercise 2 when $\{\Omega_\alpha\}$ is the family of open subsets of Ω whose closure is compact and contained in Ω.

4. Show by example that condition (ii) in Exercise 2 cannot be omitted. (Hint: Let $Pu(x) = u(x + a)$ on $\Omega = \mathbb{R}^n$, and take each Ω_α to be a set of diameter $< |a|$.)

BIBLIOGRAPHY

1. R. A. Adams, *Sobolev Spaces*, Academic Press, New York, 1975.
2. S. Agmon, The coerciveness problem for integro-differential forms, *J. d'Analyse Math.* **6** (1958), 183–223.
3. S. Agmon, *Lectures on Elliptic Boundary Value Problems*, Van Nostrand, New York, 1965.
4. The Bateman Manuscript Project (A. Erdélyi, ed.), *Higher Transcendental Functions* (3 vols.), McGraw-Hill, New York, 1953–55.
5. R. Beals, A general calculus of pseudodifferential operators, *Duke Math. J.* **42** (1975), 1–42.
6. R. Beals and C. Fefferman, On local solvability of linear partial differential equations, *Ann. of Math.* **97** (1973), 482–498.
7. L. Bers, F. John, and M. Schechter, *Partial Differential Equations*, Wiley-Interscience, New York, 1964.
8. N. N. Bogoliubov and D. V. Shirkov, *Quantum Fields*, Benjamin-Cummings, Reading, MA, 1983.
9. E. A. Coddington and N. Levinson, *Theory of Ordinary Differential Equations*, McGraw-Hill, New York, 1955.
10. R. Courant and D. Hilbert, *Methods of Mathematical Physics* (2 vols.), Wiley-Interscience, New York, 1953 and 1962.
11. S. R. Deans, *The Radon Transform and Some of its Applications*, John Wiley, New York, 1983.
12. P. Enflo, A counterexample to the approximation problem in Banach spaces, *Acta Math.* **130** (1973), 309–317.
13. E. Fabes, M. Jodeit, and N. Rivière, Potential techniques for boundary value problems on C^1 domains, *Acta Math.* **141** (1978), 165–186.
14. G. B. Folland, *Real Analysis*, John Wiley, New York, 1984.
15. G. B. Folland, *Harmonic Analysis in Phase Space*, Princeton Univ. Press, Princeton, NJ, 1989.

16. G. B. Folland, Remainder estimates in Taylor's theorem, *Amer. Math. Monthly* **97** (1990), 233–235.

17. G. B. Folland, *Fourier Analysis and its Applications*, Wadsworth and Brooks/Cole, Pacific Grove, CA, 1992.

18. G. B. Folland and J. J. Kohn, *The Neumann Problem for the Cauchy-Riemann Complex*, Princeton Univ. Press, Princeton, NJ, 1972.

19. H. W. Fowler, *A Dictionary of Modern English Usage*, Oxford Univ. Press, New York, 1965.

20. A. Friedman, *Partial Differential Equations of Parabolic Type*, Prentice–Hall, Englewood Cliffs, NJ, 1964.

21. A. Friedman, *Partial Differential Equations*, Holt, Rinehart, and Winston, New York, 1969.

22. L. Gårding, *Singularities in Linear Wave Propagation*, Springer-Verlag, Berlin, 1987.

23. P. C. Greiner and E. M. Stein, *Estimates for the $\bar{\partial}$-Neumann Problem*, Princeton Univ. Press, Princeton, NJ, 1977.

24. S. Helgason, *The Radon Transform*, Birkhäuser, Boston, 1980.

25. H. Hochstadt, *The Functions of Mathematical Physics*, Wiley-Interscience, New York, 1971; reprinted by Dover, New York, 1986.

26. L. Hörmander, *Linear Partial Differential Operators*, Springer-Verlag, Berlin, 1963.

27. L. Hörmander, *The Analysis of Linear Partial Differential Operators* (4 vols.), Springer-Verlag, Berlin, 1983–85.

28. D. Jerison and C. Kenig, Boundary value problems on Lipschitz domains, pp. 1–68 in *Studies in Partial Differential Equations* (W. Littman, ed.), Math. Assoc. of America, 1982.

29. F. John, *Plane Waves and Spherical Means Applied to Partial Differential Equations*, Wiley-Interscience, New York, 1955.

30. F. John, *Partial Differential Equations* (4th ed.), Springer-Verlag, New York, 1982.

31. O. D. Kellogg, *Foundations of Potential Theory*, Springer-Verlag, Berlin, 1929; reprinted by Dover, New York, 1954.

32. A. V. Lair, A Rellich compactness theorem for sets of finite volume, *Amer. Math. Monthly* **83** (1976), 350–351.

33. H. Lewy, An example of a smooth linear partial differential equations without solution, *Ann. of Math.* **66** (1957), 155–158.

34. J. L. Lions and E. Magenes, *Non-homogeneous Boundary Value Problems and Applications* (vol. I), Springer-Verlag, New York, 1972.

35. D. Ludwig, The Radon transform on Euclidean space, *Comm. Pure Appl. Math.* **19** (1966), 49–81.

36. S. G. Mikhlin, *Mathematical Physics: an Advanced Course*, North Holland, Amsterdam, 1970.

37. C. Miranda, *Partial Differential Equations of Elliptic Type* (2nd ed.), Springer-Verlag, New York, 1970.

38. L. Nirenberg, On elliptic partial differential equations, *Ann. Scuola Norm. Sup. Pisa* **13** (1959), 115–162.

39. J. C. Peral, L^p estimates for the wave equation, *J. Funct. Anal.* **36** (1980), 114–145.

40. M. H. Protter and H. F. Weinberger, *Maximum Principles in Differential Equations*, Prentice-Hall, Englewood Cliffs, NJ, 1967; reprinted by Springer-Verlag, New York, 1984.

41. W. Rudin, *Functional Analysis* (2nd ed.), McGraw-Hill, New York, 1991.

42. X. Saint Raymond, *Elementary Introduction to the Theory of Pseudo-differential Operators*, CRC Press, Boca Raton, FL, 1991.

43. I. J. Schoenberg, The elementary cases of Landau's problem of inequalities between derivatives, *Amer. Math. Monthly* **80** (1973), 121–158.

44. R. T. Seeley, Extension of C^∞ functions defined in a half-space, *Proc. Amer. Math. Soc.* **15** (1964), 625–626.

45. E. M. Stein, *Singular Integrals and Differentiability Properties of Functions*, Princeton Univ. Press, Princeton, NJ, 1970.

46. E. M. Stein, *Harmonic Analysis*, Princeton Univ. Press, Princeton, NJ, 1993.

47. R. S. Strichartz, *A Guide to Distribution Theory and Fourier Transforms*, CRC Press, Boca Raton, FL, 1994.

48. M. E. Taylor, *Pseudodifferential Operators*, Princeton Univ. Press, Princeton, NJ, 1981.

49. F. Treves, *Topological Vector Spaces, Distributions, and Kernels*, Academic Press, New York, 1967.

50. F. Treves, On local solvability of linear partial differential equations, *Bull. Amer. Math. Soc.* **76** (1970), 552–571.

51. F. Treves, On the existence and regularity of solutions of linear partial differential equations, pp. 33–60 in *Proc. Symp. Pure Math.* vol. XXIII, Amer. Math. Soc., Providence, RI, 1973.

52. F. Treves, *Basic Linear Partial Differential Equations*, Academic Press, New York, 1975.

53. F. Treves, *Introduction to Pseudodifferential and Fourier Integral Operators* (2 vols.), Plenum Press, New York, 1980.

54. G. Verchota, Layer potentials and regularity for the Dirichlet problem for Laplace's equations in Lipschitz domains, *J. Funct. Anal.* **59** (1984), 572–611.

INDEX OF SYMBOLS

Multi-indices and (Pseudo-)Differential Operators:

D^α, 216, 266

$p(x, D)$ (operator defined by a symbol), 267

P_a (operator defined by an amplitude), 284

$|\alpha|$, $\alpha!$, x^α (multi-index notation), 2

Δ (Laplacian), 66

Λ^s, 193

∂_j, 2

∂^α, 3

∂_ν (normal derivative), 5

$\partial_{\nu-}$, (interior normal derivative), 116

$\partial_{\nu+}$, (exterior normal derivative), 117

$\partial_{\bar{z}}$ (Cauchy-Riemann operator), 34

∇, 3

Other Operators and Functionals:

$\langle f, g \rangle$ (bilinear inner product), 2

$\langle f \mid g \rangle$ (Hermitian inner product), 2

$f * g$ (convolution), 10, 21, 22

$\hat{f}$ (Fourier transform), 14

$f^\vee$ (inverse Fourier transform), 16

$\|f\|_s$ (Sobolev norm), 190

$\|f\|_{k,\Omega}$ (Sobolev norm), 220

Miscellaneous:

$N(x)$, $N(x, y)$ (fundamental solution for Δ), 75, 76

O, o (orders of magnitude), 4

$p \sim \sum p_j$ (asymptotic expansion), 279

Γ (gamma function), 7

σ_P (symbol), 267, 289

χ_L (characteristic form), 32

Note: Symbols used only in a single section are generally not listed here.

INDEX